Biodiversity Conservation in Southeast Asia

Southeast Asia is highly diversified in terms of socio-ecosystems and biodiversity, but is undergoing dramatic environmental and social changes. These changes characterize the recent period and can be illustrated by the effects of the Green Revolution in the late 1960s and 1970s, to the globalization of trade and increasing agronomic intensification over the past decade.

Biodiversity Conservation in Southeast Asia provides theoretical overviews and challenges for applied research in living resource management, conservation ecology, health ecology and conservation planning in Southeast Asia. Five key themes are addressed: origin and evolution of Southeast Asian biodiversity; challenges in conservation biology; ecosystem services and biodiversity; managing biodiversity and living resources; policy, economics and governance of biodiversity. Detailed case studies are included from Thailand and the Lower Mekong Basin, while other chapters address cross-cutting themes applicable to the whole Southeast Asia region.

This is a valuable resource for academics and students in the areas of ecology, conservation, environmental policy and management, Southeast Asian studies and sustainable development.

Serge Morand is an evolutionary ecologist working at the French CNRS and CIRAD, Adjunct Professor at the Faculty of Veterinary Technology, Kasetsart University and Invited Professor at the Faculty of Tropical Medicine, Mahidol University, Bangkok, Thailand.

Claire Lajaunie is a doctor of law working in international environmental law and its links with regulation regarding health and ethical issues. Currently working for INSERM (French National Institute for Health and Medical Research), she conducts research in a research unit dedicated to Environmental Comparative Law (Ceric-DICE).

Rojchai Satrawaha is Associate Professor and former Deputy Director of the Walai Rukhavej Botanical Research Institute at Maha Sarakham University, Thailand.

Earthscan Conservation and Development series
Series Editor: W. M. Adams, Moran Professor of
Conservation and Development, Department
of Geography, University of Cambridge, UK

Conservation and Sustainable Development
Linking Practice and Policy in Eastern Africa
Edited by Jon Davies

Conservation and Environmental Management in Madagascar
Edited by Ivan R. Scales

Conservation and Development in Cambodia
Exploring frontiers of change in nature, state and society
Edited by Sarah Milne and Sanghamitra Mahanty

Just Conservation
Biodiversity, Well-being and Sustainability
By Adrian Martin

Biodiversity Conservation in Southeast Asia
Challenges in a changing environment
Edited by Serge Morand, Claire Lajaunie and Rojchai Satrawaha

For further information please visit the series page on the Routledge
website: http://www.routledge.com/books/series/ECCAD/

Biodiversity Conservation in Southeast Asia

Challenges in a Changing Environment

Edited by
Serge Morand,
Claire Lajaunie and
Rojchai Satrawaha

Routledge
Taylor & Francis Group

LONDON AND NEW YORK

earthscan
from Routledge

First published 2018 by Routledge

2 Park Square, Milton Park, Abingdon, Oxfordshire OX14 4RN

52 Vanderbilt Avenue, New York, NY 10017

Routledge is an imprint of the Taylor & Francis Group, an informa business

First issued in paperback 2019

British Library Cataloguing-in-Publication Data
A catalogue record for this book is available from the British Library

Library of Congress Cataloging in Publication Data
Names: Morand, S., editor. | Lajaunie, Claire, editor. | Rojchai
Satrawaha, editor.
Title: Biodiversity conservation in Southeast Asia : challenges in a
changing environment / edited by Serge Morand, Claire Lajaunie
and Rojchai Satrawaha.
Description: New York : Routledge, 2017. | Includes bibliographical
references and index.
Identifiers: LCCN 2017007652 | ISBN 9781138232044
(hardback : alk. paper) | ISBN 9781315313573 (ebook)
Subjects: LCSH: Biodiversity conservation—Southeast Asia. |
Environmental policy—Southeast Asia.
Classification: LCC QH77.S644 B56 2017 | DDC 333.950959—dc23
LC record available at https://lccn.loc.gov/2017007652

ISBN: 978-1-138-23204-4 (hbk)
ISBN: 978-0-367-33526-7 (pbk)

Typeset in Goudy
by Keystroke, Neville Lodge, Tettenhall, Wolverhampton

Contents

Illustrations

Figures

Tables

Contributors

Editors

Serge Morand is an evolutionary biologist working at the French CNRS and CIRAD, Adjunct Professor at the Faculty of Veterinary Technology, Kasetsart University and Invited Professor at the Faculty of Tropical Medicine, Mahidol University, Bangkok, Thailand. His main interests concern the links between biodiversity and health using wildlife-borne diseases as a model. He has led several collaborative projects in Southeast Asia over the last ten years.

Claire Lajaunie is a doctor of law working in international environmental law and its links with regulation regarding health and ethical issues. As a law scientist, she is currently involved in research projects in this area focusing on Southeast Asia. She taught Law in Asia and coordinated scientific programmes in Southeast Asia for the Health Department of the French Institute of Research for Development (IRD). Currently working for INSERM (French National Institute for Health and Medical Research), she conducts research in a research unit dedicated to Environmental Comparative Law (Ceric-DICE).

Rojchai Satrawaha is Associate Professor and former Deputy Director of the Walai Rukhavej Botanical Research Institute at Maha Sarakham University, Thailand. He has conducted numerous projects on freshwater biodiversity and wetland environments.

Chapter authors

Sate Ahmad is an interdisciplinary ecologist. He is working as a senior research officer at ICDDRB and as an adjunct lecturer at the Department of Environmental Science and Management, North South University, Bangladesh. His research interest lies in understanding how biodiversity and ecosystem services are affected by management decisions, and how these may affect human well-being.

Mohammed Nurul Azam is a development professional specializing in the development and running of results measurement and knowledge management systems in accordance with the DCED Standard. He has experience of working in some of the biggest M4P/market development projects in the world. He has a wide range of hard and soft skills, combined with a desire to make a difference.

Aurélie Binot is an anthropologist and agronomist working for CIRAD and based at Kasetsart University, Thailand. She coordinates OneHealth projects in Southeast Asia and implements research in the fields of collective risks management at human/animal/environment interfaces (collective action, commons, participatory approaches) and epistemology of One health and Ecohealth movements.

Julien Cappelle is a veterinarian and disease ecologist at CIRAD. Based for four years in Southeast Asia, he has been working on emerging infectious diseases at the wild–domestic–human interface, implementing OneHealth and EcoHealth approaches.

Jeremy Carew-Reid is Director General of ICEM – the International Centre for Environmental Management. He has has years' experience working in over thirty countries. With a BSc. in freshwater ecology and a Ph.D. in EIA, he has specialized in strategic environmental assessment and integrating biodiversity conservation, climate change and sustainable livelihoods with development.

Julien Claude is an assistant professor at the University of Montpellier in France, where he teaches Evolutionary Biology and Statistics. Palaeontology and evolution in Southeast Asia have been two of his main research interests for the past fifteen years.

Jean-François Cornu is Ph.D. candidate at the University of Montpellier, France. He investigates land use/land cover changes and their impacts on biodiversity and zoonotic diseases in Southeast Asia.

Pongchai Dumrongrojwatthana is a lecturer in the Department of Biology, Faculty of Science, Chulalongkorn University, Thailand. His current research fields are participatory modelling for natural resource management, ecosystem ecology, ecosystem services in urban area and geographic information systems (GIS).

Hans Keune (political scientist – Ph.D. Environmental Sciences) works on critical complexity, inter- and transdisciplinarity, action research, decision support, environment and health, ecosystem services, biodiversity and health, and OneHealth/EcoHealth. He works at the Belgian Biodiversity Platform, Research Institute for Nature and Forest (INBO) and Faculty of Medicine and Health Sciences, University of Antwerp, Belgium.

Conor Kretsch is director of Co-operation on Health and Biodiversity (COHAB), a community of individuals and organizations working together to address the gaps in awareness, policy and action on the links between biodiversity and human health and well-being. The initiative supports efforts to enhance human security through the conservation and sustainable use of biodiversity.

Chutapa Kunsook is a lecturer in the Department of Biology, Faculty of Science and Technology, Rambhai Bhani Rajabhat University, Thailand. She has been working on blue swimming crab management since 2005. Her current research fields are marine animal diversity in coastal areas and coastal resource management.

Henri Laborde is an engineer working at CNRS who specializes in remote sensing technologies and geographic information systems (GIS).

Pim Martens holds the chair for Sustainable Development at Maastricht University, Netherlands and is a professor extraordinary at Stellenbosch University, South Africa. He is also a scientivist, intending to contribute to a better, more sustainable society. He is founder of AnimalWise, a 'think and do tank' that integrates scientific knowledge and animal advocacy.

Jeffrey A. McNeely has been working in international conservation since 1968, including ten years in Thailand, three years in Indonesia, two years in Nepal, and thirty years at IUCN headquarters in Switzerland until his retirement from the position of Chief Scientist in 2009. He is author or editor of over forty books and four hundred other publications, and serves on the editorial board of a dozen journals. He currently works with the Asian Development Bank and Thailand's Department of National Parks, Wildlife and Plant Conservation.

Rachanee Nam-Matra is a lecturer in Walai Rukhavej Botanical Research Institute, Mahasakham University, Thailand. She has acquired experience in weed ecology and diversity on various projects related to agricultural practice and ecological influences. Furthermore, she has a strong interest in the interaction of Southeast Asian biodiversity and human activities.

Unnikrishnan Payyappallimana is a senior research fellow at the United Nations University–Institute for the Advanced Study of Sustainability (UNU-IAS), Tokyo and an associate fellow at the United Nations University–International Institute of Global Health (UNU-IIGH), Kuala Lumpur. His research interests are traditional medicine, traditional knowledge, biodiversity, public health and sustainable development.

Mohammed Mofizur Rahman is a multidisciplinary environmental and conservation science graduate. His research focus is environmental change in the context of health and well-being. He is attempting to communicate environmental issues with general audiences and policy-makers through the integration of knowledge from natural and social sciences.

Daniel Robinson is an associate professor in the Faculty of Arts and Social Sciences, UNSW, Australia, and a research fellow with the International Centre for Trade and Sustainable Development (ICTSD), Geneva.

Zita Sebesvari is an associate academic officer at United Nations University, Institute for Environment and Human Security, Bonn, Germany. Her work focuses on socio-ecological risk and vulnerability, ecosystem services and environmental degradation, particularly in the context of agrichemical pollution and salinity intrusion, mainly in the regional context of coastal Southeast Asia.

Joachim H. Spangenberg is an inter- and transdisciplinary researcher by education and dedication, with a Ph.D. in Economics, but an academic background Biology and Ecology. His work focuses on sustainability of socio-ecological systems, ecosystem services and the limits to valuation, and sustainable consumption. For more information, please visit http://seri.academia.edu/JoachimH Spangenberg.

Witchuda Srang-iam is an assistant professor at the Graduate School of Environmental Development Administration, National Institute of Development Administration, Thailand. She has conducted research on the political and institutional dimensions of environmental challenges, including climate change, biodiversity and ecosystem services.

Suneetha M. Subramanian works with the Institute of Advanced Studies and with the International Institute for Global Health at United Nations University. She focuses on implications to human well-being in the utilization of biodiversity, especially on issues of equity at multiple scales. Her research attempts to draw out coherences between policy goals and implementation realities in the biodiversity/ecosystem policy space.

Songtam Suksawang is currently Thailand's Director of National Parks. He is a leading scientist on a wide range of issues affecting nature conservation. His major research interests are protected areas management and biodiversity conservation, and he has led Thailand's work on conservation corridors. He has written numerous books on conservation issues, including on watershed management, ecosystem services and human dimensions of protected areas. He has participated in collaborative research and conservation programmes with UNDP, ADB, GEF and numerous bilateral development agencies.

Luke Taylor is an Environmental Scientist with ICEM – International Centre for Environmental Management. He specializes in natural resources and river basin management. He has over ten years' experience working in both Australia and the Mekong region with private and government partners conducting environmental assessments and studies. He holds Bachelor of Land and Water Science and Master of Environmental Law degrees.

Yongyut Trisurat is Professor at the Faculty of Forestry and Dean of the Department of Forest Biology, Kasetsart University (Bangkok, Thailand). He has been active in land use and species modelling for over twenty years. His current research focuses on climate change and ecosystem services in Thailand and the Lower Mekong Basin. He has published a number of peer-reviewed articles and book chapters on these subjects and recently contributed to IPBES and GEO-6 reports.

Heidi Wittmer is Deputy Head of the Department of Environmental Politics at the Helmholtz Centre for Environmental Research. She works on the governance of natural resources, especially biodiversity, the multiple values of biodiversity for human well-being, and studying and improving the interface of science and policy in this field.

Kobchai Worrapimphong is a lecturer in the Department of Agricultural Development, Faculty of Natural Resources, Prince of Songkla University, Thailand. His current research fields are coastal resource management and sustainable agricultural development.

Preliminaries

Southeast Asia, highly diversified in terms of socio-ecosystems and biodiversity, is undergoing dramatic environmental and social changes affecting biodiversity living resources and livelihoods. The climate in Southeast Asia is predicted to become dryer and to have a higher probability of extreme events, with increased climate variability affecting the monsoon. Forest cover is still declining in the region, with 0.91 percent of forest cover lost every year, and agricultural extension seems to be the major cause of deforestation of the mountainous areas of mainland and insular Southeast Asia. Although the region is recognized as a hotspot for biodiversity, and cultural diversity, it is suffering from rapid and extensive erosion of biodiversity, which urges at identifying the consequences of biodiversity decline in terms of ecosystem functions and services.

This book presents the major contributions given at an international symposium organized by Mahasarakham University (Thailand) with the support of the French CIRAD, CNRS, and IRD institutions and the French ANR program by providing overviews and challenges for generic and applied researches in living resource management, conservation ecology, health ecology, and planning in Southeast Asia.

Part I

Origin and evolution of Southeast Asian biodiversity

1 The history and diversity of Southeast Asia

Serge Morand and Claire Lajaunie

> Does Southeast Asia really exist? Many scholars, in many disciplines, writing since the 1960s have been so certain that the region is a distinctive and coherent entity that they have not bothered to define it.
>
> (Hill 2002: 1)

A remarkable Southeast Asian biodiversity

Southeast Asia is characterized by both high species richness and high level of endemism linked to its geological history. Deep evolutionary origins of flora and fauna have to be searched for in the tectonic-plate dynamics that occurred through geologic times, whereas more recent diversification origins are linked to the consequences of changing paleo-climates with sea level fluctuations (Meijaard 2003; Sodhi et al. 2004). These above processes occurring in a tropical climate favored biological speciation and led Southeast Asia to be a major biodiverse region on earth (Sodhi et al. 2004).

Southeast Asia is recognized as a major hotspot of species diversity (Myers et al. 2000; Schipper et al. 2008). As emphasized by Sodhi et al. (2010), Southeast Asia harbors the highest proportion of endemic bird and mammal species (averaged by country) in comparison to other bio-regions, and the second-highest proportion of country-endemic vascular plant species. However, Southeast Asia also has the highest proportion of threatened vascular plants, reptiles, birds and mammals, linked to forest conversion. The annual deforestation rate of Southeast Asia is one of the highest among all tropical regions and is still increasing (FAO 2006).

What is Southeast Asia? History matters

Hill (2002) is one of the scholars who has tried to define Southeast Asia both politically and geographically, emphasizing both its diversity and unity. There are eleven states in Southeast Asia inhabited by 600 million people. McGregor (2008) emphasized one main characteristic of Southeast Asia, which is its diversity: a high diversity in economical development among and within countries and a high diversity in political systems. The diversity in geography is also impressive, with landscapes varying from the high mountains of Myanmar to vast deltas, tropical rainforests and mangrove swamps (McGregor 2008). We can add the high cultural diversity, as Southeast Asia is a hotspot of language diversity. Moreover, all the biggest religions are present in Southeast Asia: Islam, Buddhism, Christianity and Hinduism, with syncretism of Taoism, Buddhism and Confucian teachings in Vietnam, and numerous community-based animist belief systems (McGregor 2008).

It may appear easy to locate Southeast Asia on the globe as a region at the interface of Asia, Australia and Oceania, the Indian Ocean and the Pacific Ocean. However, a geographical definition is more difficult to provide as such a definition should integrate the geology, biology, history and societies of the region.

As emphasized by Bankoff and Boomgaard (2007: 1) Southeast Asia has "historically been part of a worldwide network of exchange that predates 1500 and that tied the region closely to India, China, and Japan." Southeast Asia was early part of a global commodity system that consequently affected its socio-environments and its biodiversity (Figure 1.1A).

Bankoff and Boomgaard (2007: 1) added, "Historians of Southeast Asia have often ignored the question of natural resources, mainly accepting them as a given and passing on to what they identify as the central issue, trade", and underlined that:

> Southeast Asia has never been just an entrepôt, simply a "gate to China" through which goods produced elsewhere passed. It has also been an important supplier of raw and semi-processed materials. It has historically been part of a worldwide, if bounded, network of exchange that predates 1500 and that tied the region closely to India, China, and Japan . . .That this network evolved over the past five centuries into a global commodity system in no way lessens the significance of the preexisting ties and their effects on the peoples and environments of the region.

European colonialism in Southeast Asia started in the sixteenth century with the Portuguese, followed by the Spanish and soon the Dutch (Bertrand, 2011, 2014; see Figure 1.1B).

Four colonial powers were active in Southeast Asia by the beginning of the twentieth century (Booth, 2007). The British controlled Burma, the Malay peninsula, Singapore and the north of Borneo. The Dutch governed the huge Indonesian archipelago, from Sumatra to New Guinea. The French controlled the contiguous territories of Vietnam, Cambodia and Laos, a region known as French Indochina. After the defeat of Spain during the American–Spanish War of 1898, the USA imposed its administration on the Philippine islands. The Portuguese maintained a presence on East Timor. Only the Kingdom of Siam was independent and never colonized. One may also note the presence of Japan at the border of Southeast Asia (Figure 1.1C) and German colonies in New Guinea (until the end of Word War I).

It was really during and in the aftermath of the World War II that the name Southeast Asia was recognized (Booth, 2007). In the following decades, new

100-15th Century

A. Eastern Indian Ocean regional maritime networks, *c.* 100–1500 (adapted from Hall 2011)

16th Century

B. Southeast Asia in the sixteenth century (adapted from Bertrand 2011, 2014)

Figure 1.1 Maps of Southeast Asia

20th Century

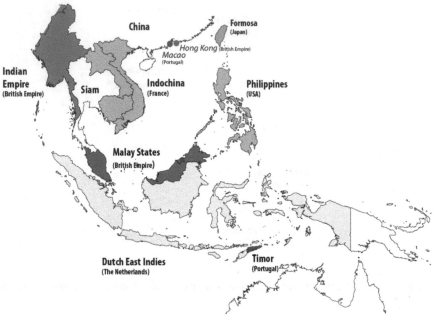

C. Southeast Asia at the start of World War II (adapted from Booth 2007)

21st Century

D. Southeast Asia at the beginning of the twenty-first century

Figure 1.1 (continued)

states gained their independence from the colonial countries. The Netherlands Indies became Indonesia. "British Malaya"—which referred to the Straits Settlements, the Federated Malay States and the Unfederated Malay States— became Malaysia, Singapore and Brunei. Cambodia, Vietnam and Laos gained independence from France. British Burma became Myanmar. The last country to gain its independence was Timor Leste—from Indonesia in 2002. Siam became Thailand in 1939. Now, Southeast Asia refers to the region covered by the ten member states of the Association of Southeast Asian Nations (ASEAN; plus Timor Leste (Figure 1.1D).

The environmental legacy of the colonial powers

During the nineteenth century, the colonial governments introduced scientific-based policies for the management of their empires. Ravi Rajan (2006: 13) showed how the settings of forest departments and other agencies "resulted in the creation of a homogeneous and assertive pancolonial community of foresters." These foresters (English, French, Dutch) were trained in the school of Germany and later the school of France, where they acquired a shared vision of forest management, a common intellectual tradition that is seen as the root and origin of environmentalism (Grove 1995; Grove et al. 1998). The German forestry school emphasized the importance of sustained yield, or sustainability (*Nachhaltigkeit*), using mathematics to achieve high timber yields while maintaining future resources. This sustainability could be attained through the regulation of local people's rights, including, for example, the complete abolition of rights in return for suitable compensation paid to the "beneficiary" (see Ravi Rajan 2006). In France, a school of forestry was established at Nancy to train foresters, including colonial ones. All these foresters shared the opinion that new forestry should preserve forests from mismanagement by local people, who were blamed for forest degradation. For instance, French foresters saw themselves as engineers concerned with the impact of deforestation on watersheds, and put forward the connection between "forest cover, healthy watersheds, and agricultural productivity" (Ravi Rajan 2006: 48).

Even more fascinating, the foresters created a coordinated network of natural, social and scientific domains (Figure 1.2; Worthington 1938, cited in Ravi Rajan 2006). This framework resembles many of today's (see also Grove 1995).

The Dutch foresters in Indonesia had quite similar concerns to those of their British and French counterparts. According to Galudra and Sirait (2009), Dutch foresters emphasized the importance of protecting upland forests to maintain a balanced hydrological system needed to avoid floods and droughts and for better agriculture productivity in lowland areas. They developed a scientific discourse which allowed the implementation of a forestry policy that targeted the customary systems of traditional land use seen as destructive and implemented forest reserves under their control. In 1865, a forestry law was introduced in Java declaring that all unclaimed land, which corresponded to forested areas, was the property of the state (Galudra and Sirait 2009).

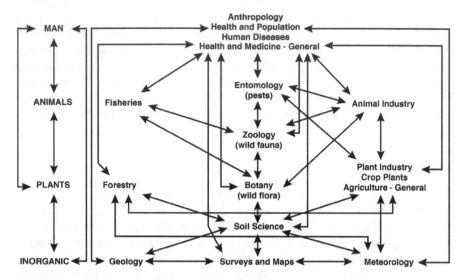

Figure 1.2 The colonial scientific network of environmental management (redrawn from Worthington 1938 in Ravi Rajam 2006)

The implementation of a forestry service in French Indochina was done later, in 1901, than in British Burma or Dutch Indonesia (Thomas 1998). In 1930, a new law made the distinction between preserved forest domain (*domaine forestier protégé*) and reservation forest domain (*domaine forestier classé*). Preserved domain allowed free access for traditional uses, also with some obligations to maintain the forest resources, timbers and other economical interests for the French colony (Maurand 1943, cited in Thomas 1998). Reservation domain is an experimental domain of the forestry service in which the forest regenerates itself. "Methodological exploitation" was the key concept in the French forestry discourse, which emphasized non-predatory colonization that can take care without exhausting the environment it exploits. Colonization was supposed not to destroy but to highlight colonial forests. However, the proportion of reserved forests never exceeded 10 percent of the global forest cover (Thomas 1998). By using the nascent and burgeoning ecological sciences of the beginning of the twentieth century, the French colonial forestry service aimed to improve then optimize the productivity of the colonial forests.

Although trained at the same forestry school, Western foresters and their forestry administrations have implemented different "ecological" concepts in Southeast Asia which have led to different outcomes and policies concerning living resources of local peoples, local rulers and Western private interests. These have without doubt influenced the post-colonial exploitation of forests, whereas in the early post-colonial years there was a reinforcement of forest management predicated by Western foresters (Bryant 1997).

Post-colonial states and the socio-ecology of Southeast Asia

The colonial states and their forest administrations used "scientific" expertise to exclude customary use of forest resources by local peasant populations in order to protect the forest resources for their timber productivity or supposed hydrological regulation. Although access to land and forests was a major factor in revolts against the colonial powers, the post-colonial governments of Southeast Asia prolonged many of the principles implemented by the colonial powers in the name of national development (Pye 2012). Colonial forestry departments were used to promote national timber industries (Bryant 1997; Pye 2005). In forestry and agriculture, coalitions between corporations and political elites pursued more profitable and permanent conversions such as teak, rubber, eucalyptus and oil palm (Pye 2012). But the role of post-colonial organizations such as the Food and Agriculture Organization of the United Nations (FAO) have greatly facilitated the construction of forestry as a "kind of empire" since World War II, paraphrasing Vandergeest and Peluso (2006a: 369), who added: "The FAO advanced the idea that forests were important not only for forestry, but also for agricultural lands, the wealth and welfare of the nation-state, and national progress in general." The transformation of colonial forests into national forests controlled by governments intended to favor profits from the forestry industry and increase the commercialization of certain timber species. It also contributed to the criminalization of people's access to the free forest products for their daily existence by forcing local populations to obtain forest products through government agencies or markets, as emphasized by Vandergeest and Peluso (2006b). These policies had dramatic impacts on indigenous populations, their mode of subsistence and their culture (Forsyth and Michaud 2011; Cairns 2015; McElwee 2016).

Post-colonial wars have dramatically impacted the natural resources and the livelihoods of local people, such as the use of the defoliant Agent Orange in Vietnam (Zierler 2011), the various communist insurgencies and counterinsurgencies in several Southeast Asian countries, and the post-war violence in Cambodia (Milne and Mahanty 2015). Under the Khmer Rouge agriculture production was collectivized and millions of people were killed. In the subsequent civil war that lasted into the 1990s, the Khmer Rouge and the Vietnam forces financed their troops through extensive logging (Le Billon 2000; Pye 2012). As emphasized by Pye (2012), this war created a peculiar political ecology of forest destruction based on conflict and cooperation between engaged parties, including the Thai state.

National development policies were based on nation-state building concepts and the science-based expertise of international organizations, which ultimately led to increased deforestation (Delang 2005) and the removal of access rights from local people subjected to ostracism by national elites (Delang 2002). Hence, as an example, "the FAO gave considerable attention to the notion that the Chao Phraya River watershed in Thailand was repeatedly deforested by the shifting cultivation practices of what they called nomadic hill tribes" (Vandergeest and Peluso 2006b: 378).

Interestingly, a recent study has shown that the conversion of rainforest ecosystems to oil palm plantations in Java may impact the hydrological system through a redistribution of precipitated water by runoff associated with reported periodic water scarcity (Merten et al. 2016). This study revisits colonial forestry science by emphasizing the protective role of forests and the eco-hydrological consequences of oil palm expansion.

Recent environmental changes

Southeast Asia is a region prone to multiple, rapid change. Such change characterizes the recent period and could be illustrated by the effects of the Green Revolution from the late 1960s and 1970s. This Green Revolution succeeded in fighting widespread famine and contributed to a drastic reduction of poverty and to economic growth across the region. For instance, from 1970 to 1995, while the population increased by more than 60 percent (68.2%; see Hazell 2009), the calorie intake per person and per day increased by more than 30 percent (33.5%; see Hazell 2009) due to high growth in agricultural production. Though it can be explained by various reasons, we can underline that the success of the Green Revolution in Southeast Asia is at least partly due to the policy-supported favorable environment (Estudillo and Otsuka 2013) for the development of agricultural research and to the strengthening of regional cooperation through ASEAN (Tongzon 2002). The impetus given by the adhesion to a regional community which aims for the acceleration of economic growth in the region in order to strengthen the foundations for a prosperous and peaceful community (Article 6, Treaty of Amity and Cooperation in Southeast Asia, Indonesia, 1976) has helped to determine the process of agricultural modernization and cooperation among ASEAN member states, a process that continues to be promoted in the post-2015 vision (ASEAN Ministers of Agriculture and Forestry 2015).

The transformation of the overall economic situation of the region, though uneven across the different countries, induced different changes regarding the demography, the agricultural intensification, land use changes, rapid urbanization and biodiversity. These changes have consequences for livelihood, management of natural resources and conservation (Milne and Mahanty 2015). The socio-economic and environmental changes have a direct influence on the emergence of Emerging Infectious Diseases (EID) in the region (Coker et al. 2011; Morand et al. 2015), which originate mostly from wildlife and biodiversity changes.

This reminds us that investigating biodiversity and its crisis in Southeast Asia has to embed the links between biodiversity, natural resources and people not in the short causal effect of the global economy but in more long-term historical perspectives.

Biodiversity conservation and development in Southeast Asia

Southeast Asia has faced dramatic socio-economic transformations in the twentieth century, with increased demographics (the overall population doubled from 1970 to 2000), and associated problems of poverty and governance (Shively and Smith 2015; see also Coxhead 2015). The environment, the landscapes, the living resources and the biodiversity have all been deeply affected.

One of the major changes has concerned the high level of deforestation. Forest clearing seems to have resulted from both the historical colonial legacy and post-colonial decisions. Large endowments of forest resources have characterized Southeast Asian nations and authoritarian or centralized governments have used these endowments to sell large forest concessions to logging interests and to promote extraction at unsustainable rates (Rola and Coxhead 2005; Shively and Smith 2015). As stressed by Dennis et al. (2008), sustainable forest management is still far from being implemented, despite decades of attempts at improvement. While most of the easily accessible timber has been removed, fast-growing plantations are favored and large-scale transformation of forests to rubber trees or palm oil plantations is still occurring, threatening emblematic species such as elephants and orangutan.

Major investment in hydroelectric dams in order to provide electricity for cities and industries threatens both the remaining forests and the aquatic ecosystems (Molle et al. 2009). Already built dams along the upper course of the Mekong and ongoing plans for new dams in Laos and Cambodia will have profound consequences for biodiversity, fisheries and human livelihoods (Vörösmarty et al. 2010; Dudgeon 2011; Ziv et al. 2012; Sunderland et al. 2013).

Aims of the book

This book intends to present the main challenges affecting biodiversity in the Southeast Asian region in regard to the rapid changes it is experiencing. But if it is obvious that there have been dramatic recent changes, we should consider the trends on a longer temporal scale both evolutionarily and historically. It is worth defining Southeast Asia as a product of history and also as a biogeographical entity. By taking different views on biodiversity in Southeast Asia, from past to actual, and from biology to the social sciences, this book attempts to complement several related, previously published studies (i.e. Sodhi et al. 2008; Sunderland et al. 2013; Morand et al. 2015). Furthermore, particular emphasis will be placed on health and infectious diseases by showing the connections between sociodynamics, ecology, biodiversity and health.

Introducing the chapters

To understand the current richness of Southeast Asian biodiversity, we should first look at the origins and evolution of the environment and its biodiversity. Studying the continental fossil record in various places in Southeast Asia can help us understand the place and the importance of Southeast Asia in the distribution, migration and origination of plants and animals; studying the complexity of paleogeographic and tectonic history (Chapter 2) allows better understanding of the evolution of biodiversity in the region and allows us to see potential future directions. The actual biodiversity can be understood in light of the Last Glacial Maximum, which occurred around 150,000 years ago (Voris 2000; Sathiamurthy and Voris 2006; Figure 1.3A) and finally shaped the biogeographic regions (Woodruff 2010; Figure 1.3B).

Southeast Asia is a hotspot of language diversity. An interesting pattern that emerged from analysis of the region's cultural diversity is its strong correlation with the total number of bird and mammal species (Figure 1.4). The higher the biodiversity in a country, measured by the numbers of bird and mammal species, the higher the human cultural diversity, measured by the number of languages spoken (Figure 1.4B). Cultural diversity therefore appears to mirror biological diversity (Maffi 2005).

Examination of cultural diversity and its linkages with traditional knowledge, whether it concerns agriculture, medicine or forest management, gives valuable insights on sustainable ways to protect and manage biodiversity as well as adaptation solutions (Chapter 3).

Southeast Asia has been dramatically affected by land use changes over the centuries (Chapter 4). These have been acknowledged in several regional and national actions to better preserve biodiversity and the services provided by ecosystems.

The conservation of biodiversity is thus one of the main challenges in conservation biology, together with the emergence of diseases. Southeast Asia is recognized as a hotspot for biodiversity (Myers et al. 2000) that also suffers from rapid and extensive erosion of that diversity (Sodhi et al. 2004; Wilcove et al. 2013). As exemplified by Schipper et al. (2011), this region is both a major hotspot of mammal diversity and a major hotspot of mammal diversity *at threat*. Moreover, there is a good relationship between the number of languages at threat and the number of bird and mammal species at threat (Figure 1.4B).

Facing constant and rapid environmental changes, climate change and natural hazards, Southeast Asian countries have to develop strategies and policies in order to actively conserve biodiversity and manage their natural resources; thus ASEAN countries all have established systems of protected areas (Figure 1.5) that will be presented in Chapter 5.

Southeast Asia as a hotspot of biodiversity is also experiencing rapid biodiversity loss which is seen as a major driver of the emergence of infectious diseases. The most important Emerging Infectious Diseases (EIDs) over the past twenty years in the region originate from wildlife. Wildlife ecology is important to consider as several ecological traits of hosts may have an impact on the circulation of infectious diseases (host population dynamics, species behavior, spatial distribution, temporal changes). Surveillance of EIDs in the region could target certain groups of wild animals, like bats or rodents.Coordination between administrative sectors, researchers and local populations is needed in wildlife monitoring and surveillance to ensure a timely response as well as in wildlife management in order to develop strategies which could benefit both public health and the conservation of endangered species (Chapter 5).

A. Southeast Asia at the Last Glacial Maximum (adapted from Sathiamurthy and Voris 2006)

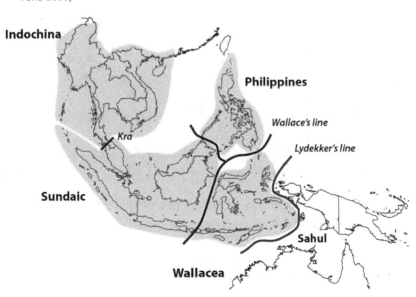

B. The biogeographic regions of Southeast Asia (adapted from Woodruff 2010)

Figure 1.3 Southeast Asia at the Last Glacial Maximum and biogeographic regions of Southeast Asia

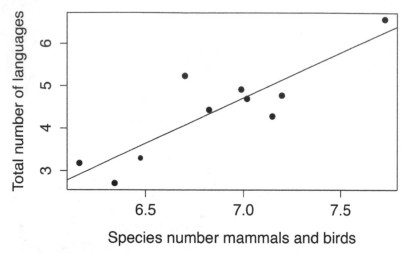

A. Relationship between the number of languages and the number of bird and mammal species per nation in Southeast Asia (adapted from Morand et al. 2014, 2015)

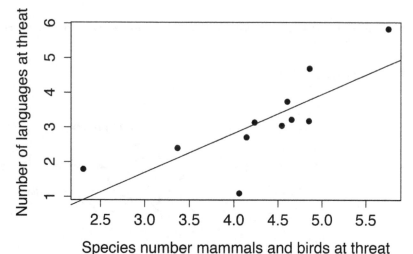

B. Relationship between the number of languages at threat and the number of bird and mammal species at threat

Figure 1.4 Relationship between the number of languages and the number of bird and mammal species per nation in Southeast Asia

The ecosystems services approach is another way to look at biodiversity changes: they can notably be studied through their role in agricultural management, through examination of the consequences of climate and land use land cover changes on watershed services and sediments or through the determination of ecosystem services necessary for health regulation.

Figure 1.5 Map of the protected areas of Southeast Asia (from UNEP-ICUN n.d.)

The agricultural practices changes (with respect to farm structure, kind of crop, farming systems) led to greater reliance on chemical inputs which modify weed management and thus change weed distribution and diversity. The practice of weed conservation is important when it comes to preserving weed diversity. It can rely on local traditional knowledge and in the end research on weed conservation is necessary to develop sustainable agriculture (Chapter 6).

Deforestation and climate change are becoming serious threats to watershed services as they can dramatically affect hydrological processes and related services. Models associating the combination of transformative land use and climate change in the Thadee watershed in southern Thailand are used to predict the evolution of water yield and water quality in order to maximize the resilience of the watershed ecosystem (Chapter 7). While a link between ecosystem health and human and animal health is clearly expressed and increasingly assumed, Chapter 8 discusses the links between an ecosystem's functions and associated ecosystem services in terms of infectious disease regulation, agricultural pest protection, resistance to biological invasion, or the impact on animal and plant communities' structure.

These issues are addressed at the ecological and social interface. The increasing threat to biodiversity in Southeast Asia leads researchers to think about future directions to manage biodiversity and living resources in a sustainable manner. The Lower Mekong Basin is undergoing total transformation of its social, economic and natural environment, implying long-term implications for its sustainability. Some 80 percent of families in the region are small-scale farmers dependent on healthy natural systems for their livelihoods and subsistence, and understanding the resilience of natural systems is crucial when it comes to evaluating the impact of climate change on various ecosystems and species. As there are shifts in the farming ecosystems in reaction to shifts in climate and linked ecological systems, it is necessary to build on more resilient and sustainable cropping systems. Thus the adaptation strategies of natural systems in the Lower Mekong Basin are examined as they would bring diversity and complexity back into the agricultural landscape (Chapter 9).

Regarding the management of biodiversity, the implementation of companion modeling to develop simulation models integrating various stakeholders' points of view and to facilitate dialogue, shared learning and collective decision-making is examined in three cases of biodiversity management in different regions of Thailand. Chapter 10 shows how the approach is suitable and useful for achieving co-learning and problem mitigation in interactive fashion.

Issues regarding the rapid changes affecting biodiversity are taken into account by policies and law at different levels. The international agreement known as the Convention on Biodiversity (1992) transformed the landscape of biodiversity politics and strengthened international environmental law. It also fostered research at the science–policy interface. Chapter 11 focuses on the shift from a single domain of governance towards 'hybrid' governance platforms of biodiversity through three modes of political practice in the creation of biodiversity markets that occur in different settings in Thailand. It makes suggestions regarding the design of policies and practices to achieve effectiveness and equity of biodiversity governance. These suggestions could be extended to other Southeast Asian countries.

The main steps leading to the elaboration of the international instruments related to health and biodiversity are presented in Chapter 12, which shows the ways those instruments have been integrated and implemented at the regional level in light of the involvement of different Southeast Asian countries in the construction of the ASEAN community.

The main biodiversity challenges in relation to human health are scrutinized by Chapter 13 from a science–society interface perspective. It specifically focuses on the need for integration and collaboration to confront complexity in a societally relevant manner using several case studies.

The lessons learned from these case studies allow us to draw conclusions regarding the need for enhanced attention to the relationship between biodiversity and human health and offer direction in relation to important challenges for society, both in Asia and internationally.

Finally, taking into consideration the fact that, in Southeast Asia, the governments have become increasingly aware of the importance of biodiversity conservation for human development and ecosystem health, Chapter 14 proceeds backwards from the present to determine the main steps leading to the elaboration of international instruments related to health and biodiversity and to discuss the ways in which they have been integrated and implemented at the regional level.

References

ASEAN Ministers of Agriculture and Forestry. 2015. Vision and Strategic Plan for ASEAN Cooperation in Food, Agriculture and Forestry (2016-2025), endorsed at the 37th AMAF, 10 September, Makati City, Philippines.

Bankoff, G. and P. Boomgaard. 2007. Introduction: natural resources and the shape of Asian history, 1500–2000. In G. Bankoff and P. Boomgaard (eds.), *A History of Natural Resources in Asia*. New York: Palgrave Macmillan.

Bertrand, R. 2011. *L'Histoire à parts égales: récits d'une rencontre Orient-Occident (XVIe– XVIIe siècle)*. Paris: Seuils.

Bertrand, R. 2014. *Le Long Remords de la conquête*. Paris: Seuil.

Booth, A. 2007. *Colonial Legacies: Economic and Social Development in East and Southeast Asia*. Honolulu: University of Hawai'i Press.

Bryant, R.L. 1997. *The Political Ecology of Forestry in Burma: 1824–1994*. Honolulu: University of Hawai'i Press.

Cairns, M.F. 2015. *Shifting Cultivation and Environmental Change: Indigenous People, Agriculture and Forest Conservation*. Abingdon: Routledge.

Coker, R.J., B.M. Hunter, J.W. Rudge, M. Liverani, and P. Hanvoravongchai. 2011. Emerging infectious diseases in Southeast Asia: regional challenges to control. *Lancet* 377: 599–609.

Coxhead, I. 2015. *Routledge Handbook of Southeast Asian Economics*. Abingdon: Routledge.

Delang, C.O. 2002. Deforestation in northern Thailand: the result of Hmong farming practices or Thai development strategies? *Society and Natural Resources* 15: 483–501.

Delang, C.O. 2005. The political ecology of deforestation in Thailand. *Geography* 90: 225–237.

Dennis, R.A., Erik Meijaard, Robert Nasi, and Lena Gustafsson. 2008. Biodiversity conservation in Southeast Asian timber concessions: a critical evaluation of policy mechanisms and guidelines. *Ecology and Society* 13: 25.

Dudgeon, D. 2011. Asian river fishes in the Anthropocene: threats and conservation challenges in an era of rapid environmental change. *Journal of Fish Biology* 79: 1487–1524.

Estudillo, J.P. and K. Otsuka. 2013. Lessons from the Asian Green Revolution in rice. In K. Otsuka and D. F. Larson (eds.), *An African Green Revolution: Finding Ways to Boost Productivity on Small Farms*. Dordrecht: Springer.

FAO. 2006. Global forest resources assessment 2005: progress towards sustainable forest management. Forestry Paper 147. Rome: United Nations Food and Agriculture Organization.

Forsyth, T. and J. Michaud. 2011. *Moving Mountains: Ethnicity and Livelihoods in Highland China, Vietnam, and Laos*. Vancouver: University of British Columbia Press.

Galudra, G. and M. Sirait. 2009. A discourse on Dutch colonial forest policy and science in Indonesia at the beginning of the 20th Century. *International Forestry Review* 11: 524–533.

Grove, R.H. 1995. *Green Imperialism: Colonial Expansion, Tropical Island Edens and the Origins of Environmentalism, 1600–1860*. Cambridge: Cambridge University Press.

Grove, R.H., V. Vinita Damodaran, and S. Sangwan. 1998. *Nature and the Orient: The Environmental History of South and Southeast Asia*. Delhi: Oxford University Press.

Hall, K.R. 2011. *History of Early Southeast Asia: Maritime Trade and Societal Development, 100–1500*. Lanham: Rowman & Littlefield.

Hazell, P.B.R. 2009. The Asian Green Revolution. IFPRI Discussion Paper 00911. Washington D.C.: International Food Policy Research Institute.

Hill, R.D. 2002. *Southeast Asia: People, Land and Economy*. Crows Nest, NSW: Allen & Unwin.

Le Billon, P. 2000. The political ecology of transition in Cambodia 1989–1999: war, peace and forest exploitation. *Development and Change* 31: 785–805.

Maffi, L. 2005. Linguistic, cultural and biological diversity? *Annual Review Anthropology* 29: 599–617.

McElwee, P.D. 2016. *Forests Are Gold: Trees, People, and Environmental Rule in Vietnam*. Seattle: University of Washington Press.

McGregor, A. 2008. *Southeast Asian Development*. Abingdon: Routledge.

Meijaard, E. 2003. Mammals of South-east Asian islands and their Late Pleistocene environments. *Journal of Biogeography* 30: 1245–1257.

Merten, J., A. Röll, T. Guillaume, A. Meijide, S. Tarigan, et al. 2016. Water scarcity and oil palm expansion: social views and environmental processes. *Ecology and Society* 21: 5.

Milne, S. and S. Mahanty. 2015. *Conservation and Development in Cambodia: Exploring Frontiers of Change in Nature, State and Society*. Abingdon: Routledge.

Molle, F., T. Foran, and M. Käkönen. 2009. *Contested Waterscapes in the Mekong Region: Hydropower, Livelihoods, and Governance*. London: Earthscan.

Morand, S., J.-P. Dujardin, R. Lefait-Rollin, and C. Apiwathnasorn. 2015. *Socio-ecological Dimensions of Infectious Diseases in Southeast Asia*. Singapore: Springer.

Morand, S., S. Jittapalapong, Y. Supputamongkol, M.T. Abdullah, and T.B. Huan. 2014a. Infectious diseases and their outbreaks in Asia-Pacific: biodiversity and its regulation loss matter. *PLoS One* 9: e90032.

Morand, S., K. Owers, and F. Bordes. 2014b. Biodiversity and emerging zoonoses. In A. Yamada, L.H. Kahn, B. Kaplan, T.P. Monath, J. Woodall, and L. Conti (eds.), *Confronting Emerging Zoonoses: The One Health Paradigm*. New York: Springer.

Myers, N., R.A. Mittermeier, C.G. Mittermeier, G.A. Da Fonseca, and J. Kent. 2000. Biodiversity hotspots for conservation priorities. *Nature* 403: 853–858.

Pye, O. 2005. *Khor Jor Kor: Forest Politics in Thailand*. Bangkok: White Lotus.

Pye, O. 2012. Editorial: changing socio-natures in South-East Asia. *Austrian Journal of South-East Asian Studies* 5: 198–207.

Ravi Rajan, S. 2006. *Modernizing Nature: Forestry and Imperial Eco-Development 1800–1950*. Oxford: Oxford University Press.

Rola, A.C. and I. Coxhead. 2005 Economic development and environmental management in the uplands of Southeast Asia: challenges for policy and institutional development. *Agricultural Economics* 32: 243–256.

Sathiamurthy, E. and H.K. Voris. 2006. Maps of Holocene sea level transgression and submerged lakes on the Sunda shelf. *Natural History Journal of Chulalongkorn University* Supplement 2: 1–43.

Schipper, J., J.S. Chanson, F. Chiozza, N.A. Cox, M. Hoffmann, et al. 2008. The status of the world's land and marine mammals: diversity, threat, and knowledge. *Science* 322: 225–230.

Shively, G. and T. Smith. 2015. Natural resources, the environment and economic development in Southeast Asia. In I. Coxhead (ed.), *Routledge Handbook of Southeast Asian Economics*. Abingdon: Routledge.

Sodhi, N.S., G. Acciaioli, M. Erb, and A.K.-J. Tan. 2008. *Biodiversity and Human Livelihoods in Protected Areas: Case Studies from the Malay Archipelago.* Cambridge: Cambridge University Press.

Sodhi, N.S., L.P. Koh, B.W. Brook, and P.K.L. Ng. 2004. Southeast Asian biodiversity: an impending disaster. *Trends in Ecology and Evolution* 19: 654–660.

Sodhi, N.S., M.R.C. Posa, T.M. Lee, D. Bickford, L.P. Koh, and B.W. Brook. 2010. The state and conservation of Southeast Asian biodiversity. *Biodiversity and Conservation* 19: 317–328.

Sunderland, T.C.H, J. Sayer, and M.-H. Hoang. 2013. *Evidence-based Conservation Lessons from the Lower Mekong.* Abingdon: Routledge.

Thomas, F. 1998. Écologie et gestion forestière dans l'Indochine française. *Revue Française d'Histoire d'Outre-Mer* 85: 59–86.

Tongzon, J.L. 2002. *The Economies of Southeast Asia, before and after the Crisis.* Cheltenham: Edward Elgar.

Vandergeest, P. and N.L. Peluso. 2006a. Empires of forestry: professional forestry and state power in Southeast Asia, Part 1. *Environment and History* 12: 31–64.

Vandergeest, P. and N.L. Peluso. 2006b. Empires of forestry: professional forestry and state power in Southeast Asia, Part 2. *Environment and History* 12: 359–393.

Voris, H.K. 2000. Maps of Pleistocene sea levels in Southeast Asia: shorelines, river systems and time durations. *Journal of Biogeography* 27: 153–167.

Vörösmarty, C.J., P.B. McIntyre, M.O. Gessner, D. Dudgeon, A. Prusevich, et al. 2010. Global threats to human water security and river biodiversity. *Nature* 467: 555–561.

Wilcove, D.S., X. Giam, D.P. Edwards, B. Fisher, and L.P. Koh. 2013. Navjot's nightmare revisited: logging, agriculture, and biodiversity in Southeast Asia. *Trends in Ecology and Evolution* 28: 531–540.

Woodruff, D.S. 2010. Biogeography and conservation in Southeast Asia: how 2.7 million years of repeated environmental fluctuations affect today's patterns and the future of the remaining refugial-phase biodiversity. *Biodiversity and Conservation* 19: 919–941.

UNEP-ICUN. n.d. World Database on Protected Areas. www.protectedplanet.net/c/world-database-on-protected-areas (accessed 7 March 2017).

Zierler, D. 2011. *The Invention of Ecocide: Agent Orange, Vietnam, and the Scientists who Changed the Way We Think about the Environment.* Athens, GA: University of Georgia Press.

Ziv, G., E. Baran, S. Nam, I. Rodriguez-Iturbe, and S.A. Levin. 2012. Trading-off fish biodiversity, food security, and hydropower in the Mekong River Basin. *Proceedings of the National Academy of Sciences of the United States of America* 109: 5609gs of.

2 The continental fossil record and the history of biodiversity in Southeast Asia

Julien Claude

Introduction

Understanding diversity can be undertaken by using inference from living forms (e.g. using DNA data to reconstruct phylogenetic trees), but the diversity of today offers just a small window on the historic succession of organisms. Although it may be sometimes scanty or just unexplored, the fossil record provides the only factual data on what was really present in a given area for a given period. The fossil record in Southeast Asia (SEA) started to be investigated during the first geological surveys of the region, especially at the end of the nineteenth century, when European geologists launched scientific expeditions. French geologists started to document continental fossil assemblages in Indochina in the late Paleozoic of Laos while Dutch geologists were the first to document late tertiary fossil continental assemblage in the former Dutch East Indies. Put into the geological context, the early paleontological discoveries soon helped to establish a stratigraphical frame for the region to provide geological maps gaining in stratigraphical precision, especially in continental sediments.

The fossil discoveries in Southeast Asia also had a rapid and important impact on our understanding of the history of life. For instance, soon after the publication of Darwin's *On the Origin of Species*, scientists started to investigate the fossil record to discover a missing link between apes and humans. Also, while it was initially welcomed with some skepticism by the scientific community (as was the discovery of Neanderthals), the discovery of *Homo erectus* (formerly *Pithecantropus erectus*) in the Solo River at Trinil, Java, quickly became a cornerstone for our understanding of human evolution and the first of the *Homo erectus* species.

Southeast Asia: a complex geological history

Paleontological expeditions accumulated more and more fossil discoveries, permitting the documentation not only of lineages but also of fossil assemblages in various places in Southeast Asia. The study of these assemblages can provide first an idea of the succession of faunas and floras in Southeast Asia, but more importantly, by comparing these assemblages to other fossil assemblages in the world, it can help us understand the place and the importance of Southeast Asia in the distribution, migration, and origination of plants and animals. Publications started to acknowledge the importance of fossils for understanding paleobiogeo-graphy as early as the 1960s. Indeed, Southeast Asia has a very complex geologi-cal history. This continental area is and has been peripheric in comparison to major continental masses, and its modern-day configuration resulted from the accretion of several microcontinents from the Paleozoic to the Cenozoic. Moreover, in the context of the shallow sea waters covering the Sunda shelf, sea level fluctuations necessarily had an influence in connecting and disconnecting land masses, favor-ing the migration of flora and fauna. Tectonic collision within Southeast Asia or closeby (e.g. the Indian/Asian plate collision) generated mountain chains and quickly modified the geomorphology and the hydrography of continental areas, also impacting the distribution and evolution of organisms. One should finally note that Southeast Asia has suffered both geological and astronomical cata-strophes: the 0.77 Ma impact resulting in the Australasian strewn-field and the eruption of the Toba super-volcano 77,000 years ago are both the subject of consider-able debate concerning extinctions and population declines in plant and animal species, including humans.

Questions regarding biodiversity today and the importance of the fossil record

Southeast Asia's biodiversity is extraordinarily rich and geographically well struc-tured today, sometimes with a high degree of endemism that is often directly linked to geological events. Whether this rich biodiversity results from the com-posite nature of Southeast Asia, its marginal position, or its long-term tropical climate is an open question. Whether that place plays a role as origination center or as biodiversity sanctuary is another. On the other hand, this biodiversity is particularly threatened by our activities today. Documenting the fossil record in this region is therefore absolutely necessary to understand the origin of living diversity, as well as its resilience to catastrophic events. It may also help us reach decisions about our conservation priorities.

The growing interest among politicians and the public in the origins of the natural world and prehistoric creatures has motivated the establishment of natural history and paleontological museums that serve both educational and recrea-tional functions in Southeast Asia. Although these are not equally developed among the various countries, they have contributed to an awareness that the paleontological record (and more generally the geological record) is, as well as a living being, part of our patrimonial heritage, so measures should be undertaken to protect it.

The remainder of this chapter will describe several key features of the conti-nental fossil record in Southeast Asia, following a stratigraphical order. Finally, there will be a short review of the educational and touristic importance of this fossil record, including an assessment of museum and geopark projects, as well as some recommendations concerning future directions.

Paleozoic

The paleogeographic and tectonic history of Southeast Asia is extremely complex, comprising several continental and tectonic subunits (more than twenty units in total), each of which has its own unique and complex history. The reader is advised to look at Metcalfe (1998, 2009, 2011) for detailed information on the Paleozoic to Mesozoic paleobiogeography of the region.

Before the emergence of land life, terranes that would eventually constitute Southeast Asia and China were mostly distributed on the northern margin of Gondwanaland, also made up of India, Australia, Antarctica, Africa and South America (400 million years ago). This northern margin was covered by a relatively shallow sea which surrounded big islands made of initial cratons. This portion of Gondwanaland would soon be partially and completely separated from main land mass by the effect of rifting and the birth of an ocean between a continental mass comprising the Indochinese block, the two Chinese blocks and the Tarim block (part of which would later become north-western China). Until the late Permian, these four blocks continued to drift northwards and eventually made contact with Pangea to form what would become Eurasia, with Siberia, Kazakhstan, and Mongolia. At the end of the Triassic, two other blocks, Sibumasu and Qiangtang, located on the southern margin of the Paleo-Tethys, followed the same trajectory and also drifted to the North. A new ocean, the Meso-tethys, opened in the south, while the Paleo-Tethys in the north disappeared.

During the Permian, the Sibumasu (western Thailand and eastern Myanmar) and Qiangtang (western Himalaya) formed the Cimmerian continent between the two seas, and an island ark system formed on the northern margin of the Paleo-Tethys. At the end of the Triassic, the Paleo-Tethys definitely closed, while the Sibumasu and Qiangtang block collided with the southern margin of mainland Asia. The same scenario played out again when the Lhasa and Western Burma plates detached from the south to reach the northern margin from the late Triassic to the late Cretaceous with the opening of the Ceno-Tethys (which became the Indian Ocean once India detached from Gondwana and finally collided with Asia during the Paleogene).

Located on the margins of large land masses or drifting in oceans, most of the paleontological history of Southeast Asia is documented by marine assemblages. However, some continental fossils, consisting mostly of plants and some invertebrates, have been discovered in different tectonical units from the Carboniferous to the Permian (300 million years ago). A few localities in north-eastern Thailand and peninsular Malaysia have yielded Carboniferous flora (and some arthropods) that can be compared with that of southern China. Permian flora is better documented: it is known in Laos, northern Vietnam, Thailand, south China, peninsular Malaysia and Sumatra (Fontaine, 2002). These floras document this very complex paleobiogeographic history. In the Carboniferous, the flora from the southern Chinese block, from Indochina (known from North-eastern Thailand and peninsular Malaysia) are different from other floras but very similar between each other (Laveine et al., 2003). They suggest that these blocks were

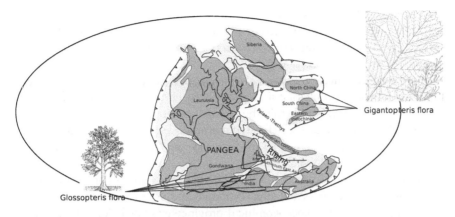

Figure 2.1 Paleobiogeographical reconstruction of land masses during the Permian. While India and eastern Southeast Asia (the Sibumasu block) harbored deciduous flora typical of high latitudes in the Southern Hemisphere, China and western Southeast Asia were covered by much more tropical and equatorial flora characterized by the presence of gigantopterids.

probably connected while they were drifting from southern Gondwana. The Permian flora from Vietnam, east and north-east Thailand, Sumatra and Laos can be defined as a particular flora (the Cathaysian *Gigantopteris* flora), while those from India and western Thailand are typically Gondwanian and found in colder climates (*Glossopteris* flora) (Xingxue and Xiuyuan, 1994). Indeed, while *Glossopteris* flora is deciduous, with growth rings, *Gigantopteris* flora indicates warmer, tropical conditions. This divergence in flora indicates that land masses occupied different positions and were not connected, and it aids our understanding of the continental drift of the Sibumasu block to the north during the Permian (Figure 2.1).

In northern Laos, in the area of Luang Prabang, a vertebrate fauna was discovered and described by Counillon (1896) and later Battail (2009). It is characterized by the presence of dicynodonts, a group of therapsids (mammalian reptiles) known throughout Pangea, which suggests that the Indochinese block was connected to the main continent by the end of the Permian.

Mesozoic

Mesozoic continental assemblages consisting of mollusks, vertebrates, and floras are particularly well documented in Thailand (Figure 2.2) from the Late Triassic to the end of the Lower Cretaceous (Buffetaut and Suteethorn, 1998; Buffetaut *et al.*, 2009). Some fossil vertebrates have been also described in the Lower Cretaceous of Laos (Hoffet, 1944) and a few have been reported from peninsular Malaya. Flora from the Mesozoic are known from Thailand, Myanmar, Laos, Singapore, Vietnam, and Sawarak (Colani, 1919; Boureau, 1950; Philippe *et al.*,

2004). Non-marine mollusks have been described in Thailand (Hayami, 1968; Kobayashi, 1968; Tumpeesuwan *et al.*, 2010) and in Laos (Hoffet, 1937), and a few arthropods have been reported in Thailand (Heggemann *et al.*, 1990) and an extremely rich micro-fauna and flora from the amber deposit of Kachin (northern Myanmar) (Grimaldi *et al.*, 2002).

The biogeographical affinities of the vertebrate assemblages have been described in detail in Buffetaut and Suteethorn (1998), Buffetaut *et al.* (2011), and Fernandez *et al.* (2009), while the paleobiogeographical importance of floras is discussed in Philippe *et al.* (2004) and that of mollusks in Sha (2010). Furthermore, palynological studies have helped to refine age and environment, especially in Thailand (Racey and Goodall, 2009).

The Triassic assemblage from the Huai Hin Lat formation in Thailand (Norian) shows affinities with European fauna (Buffetaut and Suteethorn, 1998), revealing that a large continental biogeographical province extended along the northern margin of the Meso-Tethys margin. In these ecosystems, turtles, phytosaurs, freshwater actinopterygians, lungfishes, and arthropods coexisted with ferns and sphenophytes. A similar flora was found in eastern Malaysia, in Sawarak, while a different floral assemblage was described on the island of Bintan in Indonesia (Jongmans, 1951). The differences were interpreted as resulting from taphonomical effects (Wade-Murphy and Van Konijnenburg van Cittert, 2008), and it seems that at this time the northern part of Borneo experienced similar conditions to mainland Southeast Asia, even though that continental unit split from Gondwana only at the end of late Triassic (Metcalfe, 2011). The vertebrates found in the Huai Hin Lat formation and the prosauropods and sauropod found in the Nam Phong formation (above the Hua Hin Lat formation) further demonstrate the connection of the Indochinese block to the main continent during the Triassic. In the Jurassic and Early Cretaceous, vertebrate remains from the Sibumasu and Indochinese blocks suggest a close relationship with the fauna in China and Japan (Buffetaut and Suteethorn, 1998; Buffetaut *et al.*, 2005). Patterns of provincialism are, however, evidenced (see Fernandez *et al.*, 2009; Tong *et al.*, 2009), especially during the span covered by the Sao Khua formation in Thailand. During the late Early Cretaceous, large faunal migrations are evidenced by the appearance of ceratopsian and ornithopod dinosaurs having affinities with Chinese forms (Buffetaut *et al.*, 2006). Freshwater sharks (*Thaiodus sp.*) found in the Lhasa block (now Tibet) and in the Indochinese blocks at the end of the late Early Cretaceous evidence that the Lhasa block drifted enough to the north to exchange faunal content with the main continental mass (Cappetta *et al.*, 1990).

The Late Cretaceous is documented only by the amber of Kachin in northern Myanmar and by a palynomorph found in the Mahasarakham formation in Thailand. The amber of Kachin displays an extremely rich insect, mite, and spider fauna, but also exhibits some of the first flowering plants, and some occurrences of rarely preserved animals, such as ticks and nematodes (Grimaldi *et al.*, 2002) as well as feather and skin fragments of birds and reptiles, respectively.

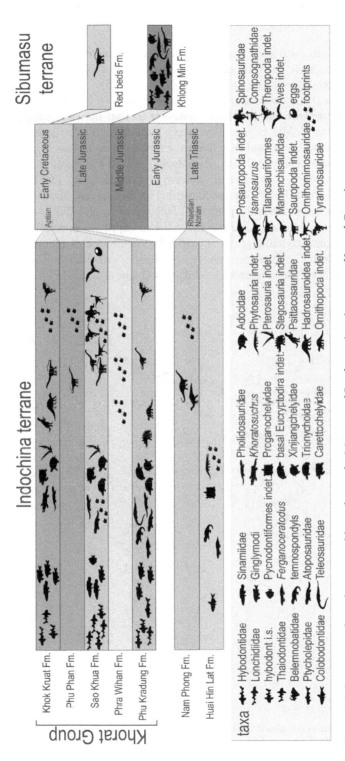

Figure 2.2 Succession of vertebrate assemblages during the Mesozoic of Southeast Asia (courtesy of Lionel Cavin)

Cenozoic

Few fossil continental assemblages are known or well described from the Paleogene of Southeast Asia, but they could potentially help to document the complex tectonic and eustatic history of the region (Figure 2.3).

Indeed, important tectonic events occurred between Australia and the Sunda shelf, and the collision with India certainly had an important impact on biodiversity. The Middle Eocene Pondaung formation in Myanmar is the oldest well-documented floral and vertebrate assemblage of Southeast Asia (Licht *et al.*, 2015; Pilgrim and Cotter, 1916). It comprises several mammals (primates, chalicotheres, creodont, artiodactyles, rodents, perissodactyles) (Tsubamoto *et al.*, 2005), reptiles (turtles, snakes, crocodiles) (Hutchison *et al.*, 2004). The mammal fauna shows some endemic patterns (Tsubamoto *et al.*, 2005) and resembles some contemporaneous Chinese fauna. The reptiles, especially the turtles, also show some endemic patterns, with the occurrence of pleurodiran species reflecting some Gondwanian influence (collision with India). The presence of dipterocarp flora also contribute to this view (Licht *et al.*, 2015).

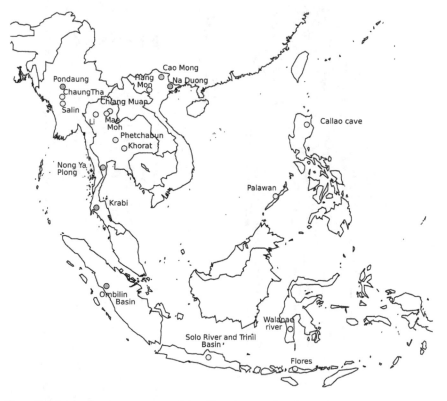

Figure 2.3 Map of major continental fossil localities in Southeast Asia

Note: Most Pleistocene karstic sites are not shown. Dark circle: Paleogene locality. Light circle: Neogene locality.

Stratigraphically, this fauna is followed by the recently described assemblage of Na Duong from the Late Eocene of Vietnam, for which a rich vertebrate fauna is described together with the flora, as well as the malacofauna living in an environment similar to the current-day swamp forest found in Indonesia (Böhme *et al.*, 2011, 2013). The lacustrine Sangkarewang formation in Sumatra (probably Late Eocene or Early Oligocene) has yielded one pelicantiform bird (Lambrecht, 1931) as well as fish fauna (more than fifteen species have been identified) (Sanders, 1934; Murray *et al.*, 2015). In Thailand, the Krabi Basin is dated close to the Eocene/Oligocene boundary. It displays a rich tropical fauna with plant rests, mollusks, fish, turtles, snakes, crocodiles, and mammals that typically fit well with the marginal position of Southeast Asia in Eurasia (Songtham and Watanasak, 1999; Claude *et al.*, 2007). A well-documented Oligocene vertebrate assemblage is known in Phetchaburi province, Thailand, and consists of mammals, fish, and freshwater turtles (Marivaux *et al.*, 2004; Tong *et al.*, 2006); a potential Oligocene assemblage may also occur in northern Vietnam (Rin Chua formation above the Na Duong formation), although the fish assemblage found there has not been studied in detail (Böhme *et al.*, 2011).

Miocene localities are more numerous, especially in northern Thailand, in small pull-apart basins where lacustrine conditions allowed the preservation of diverse vertebrate fauna and flora (Chaodumrong and Chaimanee, 2002). Similar systems are also known in Vietnam, close to the border with Laos, in the Hang Mon Basin that yielded a mammal assemblage (Ginsburg *et al.*, 1992; Covert *et al.*, 2001; Böhme *et al.*, 2011). The paleo-Irrawady and paleo-Moon river system, spanning from the Miocene to the Pleistocene, also yielded important vertebrate assemblages, some of which have been interpreted in regard to the changing hydrographic system (Claude *et al.*, 2011). Other potentially interesting continental Neogene basins occur in Malaysia and Laos, but their fossil content has not been studied.

Numerous cave deposits from the Plio-Pleistocene have yielded abundant fossil material (Chaimanee, 1998; Zeitoun *et al.*, 2010), and one must remember that several Pleistocene geological localities on Indonesia islands have yielded vertebrate remains, including those of *Homo erectus* (Dubois, 1894) and more recently *Homo floresiensis*. Recent discoveries in the Philippines archipelago have shown that the region is important for understanding human evolution (Détroit *et al.*, 2004).

Concluding remarks

Although incomplete, the fossil record in Southeast Asia is starting to be well documented and it allows better understanding of the relationship between geography, geological events, and the evolution of biodiversity. There are still a lot of uncertainties, however. In particular, the Late Cretaceous/Paleocene is not well documented, and Neogene fossil evidence is absent or not abundant in the Philippines and Indonesia. More research could shed light on the establishment and fluctuation of biogeographical faunal boundaries, such as the Wallace Line.

Figure 2.4 The Siridhorn dinosaur museum, opened in 2005, which welcomes thousands of visitors every year, Phu Kum Kao, Kalasin, Thailand

Fossils, especially but not solely those of dinosaurs and hominids, are starting to have touristic and educational importance in Southeast Asia, with museums displaying the geological patrimony to a large audience (Boonchai *et al.*, 2009; see Figure 2.4). Although some fossils are still sold in markets, the scientific importance of this heritage has started to be recognized, with the introduction of legislation and public surveys in Southeast Asia. It is mandatory for the future that international programs are launched for research and higher education purposes. The foundation of the Paleontological Research and Education Center in Mahasarakham, Thailand, some seventeen years ago has already resulted in the education of a new generation of young paleontologists.

References

Battail, B. 2009. Late Permian dicynodont fauna from Laos. In E. Buffetaut, G. Cuny, J. Le Loeuff, and V. Suteethorn (eds.), *Late Palaeozoic and Mesozoic Continental Ecosystems in SE Asia*. London: Geological Society: 33–40.

Böhme, M., M. Aigstorfer, P.-O. Antoine, E. Appel, G. Métais, L.T. Phuc, S. Schneider, F. Setzer, R. Tappert, D.N. Tran, D. Uhl, and J. Prieto. 2013. Na Duong (northern Vietnam) – an exceptional window into Eocene ecosystems from South-East Asia. *Zitteliana* A 53: 120–167.

Böhme, M., J. Prieto, S. Schneider, N.V. Hung, D.D. Quang, and D.N. Tran. 2011. The Cenozoic on-shore basins of northern Vietnam: biostratigraphy, vertebrate and invertebrate faunas. *Journal of Asian Earth Sciences* 40: 672–687.

Boonchai, N., J.P. Grote, and P. Jintasakul. 2009. Paleontological parks and museums and prominent fossil sites in Thailand and their importance in the conservation of fossils. *Carnets de Géologie/Notebooks on Geology* 3: 75–95.

Boureau, E. 1950. Contribution à l'étude paléoxylologique de l'Indochine. 1: Présence du *Xenoxylon latiporosum* (Cramer) Gothan dans le Lias du Centre-Annam. *Bulletin du Service Géologique de l'Indochine* 29: 1–16.

Brown, P., T. Sutikna, M.J. Morwood, R.P. Soejono, et al. 2004. A new small-bodied hominin from the Late Pleistocene of Flores, Indonesia. *Nature* 431: 1055–1061.

Buffetaut, E., G. Cuny, J. Le Loeuff, and V. Suteethorn. 2009. Late Palaeozoic and Mesozoic ecosystems in SE Asia. In E. Buffetaut, G. Cuny, J. Le Loeuff, and V. Suteethorn (eds.), *Late Palaeozoic and Mesozoic Continental Ecosystems in SE Asia*. London: Geological Society: 315.

Buffetaut, E. and V. Suteethorn. 1998. The biogeographical significance of the Mesozoic vertebrates from Thailand. In R. Halland and J.D. Holloway (eds.), *Biogeography and Geological Evolution of SE Asia*, Amsterdam: Backhuys: 83–90.

Buffetaut, E. and V. Suteethorn. 2011. A new iguanodontian dinosaur from the Khok Kruat formation (Early Cretaceous, Aptian) of northeastern Thailand. *Annales de Paleontologie* 97: 51–62.

Buffetaut, E., V. Suteethorn, and H. Tong. 2006. Dinosaur assemblages from Thailand: a comparison with Chinese faunas. In J.C. Lü, Y. Kobayashi, D. Huang, and Y.N. Lee (eds.), *Papers from the 2005 Heyuan International Dinosaur Symposium*. Beijing: Geological Publishing House: 19–37.

Buffetaut, E., V. Suteethorn, H. Tong, and A. Kosir. 2005. First dinosaur from the Shan-Thai block of SE Asia: a Jurassic sauropod from the southern peninsula of Thailand. *Journal of the Geological Society* 162: 481–484.

Cappetta, H., E. Buffetaut, and V. Suteethorn. 1990. A new hybodont shark from the Lower Cretaceous of Thailand. *Neues Jahrbuch für Geologie und Paläontologie, Monatshefte* 11: 659–666.

Chaimanee, Y. 1998. Plio-Pleistocene rodents of Thailand. *Thai Studies in Biodiversity* 3: 1–303.

Chaodumrong, P. and Y. Chaimanee. 2002. Tertiary sedimentary basins in Thailand. In N. Mantajit (ed.), *Proceedings of the Symposium on Geology of Thailand*. Bangkok: Department of Mineral Resources: 156–169.

Claude, J., W. Naksri, N. Boonchai, E. Buffetaut, J. Duangkrayom, C. Laojumpon, P. Jintasakul, K. Lauprasert, J. Martin, V. Suteethorn, and H. Tong. 2011. Neogene reptiles of northeastern Thailand and their paleogeographical significance. *Annales de Paléontologie* 97: 113131.

Claude, J., V. Suteethorn, and H. Tong. 2007. Turtles from the Late Eocene Early Oligocene of the Krabi Basin (Thailand). *Bulletin de la Societe Geologique de France* 178: 305316.

Colani, M. 1919. Sur quelques *Araucarioxylon* indochinois. *Bulletin du Service Géologique de l'Indochine* 6: 5–15.

Counillon, H. 1896. Documents pour servir à l'étude géologique des environs de Luang Prabang (Cochinchine). *Comptes Rendus de l'Académie des Sciences Paris* 123: 1330–1333.

Covert, H. H., M.V. Hamrick, T. Dzanh, and K.C. McKinney. 2001. Fossil mammals from the Late Miocene of Vietnam. *Journal of Vertebrate Paleontology* 21: 633–636.

Détroit, F., E. Dizon, C. Falguères, S. Hameau, W. Ronquillo, and F. Sémah. 2004. Upper Pleistocene *Homo sapiens* from the Tabon cave (Palawan, the Philippines): descriptions and dating of new discoveries. *Comptes Rendus Palevol* 3: 705–712.

Dubois, E. 1894. *Pithecanthropus erectus, eine menschenahnliche Ubergangsform aus Java.* Batavia: Landersdruckerei.

Fernandez, V., J. Claude, G. Escarguel, E. Buffetaut, and V. Suteethorn. 2009. Biogeographical affinities of Jurassic and Cretaceous continental vertebrate assemblages from SE Asia. In E. Buffetaut, G. Cuny, J. Le Loeuff, and V. Suteethorn (eds.), *Late Palaeozoic and Mesozoic Continental Ecosystems in SE Asia*. London: Geological Society: 285–300.

Fontaine, H. 2002. Permian of Southeast Asia: an overview. *Journal of Asian Earth Science* 20: 567–588.

Ginsburg, L.L., K.V. Minh, K.Q. Nam, and D.V. Thuan. 1992. Premières découvertes de vertébrés continentaux dans le Néogène du Nord du Vietnam. *Comptes Rendus de l'Académie des Sciences Paris* 314: 627–630.

Grimaldi, D.A., M.S. Engel, and P.C. Nascimbene. 2002. Fossiliferous Cretaceous amber from Myanmar (Burma): its rediscovery, biotic diversity, and paleontological significance. *American Museum Novitates* 3361: 1–72.

Hayami, I. 1968. Some non-marine bivalves from the Mesozoic Khorat Group of Thailand. *Geology and Paleontology of Southeast Asia* 4: 100–107.

Head, J.J., P.A. Holroyd, J.H. Hutchison, and R.L. Ciochon. 2005. First report of snakes (Serpentes) from the Late Middle Eocene Pondaung formation, Myanmar. *Journal of Vertebrate Paleontology* 25: 246–250.

Heggemann, H., R. Kohring, and T. Schlutert. 1990. Fossil plants and arthropods from the Phra Wihan formation, presumably Middle Jurassic of northern Thailand. *Alcheringa* 14: 311–316.

Hoffet, J.H. 1937. Les lamellibranches saumatres du Senonien de Muong Phalane (Bas-Laos). *Bulletin du Service Géologique de l' Indochine* 24: 4–25.

Hoffet, J.H. 1944. Description des ossements les plus cractéristiques appartenant à des Avipelviens du Sénonien du Bas-Laos. *Bulletin du Conseil des Recherches Scientifiques de l'Indochine*: 179–186.

Hutchison, J.H., P.A. Holroyd, and R.L. Ciochon. 2004. A preliminary report on Southeast Asia's oldest Cenozoic turtle fauna from the late Middle Eocene Pondaung formation, Myanmar. *Asiatic Herpetological Research* 10: 38–52.

Jongmans, W.J. 1951. Fossil plants of the island of Bintan (with a contribution by J.W.H. Adam). *Koninklijke Nederlandse Akademie van Wetenschappen* B54: 183–190.

Kobayashi, T. 1968. The Cretaceous non-marine pelecypod the Nam Phung Dam Site in the northeastern part of the Khorat Plateau, Thailand with a note on the Trigonioididae. *Geology and Paleontology of Southeast Asia* 4: 109–138.

Lambrecht, K. 1931. *Protoplotus beauforti* n.g. n.sp., ein Schlangen halsvogel aus dem Tertiar von W. Sumatra. *Wetensch. Meded. Dienst Mijnbouw Nederl. Indie* 17: 15–24.

Laveine, J.P., B. Ratanasthien, and S. Sithirach. 2003. The Carboniferous flora of northeastern Thailand. *Revue de Paléobiologie Genève* 22(2): 761–797.

Licht, A., A. Boura, D. de Franceschi, T. Utescher, S. Sein, and J.J. Jaeger. 2015. Late Middle Eocene fossil wood of Myanmar: implications for the landscape and the climate of the Eocene Bengal Bay. *Review of Palaeobotany and Palynology* 216: 44–54.

Marivaux, L., Y. Chaimanee, C. Yamee, P. Srisuk, and J.-J. Jaeger. 2004. Discovery of *Fallomus ladakhensis* Nanda & Sahni, 1998 (Mammalia, Rodentia, Diatomyidae) in the lignites of Nong Ya Plong (Phetchaburi Province, Thailand): systematic, biochronological and paleoenvironmental implications. *Geodiversitas* 26(3): 493–507.

Metcalfe, I. 1998. Palaeozoic and Mesozoic geological evolution of the SE Asian region: multidisciplinary constraints and implications for biogeography. In R. Hall and J.D. Holloway (eds.), *Biogeography and Geological Evolution of SE Asia*. Amsterdam: Backhuys: 25–41.

Metcalfe, I. 2009. Late Palaeozoic and Mesozoic tectonic and palaeogeographic evolution of SE Asia. In E. Buffetaut, G. Cuny, J. Le Loeuff, and V. Suteethorn (eds.), *Late Palaeozoic and Mesozoic Continental Ecosystems in SE Asia*. London: Geological Society: 7–22.

Metcalfe, I. 2011. Palaeozoic–Mesozoic history of SE Asia. In R. Hall, M.A. Cottam, and M.E.J. Wilson (eds.), *The Southeast Asian Gateway: History and Tectonics of Australia–Asia Collision*. London: Geological Society: 7–35.

Murray, A.M., Y. Zaim, Y. Rizal, Y. Aswan, G.F. Gunnel, and R.L. Ciochon. 2015. A fossil gourami (Teleostei, Anabantoidei) from probable Eocene deposits of the Ombilin Basin, Sumatra, Indonesia. *Journal of Vertebrate Palaeontology*: E906444.

Philippe, M., V. Suteethorn, P. Lutat, E. Buffetaut, L. Cavin, G. Cuny, and G. Barale. 2004. Stratigraphical and palaeobiogeographical significance of fossil woods from the Mesozoic Khorat Group of Thailand. *Geological Magazine* 141: 319–328.

Pilgrim, G.E. and G.P. Cotter. 1916. Some newly discovered Eocene mammals from Burma. *Records of the Geological Survey of India* 47: 42–77.

Racey, A. and J.G.S. Goodall 2009. Palynology and stratigraphy of the Mesozoic Khorat Group Red Bed Sequences from Thailand. In E. Buffetaut, G. Cuny, J. Le Loeuff, and V. Suteethorn (eds.), *Late Palaeozoic and Mesozoic Continental Ecosystems in SE Asia*. London: Geological Society: 41–68.

Sanders, M. 1934. Die Fossilen Fische der Alttertiaren Susswasserablagerungen aus Mittel-Sumatra. *Verhandelingen van het Geologisch-Mijnbouwkundig Genootschap voor Nederland en Koloni'n*. *Geologische Series* 11: 1–144.

Sha, J.G. 2010. Historical distribution patterns of trigonioidids (non-marine Cretaceous bivalves) in Asia and their palaeogeographic significance. *Proceedings of the Royal Society London B*: 277–283.

Songtham, W. and M. Watanasak. 1999. Palynology, age, and paleoenvironment of Krabi Basin, southern Thailand. In B. Ratanasthien and S.L. Reib (eds.), *Proceedings of the International Symposium on Shallow Tethys (ST) 5, Chiang Mai*: 426–439.

Tong, H., J. Claude, E. Buffetaut, V. Suteethorn, W. Naksri, and S. Chitsing. 2006. Fossil turtles of Thailand: an updated review. In J.C. Lü, Y. Kobayashi, D. Huang, and Y.N. Lee (eds.), *Papers from the 2005 Heyuan International Dinosaur Symposium*. Beijing: Geological Publishing House: 183–194.

Tong, H., J. Claude, V. Suteethorn, W. Naksri, and E. Buffetaut. 2009. Turtle assemblages of the Khorat Group (Late Jurassic–Early Cretaceous) of north-eastern Thailand and their palaeobiogeographical significance. In E. Buffetaut, G. Cuny, J. Le Loeuff, and V. Suteethorn (eds.), *Late Palaeozoic and Mesozoic Continental Ecosystems in SE Asia*. London: Geological Society: 141–152.

Tsubamoto, T., N. Egi, M. Takai, C. Sein, and M. Maung. 2005. Middle Eocene ungulate mammals from Myanmar: a review with description of new specimens. *Acta Palaeontologica Polonica* 50: 117–138.

Tumpeesuwan, S., Y. Sato, and S. Nakhapadungrat. 2010. A new species of *Pseudohyria* (*Matsumotoina*) (Bivalvia: Trigonioidoidea) from the Early Cretaceous Sao Khua formation, Khorat Group, northeastern Thailand. *Tropical Natural History* 10: 93–106.

Wade-Murphy, J. and J.H.A. Van Konijnenburg van Cittert. 2008. A revision of the Late Triassic Bintan flora from the Riau Archipelago (Indonesia). *Scripta Geologica* 136: 73–105.

Xingxue, L. and W. Xiuyuan. 1994. The Cathaysian and Gondwana floras: their contribution to determining the boundary between eastern Gondwana and Laurasia. *Journal of Southeast Asian Earth Sciences* 9(4): 309–317.

Zeitoun, V., A. Lenoble, F. Laudet, J. Thompson, W.J. Rink, J.-B. Mallye, and W. Chinnawut. 2010. The Cave of the Monk (Ban Fa Suai, Chiang Dao wildlife sanctuary, northern Thailand). *Quaternary International* 220: 160–173.

3 Traditional knowledge and biocultural diversity

Daniel Robinson

Introduction

This chapter acknowledges the interrelations between humans and biodiversity in Southeast Asia throughout history to the present day. Humans are entirely dependent on nature, specifically biodiversity, for a number of utilitarian purposes: foods, fibers, medicines, building materials, biotechnologies, and other things. We now know that we rely on biodiversity for important ecosystem services, such as nutrient cycling, carbon sequestration and crop pollination. Over centuries we have also developed cultural and spiritual connections with plants, animals, ecosystems, and natural sites which may have special significance for different cultural groups. This chapter specifically focuses on the traditional knowledge of biodiversity in Southeast Asia – the way humans have understood, utilized, and adapted species, plant varieties, and animals over time. It also focuses on the 'biocultural diversity' of the region, given that there are many nations and cultural and linguistic groups that have their own unique connections to biodiversity.

To start with it is worth examining some definitions. In the late 1980s and early 1990s a considerable amount of academic anthropological literature emerged on 'local', 'traditional', or 'indigenous' knowledge, particularly from postcolonial perspectives and settings (e.g. Geertz, 1983). In 1992 Johnson used the term 'traditional ecological knowledge,' explaining it as:

> A body of knowledge built by a group of people through generations living in close contact with nature. It includes a system of classification, a set of empirical observations about the local environment, and a system of self-management that governs resource use.
>
> (Johnson, 1992: 3–4)

Since that time it has become more concretely recognized by academics and policy-makers that indigenous or local knowledge makes an important contribution to the management of resources, including the use and conservation of biological resources. For example, 'traditional knowledge' is not defined by the 1992 Convention on Biological Diversity (CBD). However, in Article 8(j), it requires each party state to commit:

Subject to its national legislation, [to] respect, preserve and maintain knowledge, innovations and practices of indigenous and local communities embodying traditional lifestyles relevant for the conservation and sustainable use of biological diversity and promote their wider application with the approval and involvement of the holders of such knowledge, innovations and practices and encourage the equitable sharing of the benefits arising from the utilization of such knowledge, innovations and practices.

This article in the CBD is important because it formally acknowledges the role played by indigenous and local communities in the conservation and sustainable use of biological diversity.

All countries in Southeast Asia are party to the CBD and thus encouraged to 'respect, preserve and maintain' such knowledge. Countries such as Thailand have also opted to include mention of local community knowledge and rights relating to nature in their constitutions. Notably, the terms 'local knowledge' and 'traditional knowledge' are prevalently used in Southeast Asia (e.g. *phumpanyaa tongtin* – meaning local wisdom/knowledge – in Thailand), rather than 'indigenous knowledge', given the ambiguity over 'indigenous' identity in the region (i.e. does it refer to dominant and/or minority populations?). However, some organizations, such as the Asian Indigenous Peoples Pact (AIPP), have sought to advocate deliberately in terms of 'indigenous' knowledge and rights (particularly for ethnic minority groups), given the political momentum the indigenous rights movement has generated in recent years.

Indigenous or traditional knowledge of biodiversity can be thought of in discrete contexts, such as agricultural or medicinal knowledge, or in terms of specific biomes, mountain ecosystems, forests, or coastal and/or marine ecosystems, all of which are relevant in Southeast Asia. For the sake of brevity, this chapter will focus on agricultural and medicinal biodiversity knowledge, as well as forest ecosystems. Having said that, there is some risk in narrowly defining 'environmental knowledge' by anthropocentric use or ecosystem type because of the intrinsic value that is often placed on nature by indigenous and local communities in Southeast Asia (and elsewhere).

This brings us to our next definition: biocultural diversity:

Biocultural diversity comprises the diversity of life in all of its manifestations – biological, cultural, and linguistic – which are inter-related (and likely co-evolved) within a complex socio-ecological adaptive system.

(Maffi and Woodley, 2010: 5)

Maffi and Woodley (2010) explain further that the diversity of life (or nature) is made up not only of the diversity of plants and animal species, habitats, and ecosystems, but also the diversity of human cultures and languages. Indeed, other authors have gone to great lengths to explain the interlinkages between specific cultures and biodiversity (e.g. Posey, 1999). It is thought that linguistic and cultural diversity has important ramifications for biodiversity, because traditional languages, folklore, dance, and spiritual beliefs form a significant part of the reason why people choose to conserve biodiversity or to use it in a sustainable way. Global studies indicate that the world's most biodiverse regions and hotspots contain considerable linguistic diversity, accounting for 70 percent of all languages on earth (Gorenflo *et al.*, 2012). The biological and linguistic diversity of these areas is being simultaneously threatened. Biologists estimate annual loss of species at 1,000 times or more than historic rates, and linguists predict that 50–90 percent of the world's languages will disappear by the end of this century (Gorenflo *et al.*, 2012).

In this context, indigenous struggles to assert a range of cultural, land, and environmental rights globally have led to the emergence of 'biocultural rights' (Bavikatte, 2014). Indeed, the movement towards biocultural rights encourages the recognition of personhood in property, land, resources, objects, and even knowledge – that people often have a personal or cultural interest in these objects and knowledge (Bavikatte and Robinson, 2016). Traditional ways of life are often counterposed to 'industrialized' society and the impacts of 'globalization', which have a 'materializing effect' – turning understandings of nature in holistic forms into nature as resource and/or commodity. The reality for most indigenous and local communities in Southeast Asia is not so black and white – their lives are often a complex mix of traditional and modern. Indeed, environmental knowledge is heavily politicized in the region. For example, there are prominent national politics and academic discussions of the instrumentalist formulation of knowledge as 'local' and 'traditional' to counterpose the potential local impacts of state regulations and commercialization in countries like Thailand (Forsyth and Walker, 2008; Hirsch, 1996).

To explore the local contexts of traditional knowledge and these biocultural interconnections further, this chapter will now turn to specific examples.

Traditional agricultural knowledge

As Pfeiffer *et al.* (2006: 610) note, 'within southeast Asia, the evolution of rice-related cultural diversity – agronomic, culinary, medicinal, social, and spiritual practices involving rice – has both paralleled and influenced the development of rice genetic diversity in a two-way process'. Rice is often bred to meet specific social demands and environmental conditions, and in turn the social and cultural norms, practices, and rituals often encourage the breeding of specific landrace rice varieties.

Pfeiffer *et al.* (2006) examine the Tabo community, known ethnically and linguistically as Kemp Manggarai, of Flores Island in East Nusa Tenggara province of Indonesia. The Tado farmers previously relied on traditional landraces originating from their region as well as from Sumbawa, Java, and Sulawesi. They practiced a swidden rotational form of agriculture known as *lingko*. However, this has diminished with the widespread governmental promotion of hybrid crop lines since the 1960s (Pfeiffer *et al.*, 2006). This is a story that can be heard in many other parts of Southeast Asia. Often the farmer varieties have been bred and adapted to certain environmental conditions, and provide an important source of food security for the community. At the same time, the farmer varieties and cropping patterns were often linked to stories, folklore, or rituals, so the introduction of new crops and cropping systems can disrupt this interrelationship. Whilst there have been benefits from high-yield varieties in terms of overall increases in production, there have also been many adverse impacts, including farmer debt, crop failures due to adaptation problems, increased irrigation and use of fertilizers and pesticides, amongst other things (see Tilman *et al.*, 2002). As Pfeiffer *et al.* (2006) note, there are few Manggarai residents under the age of forty who can describe ancestral cultivation rituals or sing the songs associated with traditional upland rice landraces, and also few farming households with more than half a dozen rice varieties on their farms since the introduction of the hybrid crop lines. These authors note that some local landraces, expressing distinct genotypes, have now become locally extinct. This represents a loss of genetic diversity that might prove important for future plant breeding and food security, including resistance to pests.

Similarly, I conducted fieldwork in the province of Roi Et in northeast Thailand (Issan) in 2005 with the NGO BioThai and members of the Alternative Agriculture Network of Thailand (see Robinson, 2011). We attended a seed exchange fair where plant seeds (predominantly rice) were exchanged between local farmers and the planting and adaptation of local landraces was encouraged. Many of the rice varieties displayed (there were more than twenty) were locally adapted to specific environmental conditions on the often arid plateau of the northeast region, or bred to have specific traits such as distinct flavors, yields, or resistances (see Figure 3.1). Rituals were also explained, including the use of several varieties of rice for specific rituals: for example, Khao Dam (or black rice) is highly revered by the local Issan people. Farmers demonstrated the manual selection of rice seed for replanting and genetic improvement of their crops. They also discussed the use of some varieties for specific ceremonial use, such as in weddings and coming-of-age ceremonies, and sang songs at the seed fair about their agricultural heritage.

Nathalang (2004) and Kriengkraipetch (2004) explain much about Thai religious (animist/indigenous and Buddhist) as well as folk beliefs about rice and attitudes towards nature more generally. For example, Thai people and ethnic Tai found in neighboring Southeast Asian countries traditionally (and often to this day) believe in a rice goddess (often described as khwan khao, Mae khwan khao, or Mae Phosop), who is the spirit protector of the crop (Nathalang, 2004). More broadly, Kriengkraipetch (2004: 176) describes a number of stories, songs, and riddles which 'form a child's attitude toward nature as part of his life, since that relationship between man, animals and plants is that of cohabitants'. The farmers interviewed at the seed fair in Roi Et noted that similar external factors to the above case had both influenced and diminished cultural associations with rice varieties (e.g. ceremonies and rituals) and been drivers which impacted upon local landrace extinctions in the Issan region.

It is also worth considering the impact that traditional knowledge has had on mainstream crop varieties. For example, the Jasmine rice variety (known as Khao

Figure 3.1 An array of local rice varieties grown by farmers in the Issan region, used as food crops and also in some cases for ceremonies (D. Robinson, 2005)

Dok Mali 105), which had its origins in Chonburi province (central region of Thailand), was reputedly transferred by farmers/breeders and grown in Chachoensao province, then collected and sent to the International Rice Research Institute (IRRI) in the Philippines, before being spread throughout the Issan region, where it now thrives. This variety has therefore transferred through various knowledge domains to the point where interviewees in Thailand noted that 'its origins seem almost forgotten' (interview, Songkhran, 2006). The contributions of Issan farmers to the development of this and related varieties (a number of varieties are described under the generic umbrella of 'Jasmine' but Khao Dok Mali 105 is the most prevalent) should not be neglected either.

The attempt to genetically modify and trademark and/or patent rice varieties such as Khao Dok Mali 105 by US researchers has led to significant public outcry in Thailand (see Robinson, 2010). While the complaints about the transfer of Jasmine rice germ plasm to the US were made generally in the name of Thailand's sovereignty over biological resources, there was a more complex underlying politics of ownership between origin, source, and developers, each with their own knowledge domains and cultural associations with rice. These claims were often phrased in terms of distinctly 'national heritage' claims to Jasmine rice due to the agricultural history of the country, but also due to the cultural and customary norms associated with the importance of rice in Thailand, described above.

Again relating to the CBD, there are also global rules for 'access and benefit-sharing' (ABS) relating to research into and development of 'genetic resources' and associated traditional knowledge. The CBD rules came about at a time when global negotiations towards the World Trade Organization (WTO) were under way and there was significant political pressure to include 'intellectual property' protections within the 1994 suite of WTO agreements. This influenced the CBD negotiations such that since 1992 there has been a complex and layered set of rules of ownership associated with genetic resources and traditional knowledge, including the allowance of intellectual property rights on plant varieties, microbes, and extracts from plants and animals. However, some have argued that the CBD is contradictory in assigning state sovereign rights over biodiversity and acknowledging private intellectual property rights, whilst also acknowledging the role and knowledge of indigenous peoples and local communities relating to biodiversity (see Robinson, 2010). The 2010 'Nagoya Protocol on Access to Genetic Resources and the Fair and Equitable Sharing of Benefits Arising from Their Utilization to the CBD' has been one way of resolving some of these issues (see Robinson, 2015 for a detailed overview and case studies, including two from Thailand relating to cosmetic/medicinal plants). This stipulates that prior informed consent (PIC) is required for access to genetic resources and, where relevant, to associated knowledge. It also stipulates that mutually agreed terms (MAT) must be reached between the resource provider and the researchers, such that there is a benefit sharing agreement. The Nagoya Protocol entered into force in 2014 and there were 85 ratifications at the time of writing, including Cambodia, Laos, and Vietnam.

At the national level, governments are currently developing these ABS laws. It is yet to be seen if they will include rules for PIC and MAT of 'traditional knowledge' holders (indigenous peoples and local communities) as 'providers' of genetic resources and associated knowledge in those Southeast Asian countries that are now party to the Protocol. Globally, there has been significant concern that traditional knowledge holders will be left out of these processes and/or that the whole process of commodification of the 'resources' and knowledge through patents ignores the cultural and spiritual significance of some plant and animal species (see Robinson, 2010, 2015).

Traditional medicinal knowledge

It is not too far a step now to consider traditional medicinal knowledge, because it is often the case in Southeast Asia that certain foods are also believed – or known – to have medicinal properties (e.g. see Salguero, 2003). The use of medicinal herbs and remedies has a long history and is well documented. For example, Santasombat (2003a) details several ethnographic studies of local medical systems amongst various ethnic groups in northern Thailand, noting the importance of beliefs and practices, such as belief in spirits (*phii*). Anderson (1993) and Brun and Schumacher (1994) provide two of the more detailed ethnobotanical investigations of the many ethnic minority groups in northern Thailand, which no doubt extend beyond national borders. They include herbal descriptions of treatments for malaria, broken bones, bites and stings, and venereal diseases.

During interviews I conducted in 2005, 2006, and 2009 in Chiang Mai province in northern Thailand (Samoeng and Jom Thong districts), particularly with Karen and Hmong healers, I witnessed the continued use of many of these herbs, which are found in local village gardens or surrounding forest areas. This fieldwork documented an animist-Buddhist conservation ethic, which was often linked to indigenous spiritual beliefs for both the Karen and the Hmong healers (Robinson, 2013). The medicinal herbs were often used sparingly by traditional healers, and sold in only limited quantities. There were also very often rituals required for acknowledgement of the spirits of the plants and requests for donations to be made to both the healer and the spirits. Healers in both Karen and Hmong villages noted that despite the introduction of subsidized modern medicine and local clinics, it was still common for people to use traditional medicine either in place of or in combination with modern treatments. However, in most villages it was clear that traditional medicinal knowledge was being lost because younger generations were not interested in becoming healers. This was due to a number of factors, including concerns about the customary and ritualistic obligations imposed on traditional healers (Robinson, 2013).

Whilst it was noted that traditional healers were often involved in the conservation of medicinal herbs, conservation problems have arisen due to overuse. In Robinson and Kuanpoth (2009), we observed the potential for the medicinal herbal remedies market to explode. The 'herbal' remedy of kwao krua (*Pueraria mirifica*) has been used by traditional healers in northern and northeast Thailand since at least the 1930s, and possibly much earlier. There are different varieties of this slow-growing vine, with different treatments. For example, white kwao krua contains phyto-oestrogens and when formulated into a paste it is used by women to improve the appearance of their skin, potentially by whitening it, and even reputedly to enlarge breasts. Following a number of patent applications by researchers in Thailand and other countries, as well as concerted marketing by a number of companies, demand for creams and pastes containing this medicinal plant have greatly increased. As a result, wild stocks of the vine in forest areas have been depleted by opportunists. This has led to state-imposed restrictions under regulations of the Act on Protection and Promotion of Thai Traditional Medicinal Intelligence (1999). Commercial collection of the vine in forest areas is now restricted, but local traditional use is permitted (Robinson and Kuanpoth, 2009).

This is unfortunately becoming a fairly common occurrence in Southeast Asia. For example, Felbab-Brown (2013) explores the illegal trade in wildlife in Southeast Asia and its links to East Asian markets, noting that medicinal use is a major reason for this trade. A recent IUCN Red List shows that 10 percent of snakes endemic to China and Southeast Asia are now threatened with extinction. Notably, snakes are often used in traditional medicines and anti-venom serum, sometimes as food, and also as a source of income from the sale of their skins. IUCN (2012) notes that 'nearly 43 percent of the endemic snake species in Southeast Asia in the Endangered and Vulnerable categories are threatened by unsustainable use'. Recent Red List updates also include a number of Southeast Asian plants which are used for food and/or medicine. For example, the Tsao-ko Cardamom (Amomum tsao-ko) species is listed as 'Near Threatened' because its edible fruits have been over-harvested for trading (IUCN, 2012).

Thus, while local traditional uses may have emphasized limited or restrained use and biocultural understandings of biodiversity may have encouraged respect and cohabitation with the species that are utilized, commercial-scale exploitation, population growth, and regional or global markets have encouraged their over-exploitation. The responsibility for resolving this illegal trade is largely placed on governments and enforcement agencies, like customs officers, to ensure that the trade in certain species (e.g. those listed in the CITES Appendices I, II, and III) is regulated, restricted, or prohibited. Improved awareness and education about these impacts in the source countries is also likely to improve knowledge of what can and cannot be exploited sustainably.

Traditional knowledge and forests

There is undoubtedly much traditional knowledge relevant to the management and utilization of forest areas. Regarding mainland Southeast Asia, there have been extensive debates surrounding the extent to which this knowledge is deliberately romanticized by certain actors and groups for instrumental purposes – to attain land rights to reside in forest areas (see. e.g., Walker, 2001; Santasombat, 2004). Thus it is not possible to discuss traditional knowledge and forests without some acknowledgement of this ongoing political struggle. Whilst it is not easy to describe this process succinctly, there has over recent decades been considerable conflict between states and forest-dwelling communities regarding land rights, migration, and population growth in forest areas, conservation and co-management of forest areas, swidden or rotational agricultural practices, and the often forced shifting of opium farmers to other crops and livelihoods. As described by Dearden (1995) with respect to northern Thailand, development assistance (often in response to opium growing) has often failed to recognize the importance of the environmental context of the communities, and it has encouraged a narrowing of the economic, cultural, and ecological (biocultural) characteristics of the area from heterogeneity to homogeneity. Indeed, there are often over-generalizations about specific ethnic groups and their behaviors in relation to the environment (see Walker, 2001), as well as generalizations and stereotypes about 'hill tribes' in upland Southeast Asia which have often had negative connotations.

The largest remaining tropical forest in mainland Southeast Asia exists across the mountainous region covering parts of Cambodia, Laos, Myanmar, Thailand, Vietnam and China. Rerkasem *et al.* (2009) describe a number of specific historic adaptations to forest living: for example, the Lahu were previously famous for their big-game hunting skills (e.g. tigers and wild buffaloes), and the Karen were known for their skill handling elephants, which ended up playing an important role in the timber industry. With the loss of these species and restrictions on logging, some indigenous knowledge has likely been lost, but often these people have readily adapted their skills and knowledge towards other forest management practices. Rerkasem *et al.* (2009) provide a number of agroforestry adaptations from five mainland Southeast Asian villages, largely stemming from land use intensification (thus restricting traditional shifting rotational agriculture), shifting subsistence production to cash cropping, and opium crop eradication. The authors note practices such as fallow improvement for soil fertility and weed suppression, and the establishment of distinct and ordered forest types (conservation forests, utility forests, home gardens, managed forest patches, and cropland).

Similar findings were found during fieldwork in Karen communities in the Samoeng district of Chiang Mai (conducted in 2005, 2006, and 2009). The land organization is typically defined by a number of physical factors, such as the local environment (hills, mountains, and streams) and the size of the local population, as well as by the norms, customs, and rituals that establish ownership and deline-ate boundaries. Table 3.1 provides a broad description of the relation between the type of forest (physical factor), and the customary laws and local protection mech-anisms (cultural factors) of many mountain communities in northern Thailand. During the fieldwork, local Karen people indicated that they had delineated forest areas for different purposes, such as *Pa Ton Nam*, *Pa pra-pe-nee*, and *Pa chai soi* (see Table 3.1), and that these were protected by customary rules, shrines, and corresponding spirits, which prevented over-exploitation of the forest.

The land within these villages in Samoeng utilizes a complicated system of community and private ownership that relates to long-held customs. The Karen typically recognize private ownership of some types of land, for example household compounds, cash crop gardens, and orchards. These privately owned properties may be inherited and sold according to Karen custom, but often do not have an official land title deed. Previously in this area an animist religious leader, the Zikho, had authority in allocating communally held land to individual households on an usufruct basis. In other words the villagers had rights of communality. Through communications with the local spirit, the religious leader was able to locate the village boundary (Ganjanapan, 2000). With the stronger influence of Buddhism

Table 3.1 Community forest classification and customary laws

Type of forest	Size	Customary law	Local protection mechanism
Pa Ton Nam Catchment watershed forest	300–70,000 rai (120–28,000 acres)	Strict rules and harsh punishment against any possible violation either by community members or outside encroachers Logging is strictly forbidden	*Phii khun nam* (watershed spirits) which serve as guardians of the forest
Pa pra-pe-nee Ceremonial forest	30–300 rai (12–120 acres)	Preserved for cremation and other ritual purposes The domain of ancestral spirits, whose wrath and punishment against violators generate great fear	Located near to villages Erecting shrines of the various guardian spirits
Pa chai soi Multi-purpose forest	Large areas close to villages	Economic use: animal grazing, village wood lot, food collection, and construction materials, etc.	Open boundaries, less controlled than other areas

Source: Santasombat (2003a).

in the area, the Zikho has less authority over the allocation of land, and disputes over common land are now usually transferred to a village leader or leaders (as was the case in Samoeng).

As a counterpoint to Western exclusionary ideas about forest conservation, which tend to have national parks as their centerpiece, the role of religion has played an ever-increasing role in the politics of forest settlements and co-management. Forest ordainment ceremonies are common occurrences, where Buddhist monks bless a forest area and where the conservation ethic of the community is often put deliberately on show so as to impress upon local authorities the importance of their management of the forest (Hirsch, 1996; Fahn, 2003). These Buddhist ceremonies are actually a relatively recent 're-traditionalization' of people's connection to nature, given that the traditional beliefs in the mountain areas of northern Thailand were animist. In any case, the animist and Buddhist beliefs often become intertwined in the region, and are often seen as mutually supportive. For example, the Karen have a number of practices which blend religion and ritual, connection to their surroundings, and utilitarian purposes. One such practice involves cutting a newborn baby's umbilical cord and placing it in a bamboo basket which is attached to a tree (this special tree is known as *dei-poh-tuh*). The child then must care for the tree throughout his or her entire lifetime, and no one else from the village may cut it down. These forests are described as 'umbilical forests' (*sei dei poh*) and the belief is that the umbilical cord ties the person to the tree (and nature more broadly), with the person's spirit tied to the fate of the tree. Similarly, when a local person dies, a tall tree is chosen to be their resting place such that their spirit can climb up high and watch over the village (Trakarnsuphakorn, 1997; and interviews, Baan Mae Ka Pu, Samoeng, Chiang Mai, 2005).

Figure 3.2 Image depicting and explaining 'umbilical forests' (*sei dei poh*) in a Karen
village in Samoeng district, near Chiang Mai

Notably, it has been suggested that some authors may over-romanticize the
'wisdom' of local communities to the extent that they may be seen to 'do no wrong'.
Obviously, these conservation systems are fallible, but the underlying point that
is made by several authors (see Santasombat, 2003a, 2003b; Laungaramsri, 2001) is

that these communities have been unfairly subjected to victimization by the state, and that their existing knowledge and practices have been emphasized heavily to assert their basic human rights and through political necessity.

These challenges are not unique to northern Thailand. For example, in Yunnan province, China, a number of pressures including economic growth and demand for natural resources have impacted upon the livelihoods, environments, and practices of indigenous communities. Jianchu (2010) reports on the loss of inter-generational knowledge that relates to conservation beliefs and practices. In response there have been a number of donor- and NGO-led initiatives to sustain traditional forest knowledge, particularly for tropical rainforest, subtropical broadleaf forest, and alpine ecosystems, including an emphasis on the mainte-nance of beneficial traditional spiritual practices which link people with nature. Jianchu (2010) indicates that, to ensure knowledge is transferred, local and regional networks are being established, for example through local seed fairs, cross-farm visits, and study tours.

Conclusions

The Southeast Asian region is replete with examples of the contribution of traditional knowledge to the conservation and sustainable use of biodiversity, as well as the active development of genetic resources through plant breeding. There are also many circumstances in which there are strong cultural and spirit-ual connections to nature, which undoubtedly encourage those populations to conserve biodiversity. For these and other more intrinsic reasons, it is important to recognize the role of biocultural diversity and are efforts to maintain cultural and linguistic diversity alongside biological diversity in the region. It is also important to recognize that commodification and patenting of plants and animals as 'genetic resources' in this region, may cause cultural and spiritual offense to indigenous peoples and local communities. Because of global land rights, resource rights, and even knowledge rights struggles, the concept of 'biocultural rights' has emerged as a way to combat threats to biocultural diversity.

References

Anderson, E.F. 1993. *Plants and People of the Golden Triangle: Ethnobotany of the Hill Tribes of Northern Thailand*. Dioscorides Press, Portland, OR.

Bavikatte, S.K. 2014. *Stewarding the Earth: Rethinking Property and the Emergence of Biocultural Rights*. Oxford University Press, New Delhi.

Bavikatte, D.K. and Robinson, D.F. 2016. Putting Peoplehood at the Center of the Green Economy. In Kohli, K. and Menon, M. (eds.), *Business Interests and the Environmental Crisis*. Sage, New Delhi, pp. 207–237.

Brun, V. and Schumacher, T. 1994. *The Traditional Herbal Medicine of Northern Thailand*. White Lotus, Bangkok.

Cosananund, J. 2003. Human Rights and the War on Drugs: Problems of Conception, Consciousness and Social Responsibility. In *Thailand Human Rights Journal V1*. National Human Rights Commission of Thailand, Bangkok, pp. 59–88.

Dearden, P. 1995. Development and Biocultural Diversity in Northern Thailand. *Applied Geography* 15(4): 325–340.

Fahn, J.D. 2003. *A Land on Fire: The Environmental Consequences of the Southeast Asian Boom*. Westview, Boulder, CO.

Felbab-Brown, V. 2013. The Illegal Trade in Wildlife in Southeast Asia and its Links to East Asian Markets. In Chouvy, P.A. (ed.), *An Atlas of Trafficking in Southeast Asia*. I.B.Tauris, London, pp. 137–154.

Forsyth, T. and Walker, A. 2008. *Forest Guardians, Forest Destroyers: The Politics of Environmental Knowledge in Northern Thailand*. Silkworm Books, Chiang Mai.

Ganjanapan, A. 2000. *Local Control of Land and Forest: Cultural Dimensions of Resource Management in Northern Thailand*. Regional Centre for Social Science and Sustainable Development, Chiang Mai.

Geertz, C. 1983. *Local Knowledge: Further Essays in Interpretive Anthropology*. Basic Books, New York.

Gorenflo, L.J., Romaine, S., Mittermeier, R. A., and Walker-Painemilla, K. (2012). Co-occurrence of Linguistic and Biological Diversity in Biodiversity Hotspots and High Biodiversity Wilderness Areas. *Proceedings of the National Academy of Sciences* 109(21): 8032–8037.

Hirsch, P. 1996. Environment and Environmentalism in Thailand: Material and Ideological Bases. In Hirsch, P. (ed.), *Seeing Forests for Trees: Environment and Environmentalism in Thailand*. Silkworm Books, Chiang Mai, pp. 15–36.

IUCN (2012) Securing the Web of Life. www.iucnredlist.org/news/securing-the-web-of-life (accessed 20 February 2014).

Jianchu, X. 2010. Indigenous Knowledge, Biodiversity Conservation, and Poverty Alleviation among Ethnic Minorities in Yunnan, China. In Maffi, L. and Woodley, E. (eds.), *Biocultural Diversity Conservation: A Global Sourcebook*. Earthscan, London, pp. 45–47.

Johnson, M. 1992. *Lore: Capturing Traditional Ecological Knowledge*. IDRC, Ottawa.

Kriengkraipetch, S. 2004. Thai Folk Beliefs about Animals and Plants and Attitudes toward Nature. In Nathalang, S. (ed.), *Thai Folklore: Insights into Thai Culture*. 2nd edn. Chulalongkorn University Press, Bangkok, pp. 169–188.

Laungaramsri, P. 2001. *Redefining Nature: Karen Ecological Knowledge and the Challenge to the Modern Conservation Paradigm*. Earthworm Books, Chennai.

Maffi, L. and Woodley, E. (eds.). 2010. *Biocultural Diversity Conservation: A Global Sourcebook*. Earthscan, London.

Nathalang, S. 2004. Conflict and Compromise between the Indigenous Beliefs and Buddhism as Reflected in Thai Rice Myths. In Nathalang, S. (ed.), *Thai Folklore: Insights into Thai Culture*. 2nd edn. Chulalongkorn University Press, Bangkok, pp. 99–121.

Pfeiffer, J.M., Rice, K.J., and Mulawarman, S.D.B. 2006. Biocultural Diversity in Traditional Rice-Based Agroecosystems: Indigenous Research and Conservation of Mavo (*Oryza sativa* L.) Upland Rice Landraces of Eastern Indonesia. *Environment, Development and Sustainability* 8(4): 609–629.

Posey, D.A. (ed.). 1999. *Cultural and Spiritual Values of Biodiversity*. United Nations Environment Programme, Nairobi.

Rerkasem, K., Yimyam, N., and Rerkasem, B. 2009. Land Use Transformation in the Mountainous Southeast Asia Region and the Role of Indigenous Knowledge and Skills in Forest Management. *Forest Ecology and Management* 257: 2035–2043.

Robinson, D. 2010. *Confronting Biopiracy: Cases, Challenges and International Debates*. Routledge/Earthscan, London.

Robinson, D. 2011. Local Agricultural and Environmental Knowledge: Working with Thai Institutes and Communities. In Jennings, J., Packham R., and Woodside, D. (eds.). *Shaping Change: Natural Resource Management, Agriculture and the Role of Extension.* Australasia-Pacific Extension Network (APEN), Australia, pp. 102–108.

Robinson, D.F. 2013. Legal Geographies of Intellectual Property, 'Traditional' Knowledge and Biodiversity: Experiencing Conventions, Laws, Customary Law, and Karma in Thailand. *Geographical Research* 51: 375–386.

Robinson, D.F. 2015. *Biodiversity, Access and Benefit-Sharing: Global Case Studies.* Routledge/Earthscan, Abingdon.

Robinson, D. and Kuanpoth, J. 2009. The Traditional Medicines Predicament: A Case Study of Thailand. *Journal of World Intellectual Property* 11(5/6): 375–403.

Salguero, C.P. 2003. *A Thai Herbal: Traditional Recipes for Health and Harmony.* Silkworm Books, Chiang Mai.

Santasombat, Y. 2003a. *Biodiversity Local Knowledge and Sustainable Development.* Regional Centre for Social Science and Sustainable Development, Chiang Mai.

Santasombat, Y. 2003b. Customary Rights, Ethnicity and the Politics of Location. In *Thailand Human Rights Journal*, V1. National Human Rights Commission of Thailand, Bangkok, pp. 121–136.

Santasombat, Y. 2004. Karen Cultural Capital and the Political Economy of Symbolic Power. *Asian Ethnicity* 5(1): 105–120.

Secretariat of the Convention on Biological Diversity. (1992). *Convention on Biological Diversity.* Secretariat of the Convention on Biological Diversity, Montreal.

Songkhran, S. (former Rice Research Director, Department of Agriculture). 2006., Interview, Bangkok, December.

Tilman, D., Cassman, K.G., Matson, P.A., Naylor, R., and Polasky, S. 2002. Agricultural Sustainability and Intensive Production Practices. *Nature* 418: 671–677.

Trakarnsuphakorn, P. 1997. The Wisdom of the Karen in Natural Resource Conservation. In McCaskill, D. and Kampe, K. (eds.), *Development or Domestication? Indigenous Peoples of Southeast Asia.* Silkwork Books, Chiang Mai, pp. 205–218.

Walker, A. 2001. The 'Karen Consensus': Ethnic Politics and Resource-Use Legitimacy in Northern Thailand. *Asian Ethnicity* 2(2): 145–162.

Part II

Challenges in conservation biology

4 Landscape changes and policies for biodiversity and environment conservation in Southeast Asia

Jean-François Cornu, Claire Lajaunie, Henri Laborde, and Serge Morand

Introduction

Landscapes of Southeast Asia are among the most diverse on earth, and are characterized by an important and singular terrestrial and freshwater biodiversity (Myers *et al.*, 2000). The endogenous processes that have shaped the landscape mosaic of Southeast Asia are rooted in a geography located at the interface of Asia and Oceania, a unique geological history and a stable tropical humid climate (Sodhi *et al.*, 2004; Woodruff, 2010). However, a number of important changes and modifications in the composition and in the structure of Southeast Asian landscapes have been observed under the action of human societies. The intensity and the pace of these changes resulting from exogenous processes have increased constantly since the beginning of the anthropization of the earth (Vitousek *et al.*, 1997; Steffen *et al.*, 2011), such that a significant part of Southeast Asia is nowadays linked to used anthromes, or antropogenic biomes (Ellis and Ramankutty, 2008; Ellis *et al.*, 2010; Ellis, 2011). These changes are all the most important since Southeast Asia is among the regions of the world with the highest concentration of areas of rapid landcover changes (Lepers *et al.*, 2005), and it recapitulated in a few decades the economic and environmental transformation that took centuries in Europe (Corlett, 2013). Nowadays, the average estimations of the human influence on Southeast Asian landscapes are higher than in other tropical and subtropical areas (Sanderson *et al.*, 2002; Kehoe *et al.*, 2015).

Landscape changes and modifications are deeply affecting the effective functioning of many ecosystems of Southeast Asia and the sustainability of environmental resources like biological diversity with significant consequences expected for ecosystem services supply and extinction risk (Chapin *et al.*, 2000; Sodhi *et al.*, 2004, 2012; Barnosky *et al.*, 2012; Hooper *et al.*, 2012; Nagendra *et al.*, 2013; McGill, 2015; Newbold *et al.*, 2015). It is particularly relevant for Southeast Asia since land-use and land-cover changes have been identified as the most important drivers of biodiversity changes by the year 2100 in the tropics (Sala *et al.*, 2000; Lee and Jetz, 2008; Murphy and Romanuk, 2014). Deforestation and agricultural expansion that are occurring in Southeast Asia are putting biotas in a fragile situation (Sodhi and Brook, 2008; Wilcove et al., 2013), such that Southeast Asia is identified as a region where the expected biodiversity losses are the highest in the world for terrestrial mammals (Ceballos and Ehrlich, 2002), and a world priority for averting imminent and latent species extinctions (Cardillo et al., 2006; Duckworth et al., 2012).

To control the negative consequences of landscape changes on the environment, and in particular on biodiversity and on ecosystem services (Mace et al., 2012), regional and national initiatives and policies were proposed by scientists and stakeholders (Pereira and Cooper, 2006; Reid et al., 2010; Perrings et al., 2011; De Fries et al., 2012; Ogden et al., 2013). In Southeast Asia, most effort has been directed up to now toward biodiversity conservation (McNeely et al., 2009; Koh and Sodhi, 2010; Squires, 2014). However, special attention has also been paid recently to ecosystem services like those provided to health (Walther et al., 2016).

Taking up the challenge of biodiversity and environmental conservation in Southeast Asia through the landscape prism (Lindenmeyer et al., 2008; Chazdon et al., 2009; Sodhi et al., 2010; Sayer et al., 2013), with methods inspired from land change science (Turner et al., 2007), is acknowledged as an urgent task (Lynam et al., 2016). The design of more integrated, regional, and multi-scale policies (Sodhi et al., 2011) involving all the regional actors is advocated to better reconcile competing land uses (Smith et al., 2010) and service provision (Carrasco et al., 2016) toward sustainable landscapes (Wiens, 2013) and sustainability science (Kates et al., 2001) in Southeast Asia (Sodhi et al., 2010). As we will see, the underlying logic of the promoted policies depends on the actors involved and their ultimate objectives.

In this chapter, we will describe the Southeast Asian landscapes dynamic from the beginning of the Anthropocene (i.e., the Industrial Revolution, c. 1700). We will focus essentially on the changes in the land-use composition of Southeast Asia because the impact of humans on landscapes is inherent to the definition of land-use types. The issue of landscape changes and their future consequences will also be discussed in the broader context of an evaluation of the regional and national policies for environmental conservation.

A global picture of the Southeast Asian landscapes: biomes and anthromes

While it is quite easy to locate Southeast Asia on the earth (i.e., the region at the interface of Asia and Oceania), concisely defining the Southeast Asian region in a geographically meaningful way is highly disciplinarily dependent (e.g., biological sciences versus political sciences). Because landscapes are driven by environmental and anthropic forces, it is important to provide a consensual definition that remains faithful and relevant to ecological and political constraints. From an ecological point of view, Southeast Asia is traditionally defined as the area that straddles the Indo-Burma (or Indochina), Philippines, Sundaland (or Sundaic), and Wallacean biogeographical sub-regions, whereas membership of the Association of Southeast Asian Nations (ASEAN) is one of the most frequently used criteria to define Southeast Asia by political and legal scientists. Hopefully, these two definitions of Southeast Asia are spatially congruent, and we will use the ASEAN area to define Southeast Asia spatially (see Figure 1.1D in Chapter 1).

Southeast Asia occupies 3.02 percent of the terrestrial area of the world and is covered in its quasi totality by two major terrestrial biomes: tropical and sub-tropical moist broadleaf forests (82.47 percent of the SEA area; 18.60 percent of the biome area); and tropical and subtropical dry broadleaf forests (13.31 percent of the SEA area; 15.65 percent of the biome area). Six other biomes are also present in Southeast Asia but these are highly marginal, except for mangroves, which represent only 2.21 percent of the total Southeast Asia area but 28.44 percent of the *global* mangrove biome area.

However, Southeast Asian landscapes are strongly impacted by human influences. While the semi-natural and wild anthromes (i.e., anthropogenic biomes where the human–ecosystem interactions are weakest) occupy a higher fraction in Southeast Asia than in the tropical areas (i.e., tropical and subtropical biomes) and the whole of the terrestrial part of the earth (53.98 percent versus 37.46 percent and 45.31 percent), the wild anthromes remain marginal in Southeast Asia (4.44 percent versus 8.65 percent and 26.14 percent). Moreover, urban areas and villages occupy 2.3 percent and 17.22 percent of the area of Southeast Asia and are significantly higher than those for tropical areas and the whole of the terrestrial part of the earth (1.16 percent and 10.02 percent; 1.26 percent and 6.38 percent). This high human influence on Southeast Asia is also apparent through the average Global Human Influence Index (or HI; see Sanderson *et al.*, 2002), which is estimated at 15.60 – respectively 2 and 4 points higher than the values for the tropical area and the whole terrestrial part of the earth.

Beyond this overall regional picture, human influence is not spatially homogeneous within the region, and important differences are observed between the biogeographical sub-region of the Philippines and the Sundaland and Wallacean sub-regions. While 19.87 percent of the area of the Philippines is linked to wild and semi-natural anthromes, the urban and village anthromes reach 37.94 percent and the HI Index is estimated at 21.23. Conversely, the Sundaland and Wallacean biogeographical sub-regions are characterized by HI Indices that are lower than 14. Their proportions of wild and semi-natural anthromes are significantly higher (48.5 percent and 64.75 percent, respectively), and their territorial footprints for urban and village anthromes are low (16.29 percent and 14.97 percent, respectively). The Indo-Burma biogeographical sub-region occupies an intermediary position, resulting from the mixture of two contrasting groups of countries: Thailand, Vietnam and Cambodia; and Laos and Myanmar.

The contemporary composition of land uses in Southeast Asia

Despite important human pressure on its landscapes, in 2005 the Southeast Asian region was still mainly covered by primary and secondary land (73.6 percent), a proportion that was significantly higher than the tropical areas and the whole earth (59.5 percent and 62.3 percent, respectively). This is all the more relevant when the emphasis is placed on primary land: around 64.6 percent of the Southeast Asian region was covered by primary land (non-forested and forested areas included) in 2005, while this land-use type did not reach 25 percent in the tropical areas or the whole of the earth.

The human footprint on landscape composition in Southeast Asia is mostly linked to the presence of cropland. Around 21.9 percent of the region is covered by croplands while pasture and urban areas are only marginally present (4.0 percent and 0.6 percent, respectively). However, it must be noted that the land-use composition (i.e., the areal fraction of the territory covered by a given set of land-use types) reflects only one side of Southeast Asia, and fails to take into account the *intensity* of human influence (i.e., the land-use intensity) over the territory. Despite this conceptual limitation, the Southeast Asian region shows a highly specialized land-use profile oriented toward croplands, unlike the tropical areas and the whole of the earth, which show more balanced land-use compositions (croplands: 13.3 percent and 26.8 percent, respectively; pastures: 11.9 percent and 25.3 percent, respectively), and it is twice as urbanized as the tropical areas (0.6 percent versus 0.3 percent).

However, this overall picture of the land-use composition of Southeast Asia camouflages important geographical variations at smaller scales, like those defined by the biogeographical sub-regions or the countries, reflecting the geographical diversity (e.g., elevation, etc.) and political history of the region. Among the four biogeographical sub-regions, the Philippines shows the highest fraction of croplands and other human-dominated land uses (42.4 percent of its area). Furthermore, while its proportion of pastures does not differ from the other sub-regions, it has more than twice as much urban land as the regional average.

At the opposite end of the scale, the Indo-Burma, Sundaland, and especially Wallacean sub-regions are more preserved. The fractions of their area covered by primary and secondary land are estimated at 70.8 percent, 70.3 percent and 77.5 percent, respectively. However, unlike Sundaland and Wallacea, which are quite homogeneous, the Indo-Burma sub-region is composed of countries characterized by different land-use profiles. While Myanmar and Laos have conserved high fractions of their primary land (>75 percent), crops, pastures, and urban land are more important in Vietnam, Cambodia, and Thailand (>30 percent).

Land-use changes in Southeast Asia since the beginning of the Anthropocene

While the transition from the Holocene toward the Anthropocene, defined here by the Industrial Revolution (c. 1700; Crutzen, 2002; but see Lewis and Maslin, 2015, for discussions about the beginning of the Anthropocene), was characterized by a worldwide acceleration of the human footprint on landscapes, the overall picture of land-use composition in Southeast Asia remained almost pristine and unchanged until the middle of the nineteenth century.

Box 4.1   Land-Use Harmonization: past and future land-use and land-cover changes at the global scale

The use of landscapes by humans (i.e. land uses) and their changes through history are cornerstone parameters, like past climates, for numerous ecological studies. Land-use and land-cover changes (LULCC) that occurred in the past are often linked to "ghosts of past" and "land-use legacy" concepts that shape the present and constrain the future (Foster *et al.*, 2003; Perring *et al.*, 2016), and are important inputs when assessing the future of biodiversity and ecosystems functioning.

While remote sensing approaches are accepted as the gold standards for monitoring recent landscape changes, they cannot capture the landscape dynamics that occurred before the 1970s. Other approaches based on spatially explicit LULCC models, which in turn are based on historical data sources, demographic assumptions, and spatial allocation rules, have been designed to model land-use and land-cover changes over longer time spans (e.g., Klein Goldewijk *et al.*, 2011). These approaches can also be used to infer the future of LULCC, and are nowadays fully integrated in most of the earth system models and integrated assessment models, such as those that are used as baselines for the four representative concentration pathways (RCP) presented in the fifth IPCC report (Moss *et al.*, 2010; Van Vuuren *et al.*, 2011).

The Land-Use Harmonization Project (LUH; see Hurtt *et al.*, 2009, 2011) released, in advance of the fifth IPCC report, a harmonized and spatially distributed database (LUH1) on LULCC which covered the whole earth (spatial resolution of 0.5 × 0.5 degrees) for the period from 1500 to 2100 (temporal resolution of 1 year). Based on the method developed by Hurtt *et al.* (2006), LUH1 provides information on the five major LULC types (i.e., urban, pasture, cropland, primary vegetation, and secondary vegetation) and on LULC transitions (e.g., from secondary vegetation to cropland). A special emphasis of LUH was to harmonize past and future models in order to tackle some regional discrepancies resulting from different definitions and spatial granularity (e.g., grid versus region) used by the LULCC models. However, the modeling of large-scale LULCC over long historical periods remains challenging (e.g., Verburg *et al.*, 2011), and numerous uncertainties remain with respect to both past reconstructions (Klein Goldewijk and Verburg, 2013) and future projections (Prestele *et al.*, in press).

The percentage of primary land in Southeast Asia decreased by only 2.7 percent between 1700 and 1850 (from 98.4 percent to 95.7 percent). This is significantly lower than what was backcasted by the LUH for tropical areas and for the whole earth, where the percentages of primary land decreased by 25.0 percent and

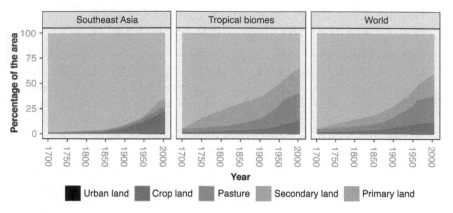

Figure 4.1 Changes in the composition of land uses since the begining of the Anthropocene (*c.* 1700) for Southeast Asia, tropical biomes and the world (LUH1)

Notes: Primary lands are areas that have never been impacted by human activies (e.g., agriculture or wood harvesting) since the beginning of the LUH backcasts. Secondary lands are areas that are recovering from previous human disturbance. The percentages of urban land are scarcely visible on the figure because of their low areal imprint.

14.1 percent, respectively. Cropland and pasture both increased significantly in this period (respectively +1.5 percent and +0.8 percent, representing 27.3 percent and 14.5 percent of the net transitions), as, to a lesser extent, did secondary land and urban land (respectively +0.4 percent and +0.1 percent, representing 7.3 percent and 1.8 percent of the net transitions). Southeast Asia differs from the tropical areas and from the whole earth, where secondary land was the land-use category that increased most during this period (respectively +19.6 percent and +9.0 percent, representing 39.2 percent and 31.9 percent of the net transitions). However, the land-use dynamics measured by annual gross transitions (i.e., the sum of gains and losses of all land-use changes) were estimated to involve 0.07 percent of the area of Southeast Asia during this period (against 0.27 percent and 0.12 percent, respectively, for the tropical areas and the whole earth), and were mostly supported by turnovers involving secondary land and cropland, on the one hand, and secondary land and pasture, on the other (52.9 percent and 34.5 percent, respectively). The loss of primary land represented only 12.4 percent of the gross transitions (3.1 percent, 3.3 percent, and 6.0 percent, respectively, toward secondary land, pasture, and cropland) and differed significantly from the estimates made for the tropical areas and the whole earth, where 31.8 percent and 41.1 percent of the gross transitions, respectively, were linked to the loss of primary land.

Changes in the composition of land uses in Southeast Asia began to accelerate after 1850, but it was around the middle of the twentieth century, the period called the Great Acceleration (Steffen *et al.*, 2011, 2015), that Southeast Asia entered in an era of more intensive change. While the percentage of Southeast

Asia covered by primary land decreased by 11.0 percent between 1850 and 1940 (compared with −16.2 percent and −20.0 percent, respectively, for tropical areas and the whole earth), the proportion of primary land in Southeast Asia shrank at a faster pace between 1940 and 2015. During this period, the loss of Southeast Asian primary land was estimated at 20.1 percent, higher than the figure for tropical areas (18.7 percent) and roughly equivalent to the figure for the whole earth (20.3 percent). Of all the land-use categories, pasture experienced the greatest increases in tropical areas and the whole earth (+20.5 percent and +19.4 percent, respectively), whereas cropland increased most in Southeast Asia (+19.3 percent). This type of land use increased by 8.5 percent between 1850 and 1940 (38.6 percent of net transitions), and covered 11.1 percent of Southeast Asia in the wake of the Great Acceleration (compared with 7.3 percent and 8.9 percent, respectively, for tropical areas and the whole earth). Southeast Asian cropland made further progress at a faster pace after 1940 (+10.8 percent), but accounted for a lower proportion of net transitions (26.7 percent) because of the rapid increase in secondary land fostered by the Great Acceleration (+7.8 percent, representing 19.4 percent of net transitions). Furthermore, the imprint of urban land increased significantly in Southeast Asia after 1940. It stood at 0.1 percent in 1850 and 0.2 percent in 1940, but had reached 0.6 percent by 2005.

This acceleration in the dynamics of the land-use system of Southeast Asia since the Great Acceleration is reflected by a significant increase in annual gross transitions. While annual gross transitions involved 0.25 percent of the area of Southeast Asia between 1850 and 1940, the average value for the period of the Great Acceleration was estimated at 0.73 percent. These estimates are lower than those provided by the LUH for tropical areas because of the comparative paucity of pasture in Southeast Asia (0.58 percent and 1.17 percent), but are at least equivalent to, and even exceed, the annual gross transitions for the whole earth since 1940 (0.30 percent and 0.57 percent). Since 1850, the loss of primary land increased its contribution to annual gross transitions and exceeded 20 percent (compared with 14.1 percent for tropical areas and 32.7 percent for the whole earth). While primary land was mostly converted to cropland before 1940 (16.9 percent), transitions toward secondary land prevailed during the Great Acceleration (13.3 percent), even though 7.2 percent of primary land continued to be converted to cropland. However, the turnovers between secondary land and cropland, and secondary land and pasture, remained the dominant drivers of land-use dynamics in Southeast Asia (76.4 percent).

The Great Acceleration was also the period when intra-regional differences began to increase significantly. While the compositions of land uses were equivalent before 1940, the Philippines biogeographical sub-region lost a significantly higher proportion of its primary land after 1940 compared to the Indo-Burma, the Sundaland, and the Wallacea subregions (−35.4 percent versus −20.3 percent, −20.2, and −19.7 percent, respectively). This trend mostly reflected a transition to cropland in all four sub-regions, but in different proportions. The Philippines sub-region was the area where cropland increased the most during the period

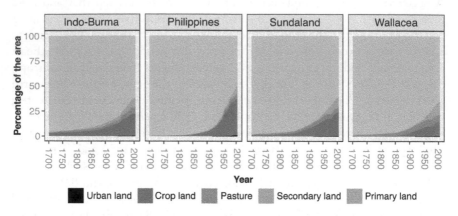

Figure 4.2 Changes in the composition of land uses since the beginning of the
Anthropocene (*c.* 1700) for the four biogeographical sub-regions of Southeast
Asia (LUH1)

1940–2005 (+23.1 percent, representing 32.6 percent of net transitions).
Secondary land also increased by 7.6 percent (10.7 percent of net transitions),
while the proportion of urban land experienced a sevenfold increase, reaching
1.4 percent of the total area of the sub-region (+1.2 percent, representing 1.7
percent of net transitions). Annual gross transitions in the Philippines sub-region
increased dramatically, and have reached 0.91 percent of the total area of the
sub-region since 1940 (compared with just 0.19 percent between 1850 and 1940).
Most of the land-use dynamics were supported by the loss of primary land oriented
toward cropland and secondary land (16.0 percent and 13.5 percent of gross
transitions, respectively), and by land-use turnovers between secondary land and
cropland (59.3 percent of gross transitions).

By contrast, the increases in cropland and the net changes linked to cropland
were lower in the Indo-Burma, Sundaland, and Wallacea sub-regions (respectively
+9.6 percent, +11.0 percent, and +8.0 percent, representing 23.6 percent, 27.3
percent, and 20.6 percent of their net transitions for the period 1940–2005).
These biogeographical sub-regions were characterized by higher net changes
linked to secondary land (>15 percent of net transitions). This is most notable in
the Sundaland and Wallacea sub-regions, where secondary land increased by
7.9 percent and 10.8 percent, respectively, but represented 19.6 percent and
27.8 percent of net transitions. Since 1940, annual gross transitions were comparable to those of the Philippines sub-region (0.99 percent and 0.89 percent), while
they were twice as high between 1850 and 1940. They were mostly supported by
transitions involving cropland and secondary land (63.9 percent and 53.3 percent
for Sundaland and Wallacea, respectively), and pasture and secondary land
(19.1 percent and 32.0 percent for Sundaland and Wallacea, respectively). The loss
of primary land represented 15.7 percent and 13.3 percent of gross transitions

in the Sundaland and Wallacea, respectively, with most of it converted into secondary land.

The profile of land-use changes for the Indo-Burma sub-region is more balanced because of its inherent heterogeneity. Secondary land and pasture increased by 6.4 percent and 3.9 percent, respectively, representing 15.7 percent and 9.6 percent of net transitions that occurred between 1940 and 2005. Annual gross transitions for Indo-Burma remained moderate (0.62 percent of the total area of the sub-region for the period 1940–2005), but it has the highest proportion of loss of primary land in gross transitions of all four biogeographical sub-regions at 25.8 percent. The transitions linked to the loss of primary land have been quite balanced between cropland and secondary land (13.2 percent and 10.1 percent, respectively). However, most of the land-use dynamics in Indo-Burma were sustained by transitions between secondary land and cropland (52.6 percent) and to a lesser extent by those between secondary land and pasture (20.7 percent).

Land-use and landscape changes in regional and national policies in Southeast Asia

Those familiar with land planning and land management may well feel that landscape is not a notion that is easily comprehended by regional and national policy-makers and legislators. Indeed, landscape is more of a geographical than a legal concept, usually associated with technical land planning rather than a legal framework. Nevertheless, we should consider that landscape was introduced into the environmental law debate jointly by the World Commission on Protected Areas and the Commission on Environmental Law of the World Conservation Union (IUCN) in 1998 through the organization of the Colloquium on Landscape Conservation Law. The central theme was the draft of the European Landscape Convention, prepared by the Council of Europe, which came into force in 2004 (Council of Europe, 2000). Central elements of the remarks and conclusions of the Colloquium help us to underline the complexity of the context of biodiversity and environmental conservation policies and to examine landscape versus land-use changes in policies in relation to the various stakeholders involved.

In Southeast Asia, in the context of the rapid changes presented in the introduction, policies related to land-uses or landscapes aim to ensure sustainable development, though the perspective adopted differs depending on the actors involved and according to their distinctive goals regarding the design of biodiversity and conservation policies. Indeed, the notion of sustainability supported by the different actors can itself be expressed according to different points of view. It can be translated in close relation to economic development and growth (Anbumozhi and Intal, 2015), with a specific focus on biodiversity conservation and the provision of ecosystem services. It can also pinpoint poverty reduction or conversely favor a more integrated approach encompassing the major environmental issues ASEAN has to face. The various logics can overlap or they may be conflicting (e.g., biodiversity conservation versus poverty reduction; Sanderson and Redford, 2003) and this issue should be taken into account to understand the perspectives of the different suggested policies and to design appropriate policy measures.

As we have seen above (see also Box 4.1), land-use and land-cover are studied through their changes and state transitions using spatially explicit land-use and land-cover changes (LULCC) models. Land-use states are studied according to five types in the LUH, which are then subdivided into distinct sub-categories. Land-cover, defining the physical state of the land, is completed by land-use (i.e., human use of the land) and can be described using land-cover classification schemes, land mapping or, as with the LUH, assumptions about the underlying potential biomass density of the different types of land.

The data and understanding of LULC dynamics (e.g. through scenario building) are integrated to draft legally binding land-use plans to be implemented by national institutions. Land-use planning has been defined by UNEP/FAO (1999: 15) as "a systematic and iterative procedure carried out in order to create an enabling environment for sustainable development of land resources which meets people's needs and demands." Landscape approaches and policies are grounded in landscape ecology and advocated by biodiversity conservation and environmental organizations (Reed et al., 2014), whereas land-use planning and management are mainly designed by institutional and intergovernmental organizations such as ASEAN through policies and legislation (for national forestry legislation in Southeast Asia, see Yasmi et al., 2010).

According to the Article I of the European Landscape Convention, landscape is "an area, as perceived by people, whose character is the result of the action and interaction of natural and/or human factors." As such, it is presented by Phillips (2000: 18) as "both natural and cultural values and features, and focuses on the relationships between these," and as "the sum of all past changes to the environment."

Through land-use planning, ASEAN intends to balance the competing interests that wish to use natural resources for food crops, industrial crops, and other purposes. For instance, in Southeast Asia, forests provide critical environmental services, contributing to the protection of land and water resources, conservation of global biodiversity, and/or climate change mitigation. ASEAN member states are committed to strengthening their cooperation and developing the appropriate regional policy framework as well as to implementing national laws and policies to tackle illegal logging and deforestation (ASEAN Ministers of Agriculture and Forestry, 2015). But the framework should be elaborated, taking into account the overall socio-economic context of the region. The rapid expansion of agriculture through land acquisition attracts regional and foreign investors (International Institute for Environment and Development, 2012) and impacts on land rights and land tenure security (Tambunan, 2016).

The ASEAN Vision and Strategic Plan for Food, Agriculture and Forestry (2016–2025) aims to integrate these sectors into the global economy in relation to the completion of the Sustainable Development Goals (SDGs). The policies are monitored and assessed with the development of key performance indicators (KPIs) by the ASEAN Ministers of Agriculture and Forestry. However, the indicators remain to be defined for effective implementation.

Research into land-use management in ASEAN showed a lack of good national coordination among government agencies and sectors. It underlined the need for good cooperation among all stakeholders to implement a good land-use management and improve the coordination between economic and agricultural development policies. It also recommended the full integration of local communities and farmers in land-use planning and insisted that local authorities must guarantee the legal rights of communities (Tambunan, 2016).

Policy-makers in ASEAN are linking land-use to food security, agriculture, and the economy: they mention a healthy environment as an element of sustainable development, not as a prerequisite for sustainability.

Landscape is difficult to comprehend as it is related to people's perception and to cultural factors, while land-use management and planning relies on quantitative datasets, GIS technologies, and mapping resources. Nevertheless, a broad landscape approach is nowadays championed by the CGIAR (Global Landscapes Forum), the CIFOR, the FAO (2012), and UNEP as it provides tools and concepts for allocating and managing land resources to achieve social, economic, and environmental objectives. This corresponds to a shift from strict conservationist considerations to a broader perspective (Sayer *et al.*, 2013) in accordance with the holistic framework of the SDGs.

The Subsidiary Body on Scientific, Technical and Technological Advice for the Convention on Biodiversity suggested focusing on the landscape level, stating that it is an important planning framework to avoid the displacement of pressures on biodiversity from one area to another (CBD SBSTTA, 2011), taking into account the interactions between different spatial scales. Ten principles to provide a normative basis for the landscape approach have been proposed by Sayer *et al.* (2013) in order to address the serious problems resulting from different land uses and promote integrated decision-making involving trade-offs between conservationist and developmental considerations (Reed *et al.*, 2014).

The landscape approach should take into account the various dimensions of diversity (biodiversity and cultural diversity) and respond to the need for integrated policies by engaging numerous stakeholders, different scales, and multiple land uses. Such an approach utilizes both local and scientific knowledge (Sayer *et al.*, 2015) to identify knowledge gaps and promote policies that are specifically designed and integrate the different types of knowledge. It is necessarily an iterative process based on establishing a dialogue among all the various stakeholders, particularly scientists and policy-makers.

Conclusion

Since the beginning of the Anthropocene, important parts of the landscape and ecosystem of Southeast Asia have been "domesticated" at a pace that has not been observed in other regions of the world. Today, Southeast Asia is a region where biodiversity is seriously threatened because of intensive land-use changes that occurred during the last century. In order to preserve the services provided by ecosystems, international calls were issued for new governance policies for biodiversity and environment conservation, and various initiatives in Southeast

Asia started to integrate the imperative to find trade-offs between short-term economic growth and long-term returns assured by the ecosystems to human welfare.

However, conservation of the environment (and, by extension, of the services provided by the ecosystems) is a complex problem that is of interest to many groups acting at different spatial and temporal scales. The design of scenarios – or "plausible alternative futures" – has proved useful when exploring the uncertainties surrounding the consequences of policies and governance decisions linked to climate change mitigation, and promises to be an essential tool for the conservation of biodiversity (e.g., Peterson *et al.*, 2003; Sala and Jackson, 2006; Visconti *et al.*, 2016), ecosystems services, and human well-being (e.g., Butler *et al.*, 2005; Butler and Oluoch Kosura, 2006; Diaz *et al.*, 2006). Although global scenarios are being developed by the Millennium Ecosystems Assessment (e.g, Raskin, 2005; Bennett *et al.*, 2005; Carpenter *et al.*, 2006; Cork *et al.*, 2006) and the Intergovernmental Science-Policy Platform on Biodiversity and Ecosystem Services (e.g., Pereira *et al.*, 2010; Kok *et al.*, in press), there is an urgent need to design regional scenarios to better capture the socio-ecological dynamics of biodiversity and ecosystem services in order to improve future policies and governance decisions.

In this context, landscapes can be the cornerstone level when designing biodiversity and environmental scenarios. Because landscapes are geographically defined, constructed by social and environmental forces, and shaped by different actors at multiple scales, tackling the problem of the alternative futures of biodiversity through the lens of landscape is one of the most promising solutions to holistically integrating socio-ecological dynamics (Kareiva *et al.*, 2007). Therefore, the analysis of land uses, which reflects the interactions between the biophysical characteristics of the landscape and human management, provides an interesting – and relatively uncomplicated – starting point for exploring socio-ecological production functions (Foley *et al.*, 2005), and for designing original and creative scenarios for biodiversity governance and environment policies. However, at present, insufficient attention is paid to changes in land-use during the design of biodiversity scenarios (Titeux *et al.*, 2016).

Studies have investigated the links between biodiversity issues, landscape changes, and regional policies in Southeast Asia. It is time to develop regional socio-ecological observatories in order to design scenarios and improve the environmental policies that are needed to maintain ecosystems services and preserve human well-being in Southeast Asia.

References

Anbumozhi, V. and P.S. Intal Jr. 2015. *Can Thinking Green and Sustainability Be an Economic Opportunity for ASEAN?* Economic Research Institute for ASEAN and East Asia Discussion Paper ERIA-DP-2015-66.

ASEAN Ministers of Agriculture and Forestry. 2015. *ASEAN Vision and Strategic Plan for Food, Agriculture and Forestry (2016–2025).* 37th AMAF, 10 September, Makati City, Philippines.

Barnosky, A.D., E.A. Hadly, J. Bascompte, E.L. Berlow, J.H. Brown, M. Fortelius, W.M. Getz, J. Harte, A. Hastings, P.A. Marquet, N.D. Martinez, A. Mooers, P. Roopnarine, G. Vermeij, J.W. Williams, R. Gillespie, J. Kitzes, C. Marshall, N. Matzke, D.P. Mindell, E. Revilla, and A.B. Smith. 2012. Approaching a state shift in earth's biosphere. *Nature* 486: 52–58.

Bennett, E.M., G.D. Peterson, and E.A. Levitt. 2005. Looking to the future of ecosystem services. *Ecosystems* 8: 125–132.

Butler, C.D., C.F. Corvalan, and H.S. Koren. 2005. Human health, well-being, and global ecological scenarios. *Ecosystems* 8: 153–162.

Butler, C.D. and W. Oluoch Kosura. 2006. Linking future ecosystem services and future human well-being. *Ecology and Society* 11: 30.

Cardillo, M., G.M. Mace, J.L. Gittleman, and A. Purvis. 2006. Latent extinction risk and the future battlegrounds of mammal conservation. *Proceedings of the National Academy of Science USA* 103: 4157–4161.

Carpenter, S.R., E.M. Bennett, and G.D. Peterson. 2006. Scenarios for ecosystem services: an overview. *Ecology and Society* 11: 29.

Carrasco, L.R., S.K. Papworth, J. Reed, W.S. Symes, A. Ickowitz, T. Clements, K.S.H. Peh, and T. Sunderland. 2016. Five challenges to reconcile agricultural land use and forest ecosystem services in Southeast Asia. *Conservation Biology* 30: 962–971.

CBD Subsidiary Body for Scientific, Technical and Technological Advice (CBD SBSTTA). 2011. *Report on How to Improve Sustainable Use of Biodiversity in a Landscape Perspective.* UNEP/CBD/SBSTTA/15/13, 7–11 November, Montreal.

Ceballos, G. and P.R. Ehrlich. 2002. Mammal population losses and the extinction crisis. *Science* 296: 904–907.

Chapin III, F.S., E.S. Zavaleta, V.T. Eviner, R.L. Naylor, P.M. Vitousek, H.L. Reynolds, D.U. Hooper, S. Lavorel, O.E. Sala, S.E. Hobbie, M.C. Mack, and S. Diaz. 2000. Consequence of changing biodiversity. *Nature* 405: 234–242.

Chazdon, R.L., C.A. Harvey, O. Komar, D.M. Griffith, B.G. Fergusson, M. Martinez Ramos, H. Morales, R. Nigh, L. Soto Pinto, M. van Breugel, and S.M. Philpott. 2009. Beyond reserves: a research agenda for conserving biodiversity in human-modified tropical landscapes. *Biotropica* 41: 142–153.

Cork, S.J., G.D. Peterson, E.M. Bennett, G. Petschel Held, and M. Zurek. 2006. Synthesis of the storylines. *Ecology and Society* 11: 11.

Corlett, R.T. 2013. Becoming Europe: Southeast Asia in the Anthropocene. *Elementa Science of the Anthropocene* 1: 16.

Council of Europe. 2000. *European Landscape Convention, adopted by the Committee of Ministers of the Council of Europe on 19 July 2000 and entered into force in 2004.* European Treaty Series No. 176. www.coe.int/en/web/conventions/full-list/-/conventions/rms/0900001680080621 (accessed 30 March 2017).

Crutzen, P.J. 2002. Geology of mankind. *Nature* 415: 23.

De Fries, R.S., E.C. Ellis, F.S. Chapin III, P.A. Matson, B.L. Turner II, A. Agrawal, P.J. Crutzen, C. Field, P. Gleick, P.M. Kareiva, E. Lambin, D. Liverman, E. P.A. Ostrom, Sanchez, and J. Syvitski. 2012. Planetary opportunities: a social contract for global change science to contribute to a sustainable future. *BioScience* 62: 603–606.

Diaz, S., J. Fargione, F.S. Chapin III, and D. Tilman. 2006. Biodiversity loss threatens human well-being. *PLoS Biology* 4: 1300–1305.

Duckworth, J.W., G. Batters, J.L. Belant, E.L. Bennett, J. Brunner, J. Burton, D.W.S. Challender, V. Cowling, N. Duplaix, J.D. Harris, S. Hedges, B. Long, S.P. Mahood, P.J.K. McGowan, W.J. McShea, W.L.R. Oliver, S. Perkin, B.M. Rawson, C.R. Shepherd,

S.N. Stuart, B.K. Talukdar, P.P. van Dijk, J.C. Vie, J.L. Walston, T. Whitten, and R. Wirth. 2012. Why South-East Asia should be the world's priority for averting imminent species extinctions, and a call to join a developing cross-institutional programme to tackle this urgent issue. *Sapiens* 5: 77–95.

Ellis, E.C. 2011. Anthropogenic transformation of the terrestrial biosphere. *Philosophical Transactions of the Royal Society* A 369: 1010–1035.

Ellis, E.C., K. Klein Goldewijk, S. Siebert, D. Lightman, and N. Ramankutty. 2010. Anthropogenic transformation of the biomes, 1700 to 2000. *Global Ecology and Biogeography* 19: 589–606.

Ellis, E.C. and N. Ramankutty. 2008. Putting people in the map: anthropogenic biomes of the world. *Frontiers in Ecology and Environment* 6: 439–447.

FAO. 2012. *Mainstreaming Climate-Smart Agriculture into a Broader Landscape Approach*. Background paper for the Second Global Conference on Agriculture, Food Security and Climate Change, 3–7 September, Hanoi, Vietnam.

Foley, J.A., R. De Fries, G.P. Asner, C. Barford, G. Bonan, S.R. Carpenter, F.S. Chapin, M.T. Coe, G.C. Daily, H.K. Gibbs, J.H. Helkowski, T. Holloway, E.A. Howard, C.J. Kucharik, C. Monfreda, J.A. Patz, I.C. Prentice, N. Ramankutty, and P.K. Snyder. 2005. Global consequences of land use. *Science* 309: 570–574.

Foster, D., F. Swanson, J. Aber, I. Burke, N. Brokaw, D. Tilman, and A. Knapp. 2003. The importance of land-use legacies to ecology and conservation. *BioScience* 53: 77–88.

Hooper, D.U., E.C. Adair, B.J. Cardinale, J.E.K. Byrnes, B.A. Hungate, K.L. Matulich, A. Gonzalez, J.E. Duffy, L. Gamfedt, and M.I. O'Connor. 2012. A global synthesis reveals biodiversity loss as a major driver of ecosystem change. *Nature* 486: 105–109.

Hurtt, G.C., L.P. Chini, S. Frolking, R.A. Betts, J. Feddema, G. Fischer, J.P. Fisk, K. Hibbard, R.A. Houghton, A. Janetos, C.D. Jones, G. Kindermann, T. Kinoshita, K. Klein Goldewijk, K. Riahi, E. Shevliakova, S. Smith, E. Stehfest, A. Thomson, A. Thornton, D.P. van Vuuren, and Y.P. Wang. 2011. Harmonization of land-use scenarios for the period 1500–2100: 600 years of global gridded annual land-use transitions, wood harvest, and resulting secondary lands. *Climatic Change* 109: 117–161.

Hurtt, G.C., L.P. Chini, S. Frolking, R. Betts, J. Feddema, G. Fischer, K. Klein Goldewijk, K. Hibbard, A. Janetos, C. Jones, G. Kindermann, T. Kinoshita, K. Riahi, E. Shevliakova, S. Smith, E. Stehfest, A. Thomson, P. Thornton, D. van Vuuren, and Y.P. Wang. 2009. Harmonisation of global land-use scenarios for the period 1500–2100 for IPCC-AR5. *iLEAPS Newsletter* 7: 6–8.

Hurtt, G.C., S. Frolking, M.G. Fearon, B. Moore, E. Shevliakovas, S. Malyshev, W. Pacala, and R.A. Houghton. 2006. The underpinnings of land-use history: three centuries of global gridded land-use transitions, wood-harvest activity, and resulting secondary lands. *Global Change Biology* 12: 1208–1229.

International Institute for Environment and Development. 2012. *Agricultural land acquisitions: a lens on Southeast Asia*. Briefing.

Kareiva, P., S. Watts, R. McDonald, and T. Boucher. 2007. Domesticated nature: shaping landscapes and ecosystems for human welfare. *Science* 316: 1866–1869.

Kates, R.W., W.C. Clark, R. Corell, J.M. Hall, C.C. Jaeger, I. Lowe, J.J. McCarthy, H.J. Schellnhuber, B. Bolin, N.M. Dickson, S. Faucheux, G.C. Gallopin, A. Grubler, B. Huntley, J. Jager, N.S. Jodha, R.E. Kasperson, A. Mabogunje, P. Matson, H. Mooney, B. Moore III, T. O'Riordan, and U. Svedin. 2001. Sustainability science. *Science* 292: 641–642.

Kehoe, L., T. Kuemmerle, C. Meyer, C. Levers, T. Vaclavik, and H. Kreft. 2015. Global patterns of agricultural land-use intensity and vertebrate diversity. *Diversity and Distribution* 21: 1308–1318.

Klein Goldewijk, K., A. Beusen, G. van Drecht, and M. de Vos. 2011. The HYDE 3.1 spatially explicit database of human-induced global land-use change over the past 12,000 years. *Global Ecology and Biogeography* 20: 73–86.

Klein Goldewijk, K. and P.H. Verburg. 2013. Uncertainties in global-scale reconstructions of historical land use: an illustration using HYDE data set. *Landscape Ecology* 25: 861–877.

Koh, L.P. and N.S. Sodhi. 2010. Conserving Southeast Asia's imperilled biodiversity: scientific, management, and policy challenges. *Biodiversity and Conservation* 19: 913–917.

Kok, M.T.J., K. Kok, G.D. Peterson, R. Hill, J. Agard, and S.R. Carpenter. In press. Biodiversity and ecosystem services require IPBES to take novel approach to scenarios. *Sustainability Science.*

Lee, T.M. and W. Jetz. 2008. Future battlegrounds for conservation under global change. *Proceedings of the Royal Society B* 275: 1261–1270.

Lepers, E., E.F. Lambin, A.C. Janetos, R. de Fries, F. Achard, N. Ramankutty, and R.J. Scholes. 2005. A synthesis of information on rapid land-cover change for the period 1981–2000. *BioScience* 55: 115–124.

Lewis, S.L. and M.A. Maslin. 2015. Defining the Anthropocene. *Nature* 519: 171–180.

Lindemayer, D., R.J. Hobbs, R. Montague Drake, J. Alexandra, A. Bennett, M. Burgman, P. Cale, A. Calhoun, V. Cramer, P. Cullen, D. Driscoll, L. Fahrig, J. Fischer, J. Franklin, Y. Haila, M. Hunter, P. Gibbons, S. Lake, G. Luck, C. MacGregor, S. McIntyre, R. MacNally, A. Manning, J. Miller, H. Mooney, R. Noss, H. Possingham, D. Saunders, F. Schmiegelow, M. Scott, D. Simberloff, T. Sisk, G. Tabor, B. Walker, J. Wiens, J. Woinarski, and E. Zavaleta. 2008. A checklist for ecological management of landscapes for conservation. *Ecology Letters* 11: 78–91.

Lynam, A.J., L. Porter, and A. Campos Arceiz. 2016. The challenge of conservation in changing tropical Southeast Asia. *Conservation Biology* 30: 931–932.

Mace, G.M., K. Norris, and A.H. Fitter. 2012. Biodiversity and ecosystem services: a multi-layered relationship. *Trends in Ecology and Evolution* 27: 19–26.

McGill, B. 2015. Land use matters. *Nature* 520: 38–39.

McNeely, J.A., P. Kapoor Vijay, L. Zhi, L. Olsvig Whittaker, K.M. Sheikh, and A.T. Smith. 2009. Conservation biology in Asia: the major policy challenges. *Conservation Biology* 23: 805–810.

Moss, R.H., J.A. Edmonds, K.A. Hibbard, M.R. Manning, S.K. Rose, D.P. van Vuuren, T.R. Carter, S. Emori, M. Kainuma, T. Kram, G.A. Meehl, J.F.B. Mitchell, N. Nakicenovic, K. Riahi, S.J. Smith, R.K. Stouffer, A.M. Thomson, J.P. Weyant, and T.J. Wilbanks. 2010. The next generation of scenarios for climate change research and assessment. *Nature* 463: 747–756.

Murphy, G.E.P. and T.N. Romanuk. 2014. A meta-analysis of declines in local species richness from human disturbances. *Ecology and Evolution* 4: 91–103.

Myers, N., R.A. Mittermeier, C.G. Mittermeier, G.A.B. da Fonseca, and J. Kent. 2000. Biodiversity hotspots for conservation priorities. *Nature* 403: 853–858.

Nagendra, H., B. Reyers, and S. Lavorel. 2013. Impacts of land change on biodiversity: making the link to ecosystem services. *Current Opinion in Environmental Sustainability* 5: 503–508.

Newbold, T., L.N. Hudson, S.L.L. Hill, S. Contu, I. Lysenko, R.A. Senior, L. Borger, D.J. Bennett, A. Choimes, B. Collen, J. Day, A. De Palma, S. Diaz, S. Escheverria Londono, M.J. Edgar, A. Feldman, M. Garon, M.L.K. Harrison, T. Alhusseini, D.J. Ingram, Y. Itescu, J. Kattge, V. Kemp, L. Kirkpatrick, M. Kleyer, D.L. Pinto Correia, C.D. Martin, S. Meiri, M. Novosolov, Y. Pan, H.R.P. Phillips, D.W. Purves, A. Robinson, J. Simpson, S.L. Tuck,

E. Weilher, H.J. White, R.M. Ewers, G.M. Mace, J.P.W. Scharlemann, and A. Purvis. 2015. Global effects of land use on local terrestrial biodiversity. *Nature* 520: 45–50.

Ogden, L., N. Heynen, U. Oslender, P. West, K.A. Kassam, and P. Robbins. 2013. Global assemblages, resilience, and earth stewardship in the Anthropocene. *Frontiers in Ecology and Environment* 11: 341–347.

Pereira, H.M. and H.D. Cooper. 2006. Towards the global monitoring of biodiversity change. *Trends in Ecology and Evolution* 21: 123–129.

Pereira, H.M., P.W. Leadley, V. Proenca, R. Alkemade, J.P.W. Scharlemann, J.F. Fernandez Manjarres, M.B. Araujo, P. Balvanera, R. Biggs, W.W.L. Cheung, L. Chini, H.D. Cooper, E.L. Gilman, S. Guenette, G.C. Hurtt, H.P. Huntington, G.M. Mace, T. Oberdorff, C. Revenga, P. Rodrigues, R.J. Scholes, U.R. Sumarila, and M. Walpole. 2010. Scenarios for global biodiversity in the 21st century. *Science* 330: 1496–1051.

Perring, M.P., P. De Frenne, L. Baeten, S.L. Maes, L. Depauw, H. Blondeel, M.M. Caron, and K. Verheyen. 2016. Global environmental change effects on ecosystems: the importance of land-use legacies. *Global Change Biology* 22: 1361–1371.

Perrings, C., A. Duraiappah, A. Larigauderie, and H. Mooney. 2011. The biodiversity and ecosystem services science–policy interface. *Science* 331: 1139–1140.

Peterson, G.D., G.S. Cumming, and S.R. Carpenter. 2003. Scenario planning: a tool for conservation in an uncertain world. *Conservation Biology* 17: 358–366.

Phillips, A. 2000. Practical considerations for the implementation of a European Landscape Convention. In IUCN Commission on Environmental Law, *Landscape Conservation Law: Present Trends and Perspectives in International and Comparative Law*. Gland and Cambridge: IUCN: 18–19.

Prestele, R., P. Alexander, M.D.A. Rounsevell, A. Arneth, K. Calvin, J. Doelman, D.A. Eitelberg, K. Engstrom, S. Fujimori, T. Hasegawa, P. Havlik, F. Humpenoder, A.K. Jain, T. Krisztin, P. Kyle, P. Meiyappan, A. Popp, R.D. Sands, R. Schaldach, J. Schungel, E. Stehfest, A. Tabeau, H. van Meijl, J. van Vliet, and P.V. Verburg. In press. Hotspots of uncertainty in land-use and land-cover change projections: a global-scale model comparison. *Global Change Biology*.

Raskin, P.D. 2005. Global scenarios: background review for the Millennium Ecosystem Assessment. *Ecosystems* 8: 133–142.

Reed J., L. Deakin, and T. Sunderland. 2014. What are "integrated landscape approaches" and how effectively have they been implemented in the tropics: a systematic map protocol. *Environmental Evidence* 4: 2.

Reid, W.V., D. Chen, L. Goldfarb, H. Hackmann, Y.T. Lee, K. Mokhele, E. Ostrom, K. Raivio, J. Rockstrom, H.J. Schellnhuber, and A.Whyte. 2010. Earth system science for global sustainability: grand challenges. *Science* 330: 916–917.

Sala, O.E. and R.B. Jackson. 2006. Determinants of biodiversity change: ecological tools for building scenarios. *Ecology* 87: 1875–1876.

Sala, O.E., F.S. Chapin Ill, J.J. Armesto, E. Berlow, J. Bloomfield, R. Dirzo, E. Huber Sanwald, L.F. Huenneke, R.B. Jackson, A. Kinzig, R. Leemans, D.M. Lodge, H.A. Mooney, M. Oesterheld, N. LeRoy Poff, M.T. Sykes, B.H. Walker, M. Walker, and D.H. Wall. 2000. Global biodiversity scenarios for the year 2100. *Science* 287: 1770–1774.

Sanderson, E.W., M. Jaiteh, M.A. Levy, K.H. Redford, A.V. Wannebo, and G. Woolmer. 2002. The human footprint and the last of the wild. *BioScience* 52: 891–904.

Sanderson, S.E. and K.H. Redford 2003. Contested relationships between biodiversity conservation and poverty alleviation. *Oryx* 37: 1.

Sayer, J., C. Margules, I. Bohnet, A. Boedhihartono, R. Pierce, A. Dale, and K. Andrews. 2015. The role of citizen science in landscape and seascape approaches to integrating conservation and cevelopment. *Land* 4: 1200–1212.

Sayer J., T. Sunderland, J. Ghazoul, J.L. Pfund, D. Sheil, E. Meijaard, M. Venter, A.K. Boedhihartono, M. Day, C. Garcia, C. van Oosten, and L.E. Buck. 2013. Ten principles for a landscape approach to reconciling agriculture, conservation, and other competing land uses. *Proceedings of National Academy of Sciences USA* 110: 8349–8356.

Smith, P., P.J. Gregory, D. van Vuuren, M. Obersteiner, P. Havlik, M. Rounsevell, J. Woods, E. Stehfest, and J. Bellarby. 2010. Competition for land. *Philosophical Transactions of the Royal Society B* 365: 2941–2957.

Sodhi, N.S. and B.W. Brook. 2008. Fragile Southeast Asian biotas. *Biological Conservation* 141: 883–884.

Sodhi, N.S., R. Butler, W.F. Laurance, and L. Gibson. 2011. Conservation successes at micro, meso and macroscales. *Trends in Ecology and Evolution* 26: 585–594.

Sodhi, N.S., L.P. Koh, B.W. Brook, and P.K.L. Ng. 2004. Southeast Asian biodiversity: an impending disaster. *Trends in Ecology and Evolution* 19: 654–660.

Sodhi, N.S., L.P. Koh, R. Clements, T.C. Wanger, J.K. Hill, K.C. Hamer, Y. Clough, T. Tscharntke, M.R.C. Posa, and T.M. Lee. 2010. Conserving Southeast Asian forest biodiversity in human-modified landscapes. *Biological Conservation* 143: 2375–2384.

Sodhi, N.S., M.R.C. Posa, K.S.H. Peh, L.P. Koh, M.C.K. Soh, T.M.L. Lee, J.S.H. Lee, T.C. Wanger, and B.W. Brook. 2012. Land use changes imperil South East Asian biodiversity. In D. Lindenmayer, S. Cunningham, and A. Young (eds.), *Land Use Intensification: Effects on Agriculture, Biodiversity and Ecological Processes*. Clayton, Australia: CSIRO Publishing: 33–39.

Squires, D. 2014. Biodiversity conservation in Asia. *Asia and the Pacific Policy Studies* 1: 144–159.

Steffen, B.W., J. Grinevald, P. Crutzen, and J. McNeill. 2011. The Anthropocene: conceptual and historical perspectives. *Philosophical Transactions of the Royal Society A* 369: 842–867.

Steffen, W., W. Broadgate, L. Deutsch, O. Gaffney, and C. Ludwig. 2015. The trajectory of the Anthropocene: the Great Acceleration. *Anthropocene Review* 2: 81–98.

Tambunan, T. 2016. *Identifying Stakeholders in the Land Use Management Process and Related Critical Factors in ASEAN*. ASEAN–Canada Research Partnership Working Paper No. 1, May.

Titeux, N., K. Henle, J.B. Mihoub, A. Regos, I.R. Geijzendorffer, W. Cramer, P.H. Verburg, and L. Brotons. 2016. Biodiversity scenarios neglect future land-use changes. *Global Change Biology* 22: 2505–2515.

Turner, B.L., E.F. Lambin, and A. Reenberg. 2007. The emergence of land change science for global environmental change and sustainability. *Proceedings of the National Academy of Sciences USA* 104: 20666–20671.

UNEP/FAO. 1999. *The Future of Our Land: Facing the Challenge*. www.fao.org/docrep/004/x3810e/x3810e00.htm (accessed 8 March 2017).

Van Vuuren, D.P., J. Edmonds, M. Kainuma, K. Riahi, A. Thomson, K. Hibbard, G.C. Hurtt, T. Kram, V. Krey, J.F. Lamarque, T. Masui, M. Meinshausen, N. Nakicenovic, S.J. Smith, and S.K. Rose. 2011. The representative concentration pathways: an overview. *Climatic Change* 109: 5–31.

Verburg, P.H., K. Neumann, and L. Nol. 2011. Challenges in using land use and land cover data for global change studies. *Global Change Biology* 17: 974–989.

Visconti, P., M. Bakkenes, D. Baisero, T. Brooks, S.H.M. Butchart, L. Joppa, R. Alkemade, M. Di Marco, L. Santini, M. Hoffmann, L. Maiorano, R.L. Pressey, A. Arponen,

L. Boitani, A.E. Reside, D.P. van Vuuren, and C. Rondinini. 2016. Projecting global biodiversity indicators under future development scenarios. *Conservation Letters* 9: 5–13.

Vitousek, P.M., H.A. Mooney, J. Lubchenco, and J.M. Melillo. 1997. Human domination of earth's ecosystems. *Science* 277: 494–499.

Walther, B.A., C. Boete, A. Binot, Y. By, J. Cappelle, J. Carrique Mas, M. Chou, N. Furey, S. Kim, C. Lajaunie, S. Lek, P. Meral, M. Neang, B.H. Tan, C. Walton, and S. Morand. 2016. Biodiversity and health: lessons and recommendations from an interdisciplinary conference to advise Southeast Asian research, society and policy. *Infection, Genetics and Evolution* 40: 29–46.

Wiens, J.A. 2013. Is landscape sustainability a useful concept in a changing world? *Landscape Ecology* 28: 1047–1052.

Wilcove, D.S., X. Giam, D.P. Edwards, B. Fisher, and L.P. Koh. 2013. Navjot's nightmare revisited: logging, agriculture, and biodiversity in Southeast Asia. *Trends in Ecology and Evolution* 28: 531–540.

Woodruff, D.S. 2010. Biogeography and conservation in Southeast Asia: how 2.7 million years of repeated environmental fluctuations affect today's patterns and the future of the remaining refugial-phase biodiversity. *Biodiversity Conservation* 19: 919–941.

Yasmi, Y., J. Broadhead, T. Enters, and C. Genge. 2010. *Forestry Policies, Legislation and Institutions in Asia and the Pacific: Trends and Emerging Needs for 2020.* Asia-Pacific Forestry Sector Outlook Study II, Working Paper No. APFSOS II/WP/2010/34.

5 How conserving biodiversity will help the ASEAN countries adapt to changing conditions

Jeffrey A. McNeely and Songtam Suksawang

Introduction: the geographical setting

The key to evolutionary success for both species and civilizations is the ability to adapt to changing conditions (Darwin, 1859). The early decades of the twenty-first century are providing multiple opportunities to adapt, with major changes affecting climate, demography, economic conditions, politics, and biodiversity, among many others. This chapter will focus on the importance of conserving biodiversity as a basis for providing the biological capacity to adapt to these changing conditions, however unpredictable the details of the changes may be.

Change is nothing new in Southeast Asia, which for convenience here will be confined to the ASEAN nations: Brunei Darussalam, Cambodia, Indonesia, Lao PDR, Malaysia, Myanmar, Philippines, Singapore, Thailand, and Vietnam. The ASEAN region has gained its biological and cultural diversity through a tumultuous environmental, social, and political history (Hall, 1968; McNeely and Sochaczewski, 1988). On the environmental front, the late Pleistocene Epoch Ice Ages that drew to a close some 11,700 years ago lowered sea levels in Southeast Asia by up to 120 meters, though over the past 17,000 years this extreme lowering covered only about 1,000 years while more modest lowering of 75 to 100 meters lasted about 9,500 years (Voris, 2000). Even this more modest lowering formed land connections among Java, Sumatra, Borneo, and the Asian mainland and enabled a relatively free interchange of species, including many that are now extinct, including three species of elephants, a large carnivore, and at least five species of grazing mammals (Hooijer, 1975).

Once the sea rose to today's higher levels, islands once again were isolated, providing conditions for rapid speciation and leading to the high levels of endemism now seen especially in the island nations of Indonesia (for example, almost half of its mammals are endemic, found only there) and the Philippines (60 percent of mammals, 35 percent of birds, 67 percent of reptiles, and 85 percent of amphibians are endemic) (Ceballos and Brown, 1995). The mainland ASEAN countries have long been connected to each other so they share many species, but with pockets of significant endemism, such as the Annamite Mountains along the borders of Lao PDR, Vietnam, and Cambodia.

More generally, their geographic histories have given ASEAN countries the highest level of country-endemic bird species (9 percent), mammals (11 percent), and vascular plants (25 percent) in the tropics (Sodhi *et al.*, 2010). This has led to ASEAN countries being included in four of the thirty-five global "biodiversity hotspots" – areas defined as having large numbers of endemic species whose habitats are facing significant anthropogenic threats (Mittermeier *et al.*, 2000). These include Indo-Burma (Myanmar, Thailand, Lao PDR, and Vietnam, as well as parts of India and southern China, with just 8.7 percent of its natural intact vegetation – NIV – remaining), the Philippines (8 percent of NIV remaining), Sundaland (a bit of southern Thailand plus Malaysia and Indonesia west of Lombok and Sulawesi, with 22.8 percent of NIV remaining), and Wallacea (Indonesia from Lombok and Sulawesi east to Halmahera and the Kai Islands, with 13.8 percent of NIV remaining) (Sloan *et al.*, 2014).

While climate change is nothing new, modern changes will affect all eco-systems (IPCC, 2014). Impacts will affect the coastline (rising sea levels, acidifi-cation, changing distribution of harvest fisheries, storm surges; see Woodruff and Woodruff, 2008), agriculture (changing rainfall patterns and temperatures, affecting seasonality and crop production), transport (threats of floods), water supply (changing rainfall patterns again), health (increasing likelihood of emerg-ing infectious diseases and invasion of harmful non-native species), and many other aspects of human well-being (Groves *et al.*, 2012). Climate change is also likely to increase the frequency and power of extreme natural events, sometimes called "natural disasters," that can include damaging storms that can lead to floods, landslides, and other disruptions (Anderson and Bausch, 2009; IPCC, 2012).

Irrespective of the long-term climate, the regional weather has remained a more immediate source of dynamism. The seasonal monsoon climate that char-acterizes the region ensures that the seasons change dramatically, with a rainy season of up to 160 days alternating with a slightly longer dry season; however, the monsoon system remains highly variable (Misra and DiNapoli, 2013). The climate change now affecting the planet is likely to affect the monsoon weather system, possibly leading to a delay of up to fifteen days in the onset of the seasonal rains and to more intensive flooding in some areas (Loo *et al.*, 2015). But the ecosystems of the region, with the great diversity of species and the genetic com-position they contain, are adapted to variable weather, including the typhoons that affect some parts of the region more than others. The Philippines is espe-cially vulnerable to these extreme events, which may become even stronger with climate change; Typhoon Haiyan in 2013 may have been the strongest storm ever measured (Emanuel, 2013).

Climate and weather are only part of the geographical story, which is enriched by active geological conditions. ASEAN is part of the "Ring of Fire," with volcanoes a regular feature of the landscape in Indonesia and the Philippines, accompanied by earthquakes and tsunamis that affect the entire region (as in 2004 in the Andaman Sea, which killed over 225,000 people, mostly in Indonesia; see Lay *et al.*, 2005). Many volcanoes in the region have had major impacts on both ecosystems and people, even beyond Southeast Asia. Examples include Mount Pinatubo in the Philippines, whose 1991 eruption had global impacts on climate (Fiocco *et al.*, 2011); Krakatau in Indonesia, whose 1883 eruption was heard

4,500 kilometers away, destroyed 165 villages in western Java, and eventually enabled the conservation of the Javan rhinoceros in the Ujung Kulon Nature Reserve (Hoogerwerf, 1970); and Mount Tambora, whose 1815 eruption was the largest volcanic eruption of the nineteenth century, devastated the islands of eastern Indonesia (Raffles, 1817/1978), contributed to Napoleon's defeat at the Battle of Waterloo, and led to 1816 being "the year without a summer" in Europe (de Boer and Sanders, 2002). Even more dramatic was the Toba eruption in northern Sumatra 73,500 years ago, the largest eruption in the past several hundred thousand years; it is thought to have significantly accelerated the shift to the glacial conditions that locked up water as ice and lowered sea levels (Rampino and Self, 1992). The next big volcanic eruption is unpredictable, as are its implications for climate and human well-being.

A brief history of humans and biodiversity in Southeast Asia

Members of the human genus *Homo* adapted to the tumultuous geographic history of Southeast Asia beginning at least 500,000 years ago, with *Homo erectus* living in Java (and China) and presumably elsewhere in the region (though probably not in the Philippines, which remained isolated by the sea; Rabett, 2012). Other hominid species, including *Homo neanderthalensis* and a little-known form known as Denisovans (possibly a subspecies of our species) were living in various parts of Southeast Asia, judging from genetic evidence still found in modern species as far east as Papua New Guinea (Reich *et al.*, 2011; Vernot *et al.*, 2016). A dwarf species known as *Homo floresiensis* is known from as early as 60,000 years ago (Sutikna *et al.*, 2016) and may have survived in eastern Indonesia as recently as 12,000 years ago, living among dwarf stegodons (*Stegodon trigonocephalus*, a relative of the elephant, now extinct) and giant monitor lizards that survive as Komodo dragons (*Varanus komodoensis*; see Audley-Charles and Hooijer, 1973; Brumm *et al.*, 2010). The arrival date of modern humans in Southeast Asia remains somewhat controversial, though early *Homo sapiens* fossils discovered in Lao PDR in 2009 have been dated at 46,000–63,000 years old (Demeter *et al.*, 2012). The different species of humans may well have interacted on occasion toward the end of the Pleistocene and even into the Holocene Epoch, which began about 11,700 years ago.

In any case, rich fossil evidence indicates that humans have evolved alongside the many other species that inhabited Southeast Asia. Their local adaptations led humans to develop numerous cultures, local knowledge, and languages that enabled them to populate the forests, shorelines, grasslands, and other ecosystem types that characterized the region. This cultural diversity mirrors the rich biological diversity of the region, with over 1,000 languages still spoken, though many by only a few older people (Maffi, 2001).

While hunting and gathering was apparently the main ecological niche of the ancestral humans, fire and other forms of ecosystem modification were widely used and contributed to some of the world's first domestication of plants and animals (Solheim, 1972), possibly linked to the improved climate beginning about 10,000 years ago (Gupta, 2004). This early domestication was enabled by the plant diversity of Southeast Asia that was accompanied by hominid experience in harvesting seeds, fruits, and vegetables from numerous local species that were suitable for domestication (rice, eggplants, coconuts, bananas, gourds, sugarcane, bamboo, yams, grapefruit, mango, and many other fruits and vegetables are endemic to the region; see Vavilov, 1951). Southeast Asia also supported, and still supports, many wild animals, such as pigs, water buffalo, cattle, chickens, and ducks, that were suitable for domestication and ultimately became features of the farmyards of the region (and indeed throughout the world, along with some of the crops that originated here). The genetic diversity of the remaining wild relatives of domestic species, found largely in the forests and wetlands, could have considerable value in enabling the domestic species to adapt to changing conditions through cross-breeding that draws on modern biotechnology (Hunter and Heywood, 2011).

The earliest agriculture was probably some form of shifting cultivation, based on burning vegetation to clear the land for planting a wide variety of crops, then abandoning the land when weeds (in other words, pioneer species that produce little of interest to humans) moved in after a few years, and subsequently moving into another forested area to clear. This form of agriculture has affected virtually all of the forests in Southeast Asia (Spencer, 1966) and helped provide an optimal habitat of low-growing vegetation for some species of large mammals, especially wild cattle (Wharton, 1968) as well as elephants, pigs, deer, and their predators. Shifting cultivation is still commonly practiced in the hilly parts of Southeast Asia, though it is declining in relative importance even as its negative impacts are increasing. With growing population and economic activity, the clouds of smoke from hilly fires are becoming increasingly harmful (Fox *et al.*, 2014; Mukherjee and Sovacool, 2014), with the cost of the damage now exceeding 15 billion dollars per year.

A major advance was the development of irrigated rice cultivation, probably based on ideas imported from India (Gesick, 1983). Growing irrigated rice led to the production of agricultural surpluses that could support civilizations that had cities, monumental architecture, a division of labor, status hierarchies, settlement patterns that include both urban and rural populations, and state religions (Childe, 1929). Coming relatively late to Southeast Asia, civilizations came to dominate the large river valleys and converted vast seasonal floodplains and lowland forests, then rich in wildlife, into biologically simplified rice fields that produced more food for humans (Bellwood, 1985).

These civilizations based on irrigated rice absorbed many of the resident ethnic groups, converted many lowland ecosystems to less diverse agricultural systems, and drew on the timber, wildlife, and medicinal plants from the surrounding forests. Archeological and historical sites of the now-departed civilizations based on Indian religions and ideas (Coedes, 1968), such as Angkor Wat, Borobudur, Sukhothai, and Pagan, draw large numbers of tourists to admire the artistic creativity of the ancient civilizations. Their descendants now form the dominant cultures of ASEAN. The impacts of modern civilization (now a singular because globalization has become dominant in virtually every country) on the most productive lands have supported wealthy cultures, at the cost of losing considerable wildlife habitats that are now all but forgotten except in protected areas (Suksawang and McNeely, 2015).

Politically and economically, the cultures of Southeast Asia have long been influenced by the states to the west (now India, Sri Lanka, and Bangladesh) and north (now China, but previously numerous dynastic entities that were themselves often at war with each other; see Sawyer, 2011; Wilkinson, 2015). Southeast Asia served as a trading link between South and East Asia as early as 2,500 years ago (Higham, 2002), helping to keep a flow of goods and ideas moving through the region (Wade, 2009). Domestic unrest and the search for new resources in what are now China and India may have led people living in these relatively advanced regions to visit Southeast Asia on a regular basis; some settled permanently, and were absorbed into the dominant local cultures.

The regular flow of people and innovations from the west and north enriched Southeast Asia, bringing new cultures, technological advances, such as new metals (Murillo-Barroso *et al.*, 2010), and new relations to natural resources. The latter may have contributed to the extinction of some species adapted to the floodplains, grasslands, and lowland forests that were converted into irrigated rice fields, especially larger species such as relatives of modern elephants (*Stegodon*, *Paleoloxodon*) and rhinoceroses (at least three now-extinct Pleistocene species) as well as numerous other species (Louys *et al.*, 2007). Even in very recent times, a swamp deer – *Recervus schomburgki* – endemic to Thailand survived until 1938 before the last one was killed, ironically in a Buddhist temple (Lekagul and McNeely, 1977). Other species that once were common in the lowlands of Thailand and Myanmar were forced to retreat to less desirable hilly habitats, where many have been reduced to small populations and are now considered "Endangered" (IUCN, 2016). These include the wild water buffalo *Bubalus bubalis*, the Javan rhinoceros *Rhinoceros sondaicus*, the Asian elephant *Elephas maximus*, the hog deer *Axis porcinus*, and the tiger *Panthera tigris*. While lowland farmers were probably not the only factor in the extinctions, they were directly responsible for the habitat change and were present as many species of mammals disappeared or moved to the hills.

In addition to the age-old trading links to South and East Asia, Southeast Asia was certainly known to Europeans by the second century CE, with trading of natural biological resources such as spices, ivory, rhinoceros horn, and medicinal plants established with at least the Romans (facilitated by traders in the Middle East who had long commercial interests with India and points east). Chinese traders were especially active in the thirteenth century, even reaching the European trading center of Venice in 1434 with a significant fleet that surely carried substantial spices and other trade goods from Southeast Asia (Menzies, 2008). This may well have served as further motivation to European traders to gain more direct access to Southeast Asia's resources.

Although the Italian merchant traveler Marco Polo stopped off in Southeast Asia in 1293 on his return to Venice after spending twenty-four years in China, having followed the inland Silk Road (Polo and Latham, 1958), regular visits to the region from Europe began between 1420 and 1436, when the Italian explorer Niccolo Da Conti visited several times (Hall, 1968). This helped stimulate numerous additional efforts to reach the Indies (including by Christopher Columbus, whose "short cut" by heading west from Europe was interrupted by his discovery of the Americas).

The Portuguese were the first Europeans to arrive in Southeast Asia in force, seeking especially spices (an early European interest in Southeast Asia's biodiversity) and gems (a continuing interest in resources valued more for their rarity and beauty than their utility). They had established Malacca as a trading port by 1509 and reached trade agreements with the Kingdom of Ayutthaya (now part of Thailand) in 1511. The Portuguese monopoly began to weaken in the early 1600s, as the Netherlands East India Company established a base in what is now Jakarta; in 1641 the latter also replaced the Portuguese in Malacca. The Spanish established themselves in the Philippines, and the French in Laos, Cambodia, and Vietnam. The British East India Company was only a minor player in the region until 1786, when it settled in Penang. Thereafter, though, the British remained in Malaysia until World War II (Hall, 1968). The British took Java temporarily from the Dutch as part of the spoils of winning the Napoleonic Wars. The colony was governed by Thomas Stamford Raffles, who also founded Singapore in 1819 as a trading post (Raffles, 1817/1978). The British also subsequently established a presence in Burma (now Myanmar), especially for the harvesting of teak, which was needed by the Royal Navy (Reid, 1988–93).

Southeast Asia was thus effectively carved up by the competing European colonial powers, primarily for enhancing their access to biological resources (Hall, 1968; Osborne, 1985). Even Thailand, which remained without a colonial occupier, was significantly affected by European trading influences and by the French taking parts of Siam's land west of the Mekong as recently as 1904–1907 (Usher, 2009). The European colonial era lasted until the late 1940s (and 1954 in Vietnam), though the transition happened at different rates in different countries, with different means and different outcomes.

While the formerly colonized ASEAN nations have been self-governing for well over half a century, the influence of the colonial powers on biodiversity has lingered. They established new forms of resource governance that have persisted, nourished the international trade that has continued to dominate the region's economies, and converted diverse nature into fewer commodities that are still found in the market place. The current approaches to forestry throughout Southeast Asia are generally based on nineteenth-century European ideas that treated forestry as above all an economic enterprise, with trees managed for the profit they could yield in terms of timber rather than as important habitats for forest-dwelling people and other species and as providers of multiple ecosystem services. Sustainability of forest harvests is given lip service through some selective logging, and some European-style teak and rubber plantations have been reasonably successful, but the result has been the steady deforestation of Southeast Asia that has only accelerated with increasing market pressures

(Zimmerman and Kormos, 2012). The last biodiversity-rich forests are increasingly confined to protected areas.

When conservation measures are adopted, the models used to protect species or establish protected areas are often based on Western examples (especially the national park model, developed in the USA) that may not always be appropriate to the conditions of Southeast Asia (Usher, 2009). New approaches to resource management are clearly required.

The ASEAN countries and biodiversity

ASEAN was founded in 1967 with just five members: Indonesia, Malaysia, the Philippines, Singapore, and Thailand. The other five members joined over subsequent years, with the current membership completed in 1999. While the ten ASEAN countries are united as a geographic, political, and economic organization, they are widely variable, ranging in size from 692 km² (Singapore) to 1.9 million km² (Indonesia); in population from 707,000 (Brunei) to over 234 million (Indonesia); in population density from 27 people per km² (Lao PDR) to 7,197 per km² (Singapore); and in annual per capita gross domestic product from US$1,740 (Myanmar) to nearly $65,000 (Singapore) (see Table 5.1). Political systems are also widely variable, posing some challenges to regional cooperation (Acharya, 2009).

Land use is similarly variable, with forest covering just 3.3 percent of Singapore but over 70 percent of Brunei (see Table 5.2). However, the forest coverage figures of the other ASEAN countries need to be viewed with considerable caution in regards to biodiversity since many governments use dubious definitions of "forest." Thailand, for example, considers forest to be any land not owned by any person under the Land Code, even if the land has few, if any, trees. This gives its Forest Department the right to manage these "forests" as it sees fit, irrespective of the traditional forest-dwelling people who often lack the necessary formal landownership documentation (Usher, 2009).

Table 5.1 Size, population and per capita GDP of ASEAN countries

Country	Size (km²)	Population (thousands)	Population per km²	Per capita GDP in US$
Brunei	5,770	707	71	53,431
Cambodia	181,035	13,396	74	2,576
Indonesia	1,904,556	234,181	122	5,214
Lao PDR	236,800	6,636	27	3,068
Malaysia	329,750	28,307	86	17,748
Myanmar	678,500	50,496	74	1,740
Philippines	300,000	94,013	313	4,682
Singapore	692	4,988	7,197	64,584
Thailand	514,000	63,525	124	9,932
Vietnam	329,560	85,790	260	4,012

Source: (of GDP) IMF, 2014 (other sources give slightly different figures, with those of the World Bank higher).

Table 5.2 Land use and population distribution in ASEAN countries

Country	Forest (%)	Agriculture (%)	Protected areas (%)	Urban population (%)	Rural population (%)
Brunei	71.8	2.2	44.0	76	24
Cambodia	56.5	32.0	26.2	20	80
Indonesia	51.7	30.1	14.7	51	49
Lao PDR	67.9	10.3	16.7	35	65
Malaysia	62.0	24.0	18.4	73	27
Myanmar	48.2	19.2	7.3	33	67
Philippines	25.9	40.6	10.9	49	51
Singapore	3.3	1.0	5.4	100	0
Thailand	37.2	41.2	18.8	34	66
Vietnam	45.0	35.0	6.5	32	68

Source: World Bank, 2014.

Note: Protected areas often overlap with forested lands.

Other ASEAN countries use similar measures essentially to nationalize their forests, which are then rented out to domestic and foreign forestry companies who pay royalties on the timber they harvest. Policies for conserving biodiversity in these timber concessions have shown very limited success (Dennis *et al.*, 2008).

"Forest" in most ASEAN countries also includes tree plantations that support so little biodiversity that they are better considered "green deserts." For example, oil palm plantations cover nearly 50,000 km² of Malaysia and over 80,000 km² of Indonesia (Obidzinski, 2013), while rubber plantations cover 35,000 km² of Indonesia, 14,000 km² of Malaysia, 26,000 km² of Thailand, and substantial parts of Cambodia and Vietnam (FAO, 2014). Especially in Indonesia, oil palm is grown on peat soils on land cleared of native forest. Tropical peat forests store over 400 tons of carbon per hectare as compared to about 230 tons per hectare for tropical rainforests (UNEP, 2014a), making them by far the most effective type of forest for storing carbon (Schrier-Uijl *et al.*, 2013).

The deforestation rates cited in Table 5.3 are net of the planting of such simplified new tree monocultures, including plantations of fruit trees, non-native *Eucalyptus*, and *Acacia*. "Forest" is therefore not necessarily a good indicator of rich biodiversity, and the relatively low rates of deforestation are not a good indicator of the status of native forests or of biodiversity conservation. Nor can it be taken to represent remaining natural forest, which is far less than these figures suggest (Sloan *et al.*, 2014). The figures cited for forest loss in Indonesia, for example, are less than half the forest loss determined by more objective assessment of remote sensing data, which found over 6 million hectares lost from 2000 to 2012, with over 40 percent of that loss coming from areas that were legally protected against such logging; in 2012 alone, Indonesia led the world in clearing 840,000 hectares of old-growth forest (Margono *et al.*, 2014).

Agriculture, widely considered the major factor in transforming natural habitats (MEA, 2005), is also variable in ASEAN but covers over 30 percent of

Table 5.3 Production of selected biological resources in ASEAN countries

Country	Fisheries harvest (000 tons)	Roundwood (000 cubic meters)	Woodfuel (000 m³)	Industrial roundwood (000 m³)	Deforestation (annual %, 2005–2010)
Brunei	2,100	119	12	107	.47
Cambodia	190,000	8304	8162	114	1.26
Indonesia	5,661,681	117,522	54,917	62,606	.71
Lao PDR	30,900	6777	5,922	855	.60
Malaysia	1,375,061	20,534	2,711	17,823	.64
Myanmar	3,329,447	43,364	38,286	5,078	.99
Philippines	2,263,206	16,002	12,144	3,858	.77
Singapore	1,618	–	–	–	.00
Thailand	1,751,037	33,680	19,080	14,600	.37
Vietnam	2,502,504	27,100	20,400	6,700	.63

Sources: Froese and Pauly, 2011; FAO, 2014; Mongabay, 2012 for deforestation rates.

the land of Cambodia, Indonesia, and Vietnam, and over 40 percent of the Philippines and Thailand. However, the expansion of agricultural land seems to have slowed considerably, with virtually all of the best agricultural land now cultivated and more people moving to the cities (half of the ASEAN countries have half of their populations living in cities).

And perhaps most important here, all ASEAN countries have allocated land to protected areas, ranging from a low of 5.4 percent of the land in mostly urban Singapore to 44 percent in mostly forested Brunei. The parties to the Convention on Biological Diversity (CBD) have agreed a target of at least 17 percent of land and inland waters as protected areas by 2020 (SCBD, 2011), a figure that has already been exceeded by Brunei, Cambodia, Malaysia, and Thailand, with Indonesia and Lao PDR close behind (Table 5.2). All except land-locked Lao PDR have also established promising systems of coastal and marine protected areas, seeking to meet the CBD target of 10 percent of marine habitats protected by 2020 (Cheung *et al.*, 2002).

ASEAN is an important part of the global economy and much of the world is drawing on the natural productivity of the region's land and seas, especially seeking fish, forest products, and agricultural crops (Table 5.3). While this is not the place for a detailed discussion of the impact of this trade on native biodiversity, it is clear that the harvests of fish and timber, and the clearing of forested land for agriculture, have had profound effects on native ecosystems (see, for example, Lenzen *et al.*, 2012; Polasky *et al.*, 2004). Trade has also led to a significant increase in invasive alien species in all ASEAN countries, to the detriment of native ecosystems (Pallewatta *et al.*, 2003). These effects seem to be worsening.

The ASEAN countries have long recognized that they need to cooperate in conserving their natural patrimony. As early as 1985 (before the concept of biodiversity was in general use), the governments drafted an ASEAN Agreement on the Conservation of Nature and Natural Resources, under which they would undertake measures

> to maintain essential ecological processes and life-support systems, to preserve genetic diversity, and to ensure the sustainable utilization of harvested natural resources under their jurisdiction in accordance with scientific principles and with a view to attaining the goal of sustainable development.
>
> (UNEP, 2014b: 3)

Unfortunately, the ASEAN Agreement on Conservation has never been formally agreed and enforced, perhaps overtaken to some extent by the global 1992 Convention on Biological Diversity, which *has* been ratified by all ASEAN countries. This latter convention covers many of the issues identified in the earlier ASEAN agreement, but with a particular focus on biodiversity, defined as "the variability among living organisms from all sources including inter alia, terrestrial, marine and other aquatic ecosystems and the ecological complexes of which they are part; this includes diversity within species, between species and of ecosystems" (CBD, 1992: 3).

This definition has now been widely adopted and will be used in this chapter. However, the only element of it that is readily measured is species, often used as a surrogate for biodiversity. Table 5.4 summarizes the species diversity of the ASEAN countries, indicating the remarkable biological richness of the region, which is based on its geographical history. The mainland ASEAN countries share many species, though the Annamite Range, which links Cambodia, Lao PDR, and Vietnam, has many remarkable and recently discovered endemic species, including the black muntjac *Muntiacus truongsonensis* (Giao et al., 1998), the giant muntjac *Megamuntiacus vuquangensis* (Schaller and Vrba, 1996), the

Table 5.4 Species diversity of ASEAN countries

Country	Plants	Mammals	Birds	Reptiles	Amphibians	Fish
Brunei	4,030	121	471	50	98	491
Cambodia	8,000	212	536	176	63	955
Indonesia	28,000	515	1536	449	242	4724
Lao PDR	9,000	178	731	186	77	585
Malaysia	23,500	210	740	250	250	1994
Myanmar	7,000	257	1061	156	156	1043
Philippines	9,250	210	612	270	270	3435
Singapore	2,282	85	375	91	91	703
Thailand	12,000	294	942	325	141	2276
Vietnam	9,628	231	889	296	162	2536

Sources: Fish data from Froese and Pauly, 2011; data for other taxa from multiple sources, not always in agreement.

Notes: Fish are both freshwater and marine (landlocked Lao PDR – all freshwater). All figures are subject to change as more fieldwork is conducted and more authoritative lists are compiled.

crested gibbon *Nomascus annamensis* (Thinh *et al.*, 2010), and a forest antelope known as the saola, *Pseudoryx nghetinhensis* (Dung *et al.*, 1993). It seems likely that the Annamite Range was a refuge that conserved species that may have been more widespread in the Pleistocene Epoch (Rabinowitz, 2001).

While the biodiversity of ASEAN is already rich in global terms, many more species remain to be discovered. A few recent examples will indicate how little is known, and how much remains to be discovered:

- Some 31 percent of the species of amphibians known from Cambodia, Lao PDR, and Vietnam in 2005 had been described since 1997 (Bain *et al.*, 2007). Many more are awaiting discovery once more sophisticated molecular studies are applied to Southeast Asian amphibians (Stuart *et al.*, 2006).
- More than 1,300 new species have been described from the Lower Mekong Basin (Thailand, Lao PDR, Cambodia, and Vietnam) since 1997, with nearly 400 in 2012–2013 alone (WWF, 2014). The latter include a giant flying squirrel from Lao PDR, *Biswamoyopterus laoensis* (Sanamxay *et al.*, 2013); a zebra-striped lizard, *Cyrtodactylus phuketensis*, from the densely populated Thai island of Phuket (Sumontha *et al.*, 2012a); a new tailorbird from Cambodia, *Orthotomus chaktomuk*, (Mahood *et al.*, 2013); a very large flying frog from Vietnam, *Rhacophorus helenae* (Rowley *et al.*, 2012); and a skydiving gecko lizard from Thailand, *Ptychozoon kaengkrachanense* (Sumontha *et al.*, 2012b).
- Many other major discoveries are appearing at a regular rate, such as the new species of snub-nosed monkey *Rhinopithecus strykeri* found in Myanmar in 2011, whose closest relatives live in China, across the Salween and Mekong rivers (Geissmann *et al.*, 2011). Another is a new species of shark, *Hemiscyllium halmahera*, found in 2013 in Indonesia, which uses its fins as legs and walks along the sea floor (Allen *et al.*, 2013).
- The "discovery" of a new tree, *Dracaena kaweesakii*, in Thailand in 2013 (Wilkin *et al.*, 2013) was the first scientific description of a tree that is commonly used in Thai horticulture and planted on the grounds of many Buddhist temples. This is a good illustration of how even fairly common and well-known species remain "unknown to science."
- Perhaps even more dramatic is the discovery of the Laotian rock rat *Laonastes aenigmamus*, which is so distinctive that it is often placed in its own family, Laonastidae (Jenkins *et al.*, 2004), though some consider it to be part of the family Diatomyidae, previously known only from 11-million-year-old fossils (Dawson *et al.*, 2006). It was found in a food market, so it was well known to local consumers as a species from the limestone habitats that are widespread in Southeast Asia and host to many endemic species (Clements *et al.*, 2006), including several others that have been recently discovered (Musser *et al.*, 2005).

Such recent discoveries, along with many others throughout the region, indicate that many species (especially of plants and invertebrates) remain unknown to modern science. This seems to justify continued support for field studies that seek to identify the biological richness of the ASEAN countries.

Especially since the entry into force of the Convention on Biological Diversity in 1993, the ASEAN countries have expanded their protected areas, investment in modern biotechnology, and conservation of genetic resources. Some of these activities have been supported by the Global Environment Facility (the funding mechanism of the CBD), and by numerous governments, international agencies, and non-governmental conservation organizations (both national and international).

Meanwhile, ASEAN has significantly expanded its own cooperation on biodiversity, including establishing the ASEAN Center for Biodiversity. This started in 1999 as a Philippine government project funded by the European Union to support the biodiversity management needs of the ASEAN member states through capacity-building, training, research, information management, and technical advice (Elliott, 2003). In 2005, it became an ASEAN program, with funding from multiple sources. The ASEAN members also agreed in 2003 to establish ASEAN Heritage Parks, which are designed to promote cooperation among the members and to recognize outstanding natural sites that are particularly rich in biodiversity. Seventeen ASEAN Heritage Parks have been established to date, with supporting activities coordinated by the ASEAN Center for Biodiversity (Ahman, 2006).

The international support for conserving biodiversity, especially through the establishment of protected areas, has increased substantially in recent years from United Nations agencies, governments, and both national and international

Table 5.5 Species on the IUCN Red List as Threatened (Critically Endangered, Endangered, and Vulnerable categories only), and Threatened as percentage of total taxa

Country	Plants	Mammals	Birds	Reptiles	Amphibians	Fish
Brunei	237 (5.8%)	34 (28%)	24 (5.1%)	7 (14%)	3 (3%)	7 (3.5%)
Cambodia	347 (23%)	37 (17.4%)	26 (4.8%)	19 (10.7%)	3 (4.7%)	40 (4.7%)
Indonesia	1172 (4.1%)	185 (35.9%)	121 (7.8%)	32 (7.1%)	32 (13.2%)	145 (14.5%)
Lao PDR	369 (4.1%)	45 (25.2%)	23 (3.1%)	16 (18.6%)	5 (6.4%)	55 (9.3%)
Malaysia	1588 (10.9%)	71 (33.8%)	45 (6%)	28 (11.2%)	47 (18.8%)	71 (11.5%)
Myanmar	574 (8.2%)	46 (17.9%)	44 (4.1%)	29 (18.6%)	0 (–)	40 (8.9%)
Philippines	660 (13.4%)	38 (18%)	74 (12.1%)	39 (14.4%)	48 (17.8%)	72 (21.8%)
Singapore	346 (15.1%)	11 (12.9%)	15 (4%)	5 (5.4%)	0 (–)	25 (19.1%)
Thailand	825 (6.8%)	57 (19.3%)	47 (4.9%)	27 (8.3%)	4 (2.8%)	96 (16.8%)
Vietnam	784 (8.1%)	54 (23.3%)	44 (4.9%)	41 (13.9%)	17 (10.4%)	73 (11.5%)

Source: IUCN, 2016.

Notes: ASEAN countries vary widely in the availability of data about category of threat under the IUCN Red List criteria. Many of the amphibian data seem dubious and inadequate, while the fish data seem more reliable, perhaps because their economic importance leads to better data collection. Percentage of fish threatened is based primarily on freshwater fish because status of marine species is poorly known.

non-governmental conservation organizations. This can be taken as an indication of how important this issue has become in a time of rapid loss of native biodiversity, and when the CBD has called for stronger support to protected areas (SCBD, 2005). Table 5.5 summarizes how many of the vertebrates and plants in the region are threatened, with mammals indicated as being of particular concern. Judging from the status of species in the ASEAN region, far more needs to be done to conserve them.

With the full establishment of the ASEAN Economic Community in 2015, economic conditions may change in unexpected ways, requiring the protected areas system to be adaptable to any new conditions that may arise. One possibility is the development of mechanisms in environmental management at the regional level that will lead to a more sustainable use of natural resources in the ASEAN region, perhaps beginning with the ASEAN Heritage Parks. The cooperation among Cambodia, Lao PDR, Thailand, and Vietnam on conservation corridors in the Greater Mekong sub-region, with support from many external development agencies, is an example that could be more widely emulated (Carew-Reid *et al.*, 2007). More generally, the establishment and effective management of protected areas is an especially promising area of ASEAN collaboration in conserving biodiversity. A regional approach can provide supranational monitoring, priority-setting, and collaborative action while avoiding duplication, filling gaps, and helping to avoid the displacement of negative impacts of logging and poaching across borders (Corlett, 2013). However, the implementation gap yawns widely.

Some benefits of biodiversity and protected areas for the ASEAN region

Given that the ASEAN region has always been geographically, culturally, and biologically dynamic, the surviving natural ecosystems may be key contributors to future forms of sustainable development. Because these systems have already shown an ability to adapt to dramatic geographic, climatic, and cultural changes, they can be an essential part of any measures designed to enable a country to adapt to the changing conditions the future will surely bring. This section will focus especially on the contributions of protected areas to such adaptations since these are where the richest species diversity is found, where "nature's toolbox" is at its fullest.

Conserving biodiversity needs broad public support, and one means of building such support is to highlight the benefits of biodiversity to people. Many will argue, rightly, that biodiversity is valuable in its own right and humans should avoid any forms of development that threaten biodiversity. The benefits biodiversity provide to nature are sufficiently well known that they need no further discussion here (see, for example, Wilson, 1986), but it is worth noting that these critically important benefits have been insufficient to ensure biodiversity conservation. This section therefore takes a strongly human-oriented perspective by focusing on the many kinds of benefits protected areas provide to people, identified through the concept of ecosystem services (Daily and Ellison, 2002; McNeely *et al.*, 2009).

As explained in the Millennium Ecosystem Assessment (a four-year study of the consequences of ecosystem change for human well-being, carried out by more than 1,360 experts from all over the world), ecosystem services are simply the benefits people receive from nature, and ultimately from ecosystem functions (MEA, 2005). Treating ecosystem functions from the anthropocentric view of ecosystem services gives them an explicitly economic dimension. If effective eco-system management is yielding economic benefits, the argument goes, then those delivering the benefits should be compensated, especially if they incur costs in delivering the benefits. Extending this to protected areas, these sites deliver many valuable ecosystem services. The services are often in the form of public goods (that is, goods whose consumption by any one person does not affect their potential for consumption for others), and need to be recognized by providing the sites with sufficient support so that they can be managed effectively and continue to provide ecosystem services through effective conservation of biodiversity.

All of the ASEAN countries have already established systems of protected areas that have biodiversity conservation as a major objective (see Table 5.2). Their conservation leaders realize that sustaining the rich diversity of native species and ecosystems depends on protected areas that ensure their continued survival, even saving some from the brink of extinction (see Table 5.5). The management effectiveness of these areas, however, remains mixed (Hockings *et al.*, 2012; Don Carlos *et al.*, 2013).

The general principle promoted here is that conserving biodiversity is linked to human well-being, and that protected areas often contain the healthiest eco-systems in the regions in which they occur. By definition, a healthy ecosystem has the ability to maintain its structure and functions over time in the face of external stress (Costanza and Mageau, 1999). In other words, the healthier the ecosystem, the higher is its likelihood of maintaining its capacity to adapt to change, espe-cially if it is linked to other ecosystems to increase its effective size. The benefits provided by conserving biodiversity through well-managed and well-located protected areas include those discussed below, but there are many others.

The Millennium Ecosystem Assessment divided ecosystem services into four broad categories (MEA, 2005), discussed below in terms of their contributions to conserving biodiversity and relevance to protected areas.

Provisioning services deliver crops, fish, timber, bamboo, rattan, medicinal plants, fresh water, firewood, fodder, wild game, and many other goods to people. The supplies of those quantified in Table 5.4 do not need to be covered further here, but many of those of greatest interest for conservation of biodiversity and protected areas are not so easy to count. For example, about a billion people globally rely on wild products harvested from nature, with wild meat providing 30–80 percent of the protein for many rural communities (Hoffman and Cawthorn, 2012). These products tend to be traded on informal markets so they are difficult to assess in economic terms, and hence are not given much attention by policy-makers. In Southeast Asia, many are harvested illegally from protected areas, judging from the considerable quantities that are traded in contravention of established law (Chouvey, 2013).

The provision of medicinal plants is of broad interest because of their direct link to human health, either when consumed directly or when included in pharmaceuticals. Many species of plants contain bioactive organic compounds that are potentially useful for pharmaceutical applications, and thousands of these are harvested for local use (Chivian and Bernstein, 2008; Miththapla, 2006). While research has been done on the status in the wild of relatively few medicinal plants (Mulliken and Crofton, 2008), they undoubtedly need the attention of protected area managers in the sites where they occur.

ASEAN countries support a rich diversity of medicinal plants (Williamson, 2003; Saralamp, 1996). Many of these are found in protected areas, and some may find their best habitats within protected areas or are even confined to them. Medicinal plants support human health in rural areas, in cities where some plant-based medicines are still in wide use, and among researchers who find medicinal plants to be useful when seeking treatments for many diseases, including cancer (Saetung et al., 2005). The role of protected areas in conserving medicinal plants could become crucial as human health faces novel threats under conditions of globalization and climate change (which create new conditions and promote rapid movement of pathogens, as in the case of SARS (severe acute respiratory syndrome), which originated in China in 2002 and soon spread to many countries in Asia, Europe, and North America; see Olsen et al., 2003). A little-known benefit to human health is that protected areas and the biodiversity they support can also reduce the emergence and transmission of infectious diseases (Keesing et al., 2010).

Protected areas in Southeast Asia also support many wild flowers and fruits of economic importance; some of these could be harvested as breeding stock without damaging the protected area (Sakai, 2002). The flowers, fruits, and nuts of many trees, such as members of the family Dipterocarpaceae, are especially important to the health of forests and depend on pollinators that increase in numbers when the trees are in flower and thereby support forest biodiversity. Some of the economically significant wild fruits from Southeast Asian forests have contributed genes to domesticated forms, including mangoes (*Mangifera*), mangosteens (*Garcinia*), rambutans (*Nepthelium*), durians (*Durio*), bananas (*Musa*), and many others. Genes from these wild species could be of considerable assistance to plant breeders. At least some of them have been planted by forest-dwelling peoples for hundreds, and perhaps thousands, of years and have contributed to their current distribution and abundance (Hunt and Rabett, 2014).

Finally, the coastal zone has been a productive habitat for people for thousands of years, with the inhabitants drawing on the multiple biological resources available there. These include shellfish, crabs, sea cucumbers, sea urchins, and a rich diversity of fish (see Table 5.4). Reef fisheries in Southeast Asia generated over US$2.4 billion per year in the early 2000s (Burke et al., 2002), and undoubtedly are worth much more today. However, destructive fishing practices have destroyed many of Southeast Asia's coral reefs, and climate change is adding more pressure to them (Bruno and Selig, 2007; Mascia et al., 2010). The supporting services provided to coral reefs and fisheries by protected areas are discussed below.

Supporting services include benefits to people from ecosystem functions, such as soil formation, photosynthesis, nutrient cycling, and fisheries production. These are public services that are difficult to quantify, but they are in a sense beyond value because all life depends on them (albeit sometimes indirectly). Considerable scientific research is seeking to enhance the productivity and efficiency of these services, especially photosynthesis (Evans and von Caemmerer, 2011), though the natural systems have already shown their value in capturing the energy of the sun.

A substantial literature has developed about the benefits of protected areas in supporting marine fisheries (Davies *et al.*, 2012). These include: enabling fish to reach adulthood within the marine protected area before they migrate into adjacent fishing grounds, where they increase fishing yields; increasing reproductive potential within the protected area, thereby seeding surrounding fished areas with eggs and larvae that enhance stock recruitment and recovery; and acting as scientific reference areas that enable assessment of trends in stock dynamics among populations of fish that are not subject to harvesting. All of these benefits depend on effective management of the marine protected area, especially to enforce no-take zones. The delivery of these services is becoming ever more urgent as the human impacts on the oceans continue to increase (Watson *et al.*, 2014).

Stimulated especially by innovative approaches to involving local fishermen in protecting fish habitats in the Sumilon and Apo islands in the Philippines since 1974 (Alcala, 2001), evidence from many parts of the world has shown that density of fish populations, size of fish and other aquatic species, and diversity of species are all significantly higher inside the no-fishing zones of protected areas than outside. Even small marine protected areas are effective, but larger reserves deliver greater benefits (Halpern, 2003). Experience has shown that almost any marine habitat can benefit from the establishment of a protected area, but success will depend on its design, management, and evaluation (Edgar *et al.*, 2014). This experience has encouraged the growing efforts of ASEAN countries to establish marine protected areas, though management effectiveness still needs to be improved (Hockings *et al.*, 2012).

Protected areas in mountains, along watercourses, on the coastline, and in the sea also help to provide resilience to extreme natural events (sometimes called "natural disasters," though they are disasters only when humans are involved; see McNeely, 2016). By keeping land in mature vegetation, protected areas can help prevent the effects of heavy rainfall, storm surges along the coast, and earthquakes (and accompanying landslides) wherever they may occur. Forested protected areas can also serve as windbreaks, protecting villages against powerful seasonal winds. Healthy mangrove forests help to prevent storm surges, but if they are to be effective in protecting against the most extreme events, such as tsunamis, they need to be at least 150 meters wide (Dahdouh-Guebas *et al.*, 2005). The healthy ecosystems that protected areas provide can also help restore floodplains after heavy rainfall, and provide emergency resources if required (Stolton *et al.*, 2008).

The diverse ecosystems supported by protected areas also help prevent outbreaks of pest insects that can damage agriculture or affect human health. For example, the diverse mosquito fauna in protected areas prevents any species from becoming dominant, thereby helping to limit the spread of malaria and other mosquito-borne diseases (Laporta, 2013).

Regulating services are especially relevant to protected areas, covering such ecological functions as watershed protection, pollination, climate regulation through carbon sequestration, and filtration of pollutants by wetlands. Some of these public goods can be quantified, such as the markets that are being developed for carbon sequestration (Hamilton *et al.*, 2011).

Much of the fresh water that provides irrigation, drinking water, and hydroelectricity for the ASEAN countries comes from protected areas, which typically have much higher rainfall than the lowlands because many are located in the hilly or mountainous areas that capture clouds (leading to their popular image as "rain towers"). The value of protected areas in providing drinking water is especially dramatic. For example, the drinking water of Jakarta comes largely from the Gunung Halimun and Gunung Gede-Pangrango national parks, and Singapore's drinking water comes largely from the forested Bukit Timah Catchment (Dudley and Stolton, 2003).

Many reservoirs throughout the ASEAN countries have their watersheds conserved by protected areas. Some governments are ensuring that some of the benefits from watershed protection are returned to the protected areas. A notable example comes from Lao PDR, where the Nakai Nam Theun National Protected Area receives funds allocated by the Watershed Management Protection Authority, which is paid US$1 million per year from the sale of electricity from the Nam Theun 2 Dam (which sells 7,000 megawatts per year to Thailand; see Whyte, 2008). So far, this seems a promising idea that is awaiting broader emulation. In any case, watershed protection and the provision of fresh water may well be the most valuable ecosystem services provided by protected areas, but many of the benefits are provided to distant users (such as residents of Jakarta, who may have little idea of where their fresh water comes from, or the urban beneficiaries of a hydroelectric dam whose forested watershed protected area is extending the life of the reservoir by slowing sedimentation rates). This complicates the provision of appropriate fees for the watershed protection values provided.

Of crucial importance in a time of rapid climate change, protected areas contribute to climate change mitigation as well as adaptation and resilience to extreme natural events (Barber *et al.*, 2004). Recent research has shown that old-growth forests, even up to 800 years old, continue to accumulate carbon at a faster rate, in both trees and the soil, than the new forests that are sometimes advocated as climate change mitigation measures (Stephenson *et al.*, 2014; Luyssaert *et al.*, 2008; Pan *et al.*, 2011). More important, much of this carbon is likely to return to the atmosphere if the old-growth forests are disturbed. This is a strong argument for conserving the old-growth forests contained within the protected areas found in the wetter parts of the ASEAN countries.

Parties to the United Nations Framework Convention on Climate Change have recognized the importance of forests in mitigating climate change and adapting to it, adopting in 2005 an initiative known as REDD (Reducing Emissions from Deforestation and Forest Degradation in Developing Countries). In 2010, REDD became REDD+ due to the addition of "conservation" (that is, the sustainable management of forests and the enhancement of forest carbon stocks) to its mandate. Since a majority of the most valuable forests in relation to climate are now found in protected areas, they have significant contributions to make to REDD+ and should be primary beneficiaries of the hundreds of millions of dollars that are spent annually on this initiative. This will require developing effective forms of forest management in protected areas that can be measured, reported, and verified. Indeed, this should become part of the standard approach to managing these important sites (UNEP, 2014a). Among ASEAN members, Cambodia, Indonesia, the Philippines, and Vietnam are REDD+ partner countries, while the others are exploring such status.

Cultural services include recreation, education, aesthetics, and spiritual values. The links between cultural and biological diversity are very strong, and reasonably well researched (Jianchu, 2000; Posey, 1999). Rural people are linked to the landscape in which they live, often with sacred overtones; many protected areas have incorporated traditional sacred natural sites (Verschuuren *et al.*, 2010).

People who have intimate links to nature, such as farmers and fishermen, invariably perceive their habitats in a way that is very different from people who live in cities. But all people, and even visitors to the ASEAN countries, cannot help but be affected by the rivers, canals, rice fields, hills, shorelines, fish, and forests that define the land. All are dependent on the products of the land, such as the rice, chickens, durians, mangoes, and other species that help define the cuisines that contribute so much to the cultures of the region.

For urban-dwellers, who comprise virtually all of Singapore and 20 percent of Cambodia (the most rural ASEAN country), visits to protected areas can enable them to reconnect to some of their historical roots to nature (Tryzna *et al.*, 2014).

Another cultural benefit of protected areas is their role in generating knowledge. Research is widely seen as a critical activity in protected areas, both to inform more effective management and to help inform the public about the wonders of nature. Since protected areas generally offer the most biodiversity-rich sites, they are especially valued as research sites by scientists seeking to provide greater benefits to society (McNeely, 1994). This research also provides support to public outreach via mass media, supporting the preparation of popular nature-based television programs and social media, and thereby helping make protected areas more attractive to the public (Reinius and Fredman, 2007) as well as building broader support for the conservation of biodiversity.

Since many protected areas still support traditional societies, greater efforts are now being made to incorporate the knowledge of the forest-dwelling peoples into protected area management. Protected areas are thereby also contributing to the conservation of such knowledge, whose value is explicitly recognized by the Convention on Biological Diversity (Articles 8j and 10c). Many are also showing that traditional ecological knowledge contributes to adaptive management that is especially valuable in times of rapid change (Berkes *et al.*, 2000).

At a psychological level, research has found that visits to protected areas can enhance a sense of well-being, and that children who are able to interact with nature are healthier and socially better balanced (Louv, 2005). These health benefits also extend to adults, which has led to an international movement called "Healthy Parks, Healthy People" (Maller *et al.*, 2009; www.hphpcentral.com). The basic message is that contact with nature is essential to human health and well-being. By providing access to nature, protected areas improve and maintain human health and well-being at both individual and community levels. And through this improvement, protected areas can reduce the burden on the health-care system through facilitating a holistic approach to health that can give people a sense of empowerment and control over their own health and well-being.

Perhaps the most important flow of cash (outside the state budget) to protected areas comes from the cultural service of tourism, which capitalizes on people's desire to visit nature. A key point is that many protected areas, even some of the smallest and most obscure ones, provide business opportunities to local people – a cultural service. Entrepreneurs can provide food and drink, accommodation, and guide services, and research has shown that tourism benefits enable people who live close to protected areas to be financially better off than people living in similar conditions farther from protected areas (Andam *et al.*, 2010). Some of the larger protected areas have attracted major investments in tourism in the surrounding lands. These large developments, too, can offer employment opportunities to local communities while also providing markets for crops and handicrafts that represent the local culture. But, of course, any such developments must give fair recognition to the original landowners, who should earn significant and continuing benefits from these commercial ventures.

In short, the ecosystem services provided by protected areas make numerous economic, social, cultural, and practical contributions to human well-being, justifying the continued efforts to enhance their effective management as a means of supporting adaptations to changing conditions (McNeely *et al.*, 2009).

Conclusions and recommendations

The problems the larger ASEAN countries are facing in their management of natural resources are well known (e.g., MacKinnon, 1997). And all indications are that they will need to continue addressing multiple social, economic, and environmental challenges that may arise outside of the individual countries but will nonetheless affect them. These issues are well recognized by the ASEAN governments and include several that will be familiar to most tropical countries:

- maintaining food and water security;
- promoting participation of local communities in natural resource management;
- ensuring jobs and sustainable livelihoods;
- supporting education and training;
- creating regional connectivity for social and economic stability; and

- maintaining the productivity of the fisheries, forestry, and agricultural sectors, all in the face of rapid climate change and while managing natural resources and the environment toward sustainability.

This is a daunting list of challenges, but biodiversity can make important contributions toward addressing them if the governments work together to conserve the ecological systems that have supported Southeast Asia's historical prosperity.

These problems affect far more than biodiversity, and arise from current patterns of social and economic development. The intersection between biodiversity and the many development goals guiding government priorities presents significant challenges to policy-makers, planners, managers, and researchers. Therefore, conserving biodiversity must contribute to the larger issues of economic and social development identified by national governments. But in the end, most of the challenges to conserving biodiversity are components of the larger challenges to ASEAN's development, calling for wider cooperation to create the conditions that will enable biodiversity to be conserved.

Arguably the most effective approach to conserving biodiversity in Southeast Asia is the establishment and effective management of protected areas, an approach that has already been broadly embraced by the ASEAN countries. This chapter has indicated many of the reasons why protected areas should be considered part of every modern approach to sustainable development, especially in the form of the numerous ecosystem services they provide.

Protected areas provide public goods benefits to virtually everyone. By establishing their systems of protected areas and managing them effectively, the ASEAN countries are contributing to national efforts to adapt to changing conditions through conserving natural habitats and the ecosystems, species, and varieties they support. Beneficiaries include rural people, urban people, researchers, international tourists, and the various businesses that are dependent on biodiversity and ecosystem services. Many of these may be in a position to increase their support for protected areas, especially in the case of the private goods that may be delivered by some ecosystem services.

But much remains to be done to enable protected areas to approach their full potential in providing ecosystem services. High-priority action points for the ASEAN countries include:

- Incorporate biodiversity and protected areas into other important sectors, including agriculture, climate change, health, rural development, tourism, and others.
- Develop innovative ways of determining the value of ecosystem services, accounting for them, and developing approaches of paying for them.
- Develop effective forms of forest management in protected areas that can be measured, reported, and verified, thereby enhancing their eligibility for funding as part of carbon sequestration.
- Develop new forms of protected area management that give greater responsibilities to the forest-dwelling peoples who have long developed ways

of living in forests that contribute to sustainable forms of national efforts to sequester carbon.

- Continue to support research in protected areas, including basic inventory and monitoring, ecology, and behavior of threatened species that need particular attention, genetic diversity of wild relatives of domestic species, sustainable approaches to managing medicinal plants, and approaches to improve resource management.
- Build management of protected areas into broader landscapes and seascapes, seeking to link protected areas to each other and to compatible land uses that will enable adaptation to changing conditions of climate and other variables. This could include transboundary links and broader ASEAN cooperation.

Such measures will help the ASEAN countries adapt to the coming social, economic, and environmental conditions.

Acknowledgements

Many thanks to Bill Jackson, Dan Navid, and Bob Dobias for their useful comments on an earlier draft of this chapter. Truong Nguyen kindly provided some very relevant papers on reptiles and amphibians.

References

Acharya, A. 2009. *Constructing a Security Community in Southeast Asia: ASEAN and the Problem of Regional Order* (2nd edition). Routledge, London.

Ahman, Y. 2006. The scope of definitions of heritage: from tangible to intangible. *International Journal of Heritage Studies* 12(3): 292–300.

Alcala, A.C. 2001. *Marine Reserves in the Philippines: Historical Development, Effects and Influence on Marine Conservation Policy.* Bookmark, Makati City.

Allen, G., et al. 2013. Hemiscyllium halmahera, a new species of bamboo shark from Indonesia. *Aqua International Journal* 19(3): 123–136.

Andam, K., P. Ferraro, K. Sims, A. Healy, and M. Holland. 2010. Protected areas reduced poverty in Costa Rica and Thailand. *Proceedings of the National Academy of Sciences USA* 107(22): 9996–10001.

Anderson, J. and C. Bausch. 2009. *Climate Change and Natural Disasters: Scientific Evidence of a Possible Relation between Recent Natural Disasters and Climate Change.* European Parliament, Brussels.

Audley-Charles, M.G. and D.A. Hooijer. 1973. Relation of Pleistocene migrations of pygmy stegodonts to island arc tectonics in eastern Indonesia. *Nature* 241: 197–198.

Bain, R.H., et al. 2007. New herpetological records from Vietnam. *Herpetological Review* 38: 107–117.

Barber, C.V., et al. (eds.). 2004. *Securing Protected Areas in the Face of Global Change: Issues and Strategies.* IUCN, Gland.

Bellwood, P. 1985. *Prehistory of the Indo-Malaysian Archipelago.* Academic Press, Sydney.

Berkes, F., J. Colding, and C. Folke. 2000. Rediscovery of traditional ecological knowledge as adaptive management. *Ecological Applications* 10(5): 1251–1262.

Brumm, A., *et al.* 2010. A new cladistics analysis of *Homo floresiensis*. Nature 464: 748–752.

Bruno, J.F. and E.R. Selig. 2007. Regional decline of coral cover in the Indo-Pacific: timing, extent, and subregional comparisons. *PLoS ONE* 2(8): e711.

Burke, L., *et al.* 2002. *Reefs at Risk in Southeast Asia.* World Resources Institute, Washington, D.C.

Carew-Reid, Jeremy, *et al.* (eds.). 2007. *Biodiversity Conservation Corridors Initiative.* Asian Development Bank, Manila.

CBD. 1992. *The United Nations Convention on Biological Diversity.* United Nations Environment Programme, Nairobi.

Ceballos, G. and J. Brown. 1995. Global patterns of mammalian diversity, endemism, and endangerment. *Conservation Biology* 9(3): 559–568.

Cheung, C.P., *et al.* 2002. *Marine Protected Areas in Southeast Asia.* ASEAN Regional Center for Biodiversity Conservation, Los Banos.

Childe, V.G. 1929. *The Most Ancient East: The Oriental Prelude to European History.* Alfred A. Knopf, New York.

Chivian, E. and A. Bernstein. 2008. *Sustaining Life: How Human Health Depends on Biodiversity.* Oxford University Press, Oxford.

Chouvey, P.-A. (ed.). 2013. *An Atlas of Trafficking in Southeast Asia.* I.B.Taurus, London.

Clements, R., *et al.* 2006. Limestone karsts of Southeast Asia: imperiled arks of biodiversity. *BioScience* 56(9): 733–742.

Coedes, G.. 1968. *The Indianized States of South-East Asia.* University of Hawai'i Press, Honolulu.

Corlett, R. T. 2013. Becoming Europe: Southeast Asia in the Anthropocene. *Elementa: Science of the Anthropocene* 1: 16.

Costanza, R. and M. Mageau. 1999. What is a healthy ecosystem? *Aquatic Ecology* 33: 105–115.

Dahdouh-Guebas, F., *et al.* 2005. How effective were mangroves as a defence against the recent tsunami? *Current Biology* 15(12): 1337–1338.

Daily, G.C. and K. Ellison. 2002. *The New Economy of Nature: The Quest to Make Conservation Profitable.* Island Press, Washington, D.C.

Davies, T.K., *et al.* 2012. *A Review of the Conservation Benefits of Marine Protected Areas for Pelagic Species Associated with Fisheries.* ISSF Technical Report 2012-02. International Seafood Sustainability Foundation, McLean, VA.

Dawson, M.R., *et al.* 2006. *Laonastes* and the "Lazarus effect" in recent mammals. *Science* 311: 1456–1458.

De Boer, J.Z. and D.T. Sanders. 2002. *Volcanoes in Human History: The Far-Reaching Effects of Major Eruptions.* Princeton University Press, New York.

Demeter, Fabrice, *et al.* 2012. Anatomically modern human in Southeast Asia (Laos) by 46 ka. *Proceedings of the National Academy of Sciences USA* 109(36): 14375–14380.

Dennis, R.A., *et al.* 2008. Biodiversity conservation in Southeast Asian timber concessions: a critical evaluation of policy mechanisms and guidelines. *Ecology and Society* 13(1): 25.

Don Carlos, A., *et al.* 2013. Building capacity to enhance protected area management effectiveness: a current needs assessment for the Asian context. *George Wright Forum* 30(2): 154–162.

Dudley, N. and S. Stolton. 2003. *Running Pure: The Importance of Forest Protected Areas to Drinking Water.* World Bank, Washington, D.C.

Dugan, P.J., *et al.* 2010. Fish migration, dams, and loss of ecosystem services in the Mekong Basin. *Ambio* 39: 344–348.

Dung, D.D., *et al.* 1993. A new species of living bovid from Viet Nam. *Nature* 363: 443–445.

Edgar, G.J., *et al.* 2014. Global conservation outcomes depend on marine protected areas with five key features. *Nature* 506: 216–220.

Elliott, L. 2003. ASEAN and environmental cooperation: norms, interests and identity. *Pacific Review* 16(1): 29–52.

Emanuel, K.A. 2013. Downscaling CMIP5 climate models shows increased tropical cyclone activity over the 21st century. *Proceedings of the National Academy of Science USA* 110(30): 12219–12224.

Evans, J.R. and S. von Caemmerer. 2011. Enhancing photosynthesis. *Plant Physiology* 155(1): 19.

FAO. 2014. FAO *Statistical Yearbook: Asia and the Pacific Food and Agriculture*. Food and Agriculture Organization of the United Nations, Rome.

Fiocco, G., D. Fu'a, and G. Visconti. 2011. *The Mount Pinatubo Eruption: Effects on the Atmosphere and Climate*. Springer, London.

Fox, Jefferson, *et al.* 2014. Swidden, rubber, and carbon: can REDD+ work for people and the environment in montane mainland Southeast Asia? *Global Environmental Change* 29: 313–326.

Froese, R. and D. Pauly (eds.). 2011. *FishBase*. www.fishbase.org (accessed 30 March 2017).

Geissmann, T., *et al.* 2011. A new species of snub-nosed monkey, genus *Rhinopithecus*, from northern Kachin State, northeastern Myanmar. *American Journal of Primatology* 73(1): 96–107.

Gesick, L. 1983. *Centers, Symbols, and Hierarchies: Essays on the Classical States of Southeast Asia*. Yale University Press, New Haven.

Giao, P.M., *et al.* 1998. Description of *Muntiacus truongsonensis*, a new species of muntjac from central Viet Nam, and implications for conservation. *Animal Conservation* 1(1): 61–68.

Groves, C., *et al.* 2012. Incorporating climate change into systematic conservation planning. *Biodiversity Conservation* 21: 1651–1671.

Gupta, A.K. 2004. Origin of agriculture and domestication of plants and animas linked to early Holocene climate amelioration. *Current Science* 87(1): 54–59.

Hall, D.G.E. 1968. *A History of South-East Asia* (3rd edition). Macmillan, London.

Halpern, B.J. 2003. The impact of marine reserves: do reserves work and does reserve size matter? *Ecological Applications* 13(1): S117–S137.

Hamilton, K., *et al.* 2011. *Building Bridges: State of the Voluntary Carbon Markets 2010*. Katoomba Group, Washington, D.C.

Higham, C. 2002. *Early Cultures of Mainland Southeast Asia*. River Books, Bangkok.

Hirsch, P. (ed.). 1996. *Seeing Forests for Trees: Environment and Environmentalism in Thailand*. Silkworm Books, Chiang Mai.

Hockings, M., P. Shadie, G. Vincent, and S. Suksawang. 2012. *Evaluating the Management Effectiveness of Thailand's Marine and Coastal Protected Areas*. IUCN, Gland.

Hoffman, L.C. and D.-M. Cawthorn. 2012. What is the role and contribution of meat from wildlife in providing high quality protein for consumption. *Animal Frontiers* 2(4): 40–53.

Hoogerwerf, A. 1970. *Udjung Kulon: The Land of the Last Javan Rhinoceros*. Brill, Amsterdam.

Hooijer, D.A. 1975. Quaternary mammal west and east of Wallace's line. *Netherlands Journal of Zoology* 25(1): 46–56.

Hunt, C.O. and R.J. Rabett. 2014. Holocene landscape intervention and plant food production strategies in island and mainland Southeast Asia. *Journal of Archeological Science* 51: 22–33.

Hunter, D. and V. Heywood. 2011. *Crop Wild Relatives: A Manual of In Situ Conservation.* Earthscan, London.

International Monetary Fund (IMF). 2014. *World Economic Outlook Database.* www.imf. com/thailand/ (accessed 30 March 2017).

IPCC. 2012. *Managing the Risks of Extreme Events and Disasters in Advance Climate Change Adaptation.* Cambridge University Press, Cambridge.

IPCC. 2014. *Climate Change 2014: Impacts, Adaptation, and Vulnerability.* Intergovernmental Panel on Climate Change, Geneva.

IUCN. 2016. *The IUCN Red List of Threatened Species.* www.iucnredlist.org (accessed 8 March 2017).

Jenkins, P.D., *et al.* 2004. Morphological and molecular investigations of a new family, genus and species of rodent from Lao PDR. *Systematics and Biodiversity* 2(4): 419–454.

Jianchu, X. 2000. *Links between Culture and Biodiversity.* Yunnan Science and Technology Press, Kunming.

Keesing, F. *et al.* 2010. Impacts of biodiversity on the emergence and transmission of infectious diseases. *Nature* 468: 647–652.

Laporta, G.Z. 2013. Biodiversity can help prevent malaria outbreaks in tropical forests. *PLoS Neglected Tropical Diseases* 7(3): e2139.

Lay, T., *et al.* 2005. The great Sumatra–Andaman Earthquake of 26 December 2004. *Science* 308: 1127–1133.

Lekagul, B. and J.A. McNeely. 1977. *Mammals of Thailand.* Sahakarn Bhaet, Bangkok.

Lenzen, M., *et al.* 2012. International trade drives biodiversity threats in developing nations. *Nature* 486: 109–112.

Loo, Y.Y., L. Billa, and A. Singh. 2015. Effect of climate change on seasonal monsoon in Asia and its impact on the variability of monsoon rainfall in Southeast Asia. *Geoscience Frontiers* 6(6): 817–823.

Louv, R.. 2005. *The Last Child in the Woods: Saving Our Children from Nature Deficit Disorder.* Algonquin Books, New York.

Louys, J., D. Curnoe, and H. Tong. 2007. Characteristics of Pleistocene megafauna extinctions in Southeast Asia. *Palaeogeography, Palaeoclimatology, Palaeoecology* 243: 152–173.

Luyssaert, S., *et al.* 2008. Old-growth forests as global carbon sinks. *Nature* 455: 213–215.

MacKinnon, J. 1997. *Protected Areas Systems Review of the Indo-Malayan Realm.* Asian Bureau for Conservation, Hong Kong.

Maffi, L. 2001. *On Biocultural Diversity: Linking Language, Knowledge and the Environment.* Smithsonian Institution Press, Washington, D.C.

Mahood, S.P., *et al.* 2013. A new species of lowland tailorbird from the Mekong flood plain of Cambodia. *Forktail* 29: 1–14.

Maller, C., *et al.* 2009. Healthy Parks, Healthy People: the health benefits of contact with nature in a park context. *George Wright Forum* 26(2): 51–83.

Margono, B.A., *et al.* 2014. Primary forest cover loss in Indonesia over 2000–2012. *Nature Climate Change* 4: 730–735.

Mascia, M., *et al.* 2010. Impacts of marine protected areas on fishing communities. *Conservation Biology* 24(5): 1424–1449.

McNeely, J.A. 1994. Protected areas for the 21st century: working to provide benefits to society. *Biodiversity and Conservation* 3: 390–405.

McNeely, J.A. 2016. Protected areas, biodiversity, and the risks of climate change. In F. Renaud, K. Sudmeier–Rieux, M. Estrella, and U. Nehren (eds.), *Ecosystem–Based Disaster Risk Reduction in Practice*. Springer International Publishing, Geneva: 379–397.

McNeely, J.A. and P.S. Sochaczewski. 1988. *Soul of the Tiger*. University of Hawai'i Press, Honolulu.

McNeely, J.A., *et al.* 2009. *The Wealth of Nature: Ecosystem Services, Biodiversity, and Human Well-being*. Conservation International, Washington, D.C.

MEA. 2005. *Millennium Ecosystem Assessment*. Island Press, Washington, D.C.

Menzies, G. 2008. *1434: The Year a Magnificent Chinese Fleet Sailed to Italy and Ignited the Renaissance*. Harper, London.

Misra, V. and S. DiNapoli. 2013. The variability of the Southeast Asian summer monsoon. *International Journal of Climatology* 43(3): 893–901.

Miththapala, S. 2006. *Conserving Medicinal Species: Securing a Healthy Future*. IUCN, Gland.

Mittermeier, R.A., N. Myers, and C.G. Mittermeier. 2000. *Hotspots: Earth's Biologically Richest and Most Endangered Terrestrial Ecoregions*. Conservation International, Washington, D.C.

Mongabay. 2012. *Deforestation*. www.mongabay.com/deforestation (accessed 30 March 2017).

Mukherjee, I. and B. K. Sovacool. 2014. Palm oil-based biofuels and sustainability in Southeast Asia: a review of Indonesia, Malaysia, and Thailand. *Renewable and Sustainable Energy Reviews* 37: 1–12.

Mulliken, T. and P. Crofton. 2008. *Review of the Status, Harvest, Trade, and Management of Seven Asian CITES-Listed Medicinal and Aromatic Plant Species*. Federal Agency for Nature Conservation, Bonn.

Murillo-Barroso, M., *et al.* 2010. Khao Sam Kaeo: an archaeometallurgical crossroads for trans-asiatic technological traditions. *Journal of Archaeological Science* 37: 1761–1772.

Musser, Guy, *et al.* 2005. Description of a new genus and species of rodent from the Khammouan Limstone National Biodiversity Conservation Area in Lao PDR. *American Museum Novitates* 3497: 1–31.

Obidzinski, K. 2013. *Fact File: Indonesia World Leader in Palm Oil Production*. CIFOR Forest News No. 8, Bogor.

Olsen, S.J., *et al.* 2003. Transmission of the severe acute respiratory syndrome on aircraft. *New England Journal of Medicine* 349(25): 2416–2422.

Osborne, M. 1985. *Southeast Asia: An Illustrated Introductory History*. George Allen and Unwin, Sydney.

Pallewatta, N., *et al.* 2003. *Invasive Alien Species in South-Southeast Asia: National Reports and Directory of Resources*. Global Invasive Species Programme, Cape Town.

Pan, Y., *et al.* 2011. A large and persistent carbon sink in the world's forests. *Science* 333: 988–993.

Polasky, S., *et al.* 2004. On trade, land-use, and biodiversity. *Journal of Environmental Economics and Management* 48(2): 911–925.

Polo, M. and R. Latham. 1958. *The Travels*. Penguin Classics, London.

Posey, D.A. 1999. *Cultural and Spiritual Values of Biodiversity*. United Nations Environment Programme, Nairobi.

Rabett, R.J. 2012. *Human Adaptation in the Asian Palaeolithic: Hominin Dispersal and Behaviour during the Late Quaternary*. Cambridge University Press, Cambridge.

Rabinowitz, A. 2001. *Beyond the Last Village: A Journey of Discovery in Asia's Forbidden Wilderness*. Island Press, Washington, D.C.

Raffles, T.S. 1817/1978. *History of Java*. Oxford University Press, Oxford.

Rampino, M. and S. Self. 1992. Volcanic winter and accelerated glaciation following the Toba super-eruption. *Nature* 359: 50–52.

Reich, D., *et al.* 2011. Denisova admixture and the first modern human dispersal into Southeast Asia and Oceania. *American Journal of Human Genetics* 89(4): 516–528.

Reid, A. 1988–93. *Southeast Asia in the Age of Commerce*. Yale University Press, New Haven.

Reinius, S.W. and P. Fredman. 2007. Protected areas as attractions. *Annals of Tourism Research* 34(4): 839–854.

Rowley, J.J.L., *et al.* 2012. A new species of large flying frog (Rhacophoridae: *Rhacophorus*) from lowland forests in southern Viet Nam. *Journal of Herpetology* 46: 480–487.

Saetung, A., *et al.* 2005. Cytotoxic activity of Thai medicinal plants for cancer treatment. *Songkhlanakarin Journal of Science and Technology* 27 (S2): 469–478.

Sakai, S. 2002. General flowering in lowland mixed dipterocarp forests of Southeast Asia. *Biological Journal of the Linnean Society* 75: 233–247.

Sanamxay, D. *et al.* 2013. Rediscovery of *Biswamoyopterus* in Asia, with the description of a new species from Lao PDR. *Zootaxa* 3686(4): 471–481.

Saralamp, P. 1996. *Medicinal Plants in Thailand*. Mahidol University, Bangkok.

Sawyer, R.D. 2011. *Ancient Chinese Warfare*. Basic Books, New York.

SCBD. 2005. *Handbook of the Convention on Biological Diversity* (3rd edition). Secretariat of the Convention on Biological Diversity, Montreal.

SCBD. 2011. *Aichi Biodiversity Targets*. Secretariat of the Convention on Biological Diversity, Montreal.

Schaller, G.B. and E.S. Vrba. 1996. Description of the giant muntjac (*Megamuntiacus vuquangensis*) in Laos. *Journal of Mammalogy* 77(3): 675–685.

Schrier-Uijl, A.P., *et al.* 2013. Environmental and social impacts of oil palm cultivation on tropical peat: a scientific review. In *Reports from the Technical Panels of the 2nd Greenhouse Gas Working Group of the Roundtable on Sustainable Palm Oil*. Jakarta, Indonesia: 131–168.

Sloan, S., *et al.* 2014. Remaining natural vegetation in the global biodiversity hotspots. *Biological Conservation* 177: 12–24.

Sodhi, N.S., *et al.* 2010. The state and conservation of Southeast Asian biodiversity. *Biodiversity Conservation* 19: 317–328.

Solheim, W.G. 1970. Northern Thailand, Southeast Asia, and world prehistory. *Asian Perspectives* 13: 145–162.

Spencer, J.E. 1966. *Shifting Cultivation in Southeast Asia*. University of California Press, Berkeley.

Stephenson, N.L., *et al.* 2014. Rate of tree carbon accumulation increases continuously with tree size. *Nature* 507: 90–93.

Stolton, S., *et al.* 2008. *Natural Security: Protected Areas and Hazard Mitigation*. WWF, Gland.

Stolton, S. and N. Dudley. 2010. *Arguments for Protected Areas: Multiple Benefits for Conservation and Use*. Earthscan, London.

Stuart, B.L., *et al.* 2006. High level of cryptic species diversity revealed by sympatric lineages of Southeast Asia forest frogs. *Biology Letters* 2: 470–474.

Suksawang, S. and J.A. McNeely. 2015. *Parks for Life: Why We Love Thailand's National Parks*. Department of National Parks, Wildlife and Plant Conservation, Bangkok.

Sumontha, M., *et al.* 2012a. A new species of parachute gecko from Kaeng Krachan National Park, western Thailand. *Zootaxa* 3513: 68–78.

Sumontha, M., *et al.* 2012b. A new forest-dwelling gecko from Phuket Island, southern Thailand, related to *Cyrtodactylus macrotuberculatus*. *Zootaxa* 3522: 61–72.

Sutikna, T., *et al.* 2016. Revised stratigraphy and chronology for *Homo floresiensis* at Liang Bua in Indonesia. *Nature* 532: 366–369.

Thinh, V.N., *et al.* 2010. A new species of crested gibbon from the central Annamite mountain range. *Vietnamese Journal of Primatology* 4: 1–12.

Tryzna, T. 2014. *Urban Protected Areas: Profiles and Best Practice Guidelines.* IUCN, Gland.

UNEP. 2014a. *Building Natural Capital: How REDD+ Can Support a Green Economy.* United Nations Environment Programme, Nairobi.

UNEP. 2014b. *UNEP Register of International Treaties and Other Agreements in the Field of the Environment.* United Nations Environment Programme, Nairobi.

Usher, A.D. 2009. *Thai Forestry: A Critical History.* Silkworm Books, Chiang Mai.

Vavilov, N. 1951. The origin, variation, immunity and breeding of cultivated plants (translated by K.S. Chester). *Chronica Botanica* 13: 1–366.

Vernot, B., *et al.* 2016. Excavating Neanderthal and Denisovan DNA from the genomes of Melanesian individuals. *Science* 17 March. http://science.sciencemag.org/content/early/2016/03/16/science.aad9416.full (accessed 8 March 2017).

Verschuuren, B., R. Wild, J. McNeely, and G. Oviedo. 2010. *Sacred Natural Sites: Conserving Nature and Culture.* Earthscan, London.

Voris, H.K. 2000. Maps of Pleistocene sea levels in Southeast Asia: shorelines, river systems and time durations. *Journal of Biogeography* 27: 1153–1167.

Wade, G. 2009. An early age of commerce in Southeast Asia, 900–1300 CE. *Journal of Southeast Asian Studies* 40(2): 221–265.

Watson, J., *et al.* 2014. The performance and potential of protected areas. *Nature* 515: 67–73.

Wharton, C. 1968. Man, fire, and wild cattle in Southeast Asia. *Proceedings of the Annual Tall Timbers Fire Ecology Conference* 8: 107–167

Whyte, W. 2008. *Evaluating Dam Sustainability: The Challenges of Lao Nam Theun 2.* n.p.

Wilkin, Paul, *et al.* 2013. A new species from Thailand and Burma, *Dracaena kaweesakii.* *PhytoKeys* 26: 101–112.

Wilkinson, E. 2015. *Chinese History: A New Manual.* Harvard University Press, Cambridge, MA.

Williamson, E.M. 2003. *Medicinal Plants of Southeast Asia* (2nd edition). Prentice Hall Asia, Selangor.

Wilson, E.O. 1986. *Biodiversity.* National Academy Press, Washington, D.C.

Woodruff, D.S. and K.A. Woodruff. 2008. Paleogeography, global sea level changes, and the future coastline of Thailand. *Natural History Bulletin of the Siam Society* 56(1): 1–24.

World Bank. 2014. *The World Bank Carbon Fund and Facilities.* World Bank, Washington, D.C.

WWF. 2014. *Mysterious Mekong: New Species Discoveries 2012–2013.* Worldwide Fund for Nature, Gland.

Zimmerman, B.L. and C.F. Kormos. 2012. Prospects for sustainable logging in tropical forests. *BioScience* 62: 479–487.

6 Wildlife and emerging diseases in Southeast Asia

Julien Cappelle and Serge Morand

Emerging diseases in a changing Southeast Asia

Global importance of Emerging Infectious Diseases

The concept of Emerging Infectious Diseases (EIDs) has been defined by Morse (1995: 7) as "infections that have newly appeared in the population or have existed previously but are rapidly increasing in incidence or geographic range." As emphasized by the historian McNeill (1976), humanity has been plagued since the emergence of the first cities, kingdoms, and trades. Several pandemics have had tremendous impacts on human demography, such as the bubonic plague, caused by *Yersinia pestis* bacilli and mainly transmitted to humans through small rodent fleas, which emerged in China in the 1330s, spread to Europe via the Silk Road, and eventually killed an estimated 30 percent of the European population (Schmid *et al.*, 2015). Just after the end of the World War I, in 1918–1919, Spanish flu, caused by an H1N1 influenza virus that likely originated among wild birds, killed over 50 million people (Worobey *et al.*, 2014). Acquired immune deficiency syndrome (AIDS) was identified in 1981 and is caused by the human immunodeficiency virus (HIV), which was likely spread to humans from chimpanzees (*Pan troglodytes*) in Africa (Gao *et al.*, 1999). AIDS may become the deadliest EID in human history, with more than 30 million deaths already related to the disease and an estimated 35 million people living with the virus in 2012 (UNAIDS, 2013). At the end of 2015, the transmission chain of the largest Ebola outbreak ever observed was declared over with a total of more than 28,000 human cases, including more than 11,000 deaths, mostly in Guinea, Sierra Leone, and Liberia (WHO, 2016). This outbreak affected an unprecedentedly high number of people, including the first cases outside Africa, showing the potential for known emerging pathogens to cause major outbreaks at regional and global scales.

It should be noted that all of these dramatic emergences originated among animals.

EIDs have also caused significant damage among domestic and wild animal populations, such as the rinderpest in wild ungulates in the Serengeti (Sinclair *et al.*, 2007), the canine distemper virus, which decimated African wild dog populations (Daszak *et al.*, 2000), and the fungus *Chytridiomycosis*, which is threatening amphibians globally (Kilpatrick *et al.*, 2010). Avian plagues, like avian influenza and the Newcastle disease, can lead to very high mortality rates (70 to 100 percent) in domestic poultry. The latter was spread globally through panzootics in the twentieth century and it remains a threat to food safety in most developing African and Asian countries, where chickens are an important source of protein and income (Alders and Spradbrow, 2001; Alexander *et al.*, 2012; Miguel *et al.*, 2013). The highly pathogenic H5N1 avian influenza virus has been responsible for the culling of hundreds of millions of poultry worldwide, and led to a major epizootic in wild migratory birds in Qinghai Lake, China, in 2005 (Liu *et al.*, 2005). Like the majority of EIDs in humans, H5N1 is a zoonotic disease. It caused 378 deaths in a reported 637 human cases between 1998 and 2013.

It is estimated that 60 percent of EIDs in humans are zoonotic, of which nearly 70 percent originate among wildlife (Jones *et al.*, 2008; Woolhouse and Gowtage-Sequeria, 2005), including AIDS, dengue fever and yellow fever, related to non-human primates, the bubonic plague and hantaviruses, related to rodents, avian influenza, related to wild birds, and SARS coronavirus and Ebola, related to bats. However, several evolutionary steps are required before an animal pathogen can spread to humans and be transmitted easily through human-to-human contact (Wolfe *et al.*, 2007).

While, the fear of new deadly pandemics faded at the start of the 1960s thanks to medical advances including antibiotics and vaccination, we have faced a dramatic increase in new infectious diseases over the last few decades (Jones *et al.*, 2008). A variety of factors, such as pathogen evolution, wildlife richness, international trade, human demographics and behaviors, agriculture intensification, land-use modifications, and habitat destruction, all help to explain the increase in infectious disease emergences over recent years (Horby *et al.*, 2013; Coker *et al.*, 2011; Keesing *et al.*, 2010). This increase has coincided with important environmental changes, with ongoing climate change being the main marker of this global change. Some scholars have argued that the new EIDs are linked to the biodiversity crisis (Chivian, 2003; Keesing *et al.*, 2010).

Southeast Asia is recognized as a hotspot for biodiversity (Myers *et al.*, 2000), but as emphasized by Sodhi *et al.* (2004), this is suffering from rapid erosion (see also Wilcove *et al.*, 2013). Southeast Asia is also a hotspot for EIDs, as illustrated by the emergence of Nipah virus and new cholera and dengue variants, among others (Coker *et al.*, 2011; Horby *et al.*, 2013). As biodiversity loss is highlighted as a major driver of the emergence of infectious diseases, Southeast Asia appears to be a region that should be investigated for any potential influence of biodiversity, and especially its loss, on patterns of emergence (Morand *et al.*, 2014).

Changing Southeast Asia as a hotspot for EIDs

Some of the most important EIDs of the past twenty years emerged in Southeast Asia (SEA) and originated from wildlife. Severe acute respiratory syndrome

(SARS), which emerged in southern China and led to more than 8,000 human cases and nearly 800 deaths in 37 countries in early 2003, was later related to wild bats' coronaviruses, with a potential transmission through the palm civet (*Paguma sp.*), though direct bat–human transmission was also plausible (Ge *et al.*, 2013). Nipah virus, which emerged in Malaysia in 1998 and led to the culling of 1.1 million pigs and 265 human cases with 105 deaths, originated among flying foxes (*Pteropus spp.*) and likely spread to pigs through contaminated fruit (Chua, 2003). The H5N1 avian influenza virus, which has been circulating in Asia, Europe, and Africa since its emergence in southern China in 1997 and led to the culling of hundreds of millions of domestic poultry (WHO, 2013), is a recombinant of other influenza viruses circulating in wild migratory birds (Mukhtar *et al.*, 2007).

Human demographics and behavior

Southeast Asia has been undergoing important demographic and socio-economic changes in the past twenty-five years. The total population (without southern China) increased from 450 to 600 million (33 percent) between 1990 and 2012 (http://faostat.fao.org/). This has been associated with dramatic increases in movement and trade in the region. The value of agricultural product exports for all countries in the region increased tenfold between 1990 and 2012 (see Figure 6.1). Visa-free and trade policies in the ASEAN countries, along with exponential increases in regional roads and budget airlines, enhanced regional connectivity, facilitating the global spread of infectious diseases like SARS (Horby *et al.*, 2013). Wildlife trade may also facilitate the emergence and spread of infectious diseases through the movement of animals (Edmunds *et al.*, 2011). Control measures could be targeted on markets that may act as disease transmission hubs between different wild species, domestic animals, and humans (Karesh *et al.*, 2005).

Rodents and bats have been important reservoir and important new zoonotic EIDs originated from wild rodents or bats species. Improvement of detection methods and research for reservoir of important EIDs, such as the Nipah virus, led to the discovery of new pathogens. Following the emergence of Nipah in Malaysia in 1998, Tioman and Pulau viruses were detected in fruit bats. Two closely related orthoreoviruses (Melaka and Kampar viruses) isolated in infected patients likely originated from flying foxes (Chua et al., 2008). Along with these technical consideration, socio-economic and environmental factors also played major roles in the emergence of the infectious diseases from wildlife in SEA.

Livestock and agriculture intensification

Demographic pressure and international trade development have dramatically increased the demand for livestock and agricultural products. The poultry population in SEA increased from 1 to nearly 3 billion between 1990 and 2012 (http://faostat.fao.org/), while the pig population increased from 40 to 70 million during

Table 6.1 Selection of zoonotic EIDs that emerged or re-emerged in the past twenty years in Southeast Asia

Pathogen	Location	Role of wildlife	Reference
HPAI H5N1 virus	Cambodia, China, Laos, Mynamar, Thailand, Vietnam	Potential spread by wild birds	(Cappelle et al., 2014)
SARS coronavirus	Emergence in south China	Horseshoe bats reservoir	(Ge et al., 2013)
Reston Ebola virus	Philippines	Bat reservoir(?)	(Miranda and Miranda, 2011)
Japanese encephalitis	Southeast Asia	Wild birds reservoir	(Impoinvil et al., 2012)
Leptospirosis	Southeast Asia	Rodents may play a role in transmission	(Victoriano et al., 2009)
Hantavirus	Southeast Asia	Rodent reservoir	(Blasdell et al., 2011)
Bordetella sp.	Laos	Rodent reservoir	(Jiyipong et al., 2013)
Nipah virus	Malaysia	Flying fox reservoir	(Chua, 2003)
Tioman virus	Malaysia	Detected in flying foxes	(Chua et al., 2001)
Pulau virus	Malaysia	Detected in flying foxes	(Pritchard et al., 2006)
Melaka virus	Malaysia	Flying fox likely reservoir	(Chua et al., 2007)
Kampar virus	Malaysia	Flying fox likely reservoir	(Chua et al., 2008)

the same period (Figure 6.1). Although the development of industrial farming will likely reduce the risk of pathogen transmission through improved biosecurity levels, backyard and semi-industrial farming are still dominant in most countries of the region. Animal density increase in these low-biosecurity settings may facilitate the emergence and spread of EIDs through increased contact rates between animals with a higher susceptibility to disease due to these farming systems (Caron et al., 2012). Increasing overuse and misuse of antimicrobial drugs will likely lead to outbreaks of multiresistant pathogens (Acar et al., 2012).

The development of integrated farming systems regrouping different animal and plant species may also facilitate interspecies transmission. The intensi-fication of rice production in SEA has been associated with an increased risk of HPAI H5N1 outbreak due to the free-grazing duck-farming system associated with rice harvesting (Gilbert et al., 2008). The intensification of pig production associated with fruit tree cultivation was likely a major factor in the emergence of the Nipah virus in Malaysia (Pulliam et al., 2012). This risk of crossover trans-mission between species might be further amplified when livestock and agriculture intensification take place in recently deforested areas, creating new interfaces between wild and domestic animals.

Environment and land-use change

Demographic changes combined with livestock and agricultural intensification are leading to major modifications of the environment and land use in the region.

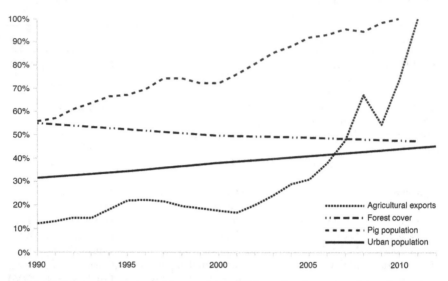

Figure 6.1 Evolution of selected factors driving EIDs in Southeast Asia (eleven countries
    – southern China not included), 1990–2011

Source: http://faostat.fao.org/.

Notes: Urban population and forest cover are expressed as percentages of total country population
and area, respectively. Agricultural exports and pig population are expressed as percentages of their
2011 values.

Urbanization has been a major change in SEA with a doubling of the urban
population from 140 to 280 million between 1990 and 2012. Urban-dwellers
represented nearly 50 percent of the total population in 2014. Cities can act as
important hubs in disease emergence and spread by bringing animals and humans
from various origins into close contact (Lam *et al.*, 2013). Certain landscapes are
associated with increased pathogen transmission: in SEA the development of
rice fields has been associated with an increased risk of Japanese encephalitis
transmission by bringing wild waterbirds, the reservoir of the virus, and vectors
breeding in the paddy fields into close contact (Le Flohic *et al.*, 2013).

   The forest cover in SEA has decreased by nearly 20 percent over the last few
decades (from 55 percent of total SEA area in 1990 to 47 percent in 2012).
Deforestation is often associated with habitat fragmentation and destruction,
which impact on wildlife populations. The distribution range of bats seems to
be particularly impacted by human activities in SEA, which may explain the
diversity of parasites and viruses harbored by them (Gay *et al.*, 2014). Modifications
in wildlife and parasite communities can facilitate the emergence and spread
of infectious diseases, as we elucidate in the next section of this chapter.

Factors driving EIDs in wildlife

Wildlife biodiversity and EIDs

At the global level, high wildlife biodiversity, measured in terms of high species richness of vertebrate hosts, is correlated with high richness of human pathogens (Dunn *et al.*, 2010). In a global analysis of 335 EID events in humans between 1940 and 2004, Jones *et al.* (2008) showed a positive correlation between wildlife host richness and the number of zoonotic EIDs originating among wildlife. Since such EIDs are the most prevalent (zoonoses accounted for 60.3 percent of all EIDs and 72 percent of these that originate in wildlife), there is a relatively higher risk of an EID event in areas with high wildlife biodiversity (Jones *et al.*, 2008). Hence, as SEA is a hotspot of biodiversity, it has a higher risk of EID events. When combined with the factors reviewed earlier in this chapter, this means that SEA appears to be a major hotspot for infectious disease emergence. Interestingly, at the SEA level, there also seems to be a positive correlation between bird and mammal richnes and infectious disease richness (controlling for potential confounding factors, such as area size, GPD per capita, disease surveys, and population size), indicating that areas with higher biodiversity within SEA are more at risk of infectious disease emergence than those with low biodiversity.

However, at the local level, higher biodiversity has been related to lower pathogen prevalence. This is called the "dilution effect," because it hypothesizes a dilution of infectious contacts between competent hosts when non-competent hosts are added to a community (increasing biodiversity), leading to more non-infectious contacts. This was popularized in the 2000s with studies into Lyme disease in the USA, where reduced biodiversity due to the fragmentation of forest habitat was associated with increased infection prevalence in tick nymphs (*Ixodes spp.*), the vector responsible for transmission of the disease (Blasdell *et al.*, 2012; Ostfeld and Keesing, 2012). Small patches of forest could not support large-mammal populations, like ungulate species, and therefore had lower biodiversity than larger forest patches. The white-footed mouse (*Peromyscus leucopus*), which is a more competent host for Lyme disease, was more abundant in the small forest patches, leading to increased density and infection prevalence in tick nymphs (Allan *et al.*, 2003). This dilution effect has been proposed for other host–pathogen systems, such as wild birds and West Nile virus (Ezenwa *et al.*, 2006; Swaddle and Calos, 2008), amphibians and *Batrachochytrium dendrobatidis* (Searle *et al.*, 2011), and rodents and hantaviruses (Carver *et al.*, 2011). In SEA, Blasdell *et al.* (2012) found a decrease in hantavirus prevalence in rodents with an increase of rodent species richness in several sites in Cambodia, Laos, and Thailand. Despite these studies, the mechanisms underlying the effects of species diversity on disease risk remain unclear and multiple and complex interactions within host and pathogen communities seem to play a role (Keesing *et al.*, 2006; Bordes *et al.*, 2015).

The relationship between high biodiversity at the local level and low risk of EIDs does not seem to be generalized or global. Depending on the resilience of hosts of interest for a given infectious disease, a decline in biodiversity may lead to either an increased or a decreased risk of transmission. For example, deforestation, and associated biodiversity loss, is associated with a decreased risk of malaria in Brazil because of the removal of the competent vectors (Hahn *et al.*, 2014). Thus, changes in community composition seem to be more important than biodiversity itself with respect to the risk of emergence and circulation of infectious disease (Randolph and Dobson, 2012). Pathogens with direct or indirect transmission, frequency- or density-dependent transmission, and variance in their host and vector competence may lead to either dilution or amplification effects due to biodiversity modification (Roche and Guégan, 2011). A meta-analysis suggested that the relationship between biodiversity and risk of zoonotic pathogen transmission is idiosyncratic and that a good understanding of the ecological and epidemiological characteristics of a particular host community–pathogen system is needed to evaluate the impact of biodiversity change (Salkeld *et al.*, 2013).

The universality of the dilution effect and the mechanisms driving the relationship between biodiversity and infectious disease emergence is a hot topic in the scientific community, leading to considerable debate among authors (Ostfeld, 2013; Lafferty and Wood, 2013; Wood *et al.*, 2016).

Wildlife ecology and disease emergence

Biodiversity of hosts may not always be a major factor in the likelihood of pathogen emergence and circulation in a particular ecosystem. Along with the community's composition, one could look at the ecological characteristics of the host species in the system as several ecological traits of hosts may have an impact on the circulation of infectious diseases.

Demographic factors and the population dynamics of hosts can influence pathogen circulation and the risk of disease emergence. Host abundance is an important factor limiting the maximum number of susceptible individuals in a population. Some pathogens, in particular those with density-dependent trans-mission, cannot be maintained in a population whose size is lower than a specific threshold (Anderson and May, 1979). Variations in host abundance induced by host population dynamics are therefore important to understand infectious disease dynamics in wildlife. A high turnover of new susceptible individuals can facilitate the maintenance of pathogens in a population. Species lifespan and reproduction strategies may then influence the dynamics of infectious diseases in a population (Anderson and May, 1979).

The spatial distribution of a population is another factor influencing the circula-tion and spread of infectious diseases. At the local scale, home range and habitat selection of a species will determine the areas at risk of infection for other species; generalist and widely spread species may then come into contact with a wide diver-sity of other species and pathogens (Woolhouse and Gowtage-Sequeria, 2005). The regional and migratory movements of a species within its distribution area have a direct effect on its capacity to spread infectious diseases. Eco-epidemiological studies of the spatial distribution of hosts can then help to identify areas with an increased risk of pathogen circulation and emergence (Cappelle *et al.*, 2010).

Species behavior can modify its exposure to pathogens. Gregarious species have high densities which can promote the transmission of density-dependent pathogens by increasing the contact rate between individuals (Gaidet *et al.*, 2011). Social interactions between individuals can facilitate the transmission of infectious diseases in a population, especially when there is close contact between individuals (Zhou, 2009). Diet and foraging behavior are important causes of exposure to pathogens that can persist in the environment, such as avian influenza viruses, which can infect new hosts through oral–fecal transmission (Gaidet *et al.*, 2011).

Physiological factors can modify hosts' resistance to infection. Host immunity is reduced for certain mammal species by the photoperiod or during the critical period of transition between maternal immunity and personal immunity (Dowell, 2001). Immunodeficiency can be induced physiologically when hosts' resources are allocated in priority to other functions, such as reproduction or migration (Gylfe *et al.*, 2000). This can facilitate the infection or excretion of a latent virus in immunodeficient individuals, as suggested for the Nipah virus in pregnant females (Sohayati *et al.*, 2011). Immunodeficiency can also be induced by pathogens such as SIV in monkeys, the probable origin of HIV, the virus responsible for AIDS in humans.

Temporal changes in these different factors can induce seasonality in infectious disease circulation patterns (Altizer *et al.*, 2006). Species with seasonal synchronized reproduction will have a generation of juveniles entering the population after breeding. This rapid arrival of susceptible individuals in the population can amplify the circulation of different pathogens and lead to a seasonal peak of circulation, as observed for avian influenza viruses in wild waterfowl and for various RNA viruses in bat colonies (Munster *et al.*, 2007; Drexler *et al.*, 2011). Seasonal variation of temperature or rainfall will also modify the transmission of infectious diseases: vector-borne diseases are particularly sensitive to these modifications as they have a direct impact on the vectors' population dynamics (Lambin *et al.*, 2010).

Even with limited or no information on the circulation of pathogens in wildlife species, combining ecological studies that describe the different factors discussed above allows a first assessment of the risk of emergence and circulation of infectious diseases in a population. A risk index can be calculated in a community based on the ecological characteristics of its species, and this can be used to orientate management and surveillance strategies (Caron *et al.*, 2010). In Southeast Asia, certain groups of wild animals may be targeted for EID surveillance. Several bat and rodent species (see Boxes 6.1 and 6.2) exhibit ecological characteristics that facilitate the maintenance of infectious agents that may emerge in other animal species or humans, as suggested by the number of bat-borne and rodent-borne EIDs already identified in the region (see Table 6.1).

Box 6.1 Bats as an EID reservoir

The recent emergence of deadly EIDs in humans has indicated that bats may be a major reservoir for future Emerging Infectious Diseases (Calisher *et al.*, 2006; Wibbelt *et al.*, 2010). Though bats only harbored only 4.5 percent of the 335 EIDs in Jones *et al.*'s (2008) study whilst rodents harbored 30 percent, when corrected for species diversity and research efforts, the viral diversity of bats was estimated to be 2.5 times higher than that of rodents (Luis *et al.*, 2013). Such capacity to harbor numerous potential EIDs may be related to bats' ecological characteristics. However, when compared with other mammal groups on which fewer pathogen studies have been conducted, bats do not appear to be truly "special" as reservoirs; rather, they seem to be worthy of study because they harbor deadly known Emerging Infectious Diseases, just as primates and rodents do (Olival *et al.*, 2015).

With more than 1,100 species, bats are the second-most diverse order of mammals after rodents (Simmons, 2005) and may therefore harbor a wide diversity of pathogens. They are ubiquitous and are the only mammal group capable of self-powered flight, which allows them to interact with a large number of species in diverse habitats and potentially to spread infectious agents locally. The migratory behavior of certain species also facilitates their ability to spread infectious agents over long distances (Epstein *et al.*, 2009). Some bat species roost in high densities, which may promote infectious agent circulation and maintenance. Synchronized reproduction in such colonies can lead to amplification of various agents, including RNA viruses (Drexler *et al.*, 2011). Finally, specificities of bats' immunological systems may play a role in their capacity to harbor infectious agents that are highly pathogenic for other mammals asymptomatically (Wibbelt *et al.*, 2010). In particular, a study comparing the genomes of two bat species with other mammalian species suggests the possibility that flight-induced adaptations have had inadvertent effects on bats' immune function (Zhang *et al.*, 2013). However, little data is available on this topic and research into bats' immunology is only just beginning.

Assessing the relative importance of these factors is challenging because little data is available for most bat species. A study of thirty-three bat species showed that sampling effort, endangered status, and a measure of population genetic structure were significantly correlated with viral richness, whereas colony size was not (Olival *et al.*, 2012). Another study of twenty bat species of SEA also showed that sampling effort and habitat fragmentation (which was positively correlated to endangered status) were significantly correlated with viral richness. However, these researchers also concluded that colony size is a significant factor in viral richness (Gay *et al.*, 2014).

Overall, bats appear to have the potential to be an important reservoir for future major EIDs, especially in altered ecosystems. However, predicting the next emergence in detail is not yet within reach. A better understanding of bats' ecology and immunology is needed, as well as integrated surveillance and management strategies.

Box 6.2 Rodents as an EID reservoir

Rodents are the most abundant and diversified group of mammals on earth, with more than 2,700 species recognized worldwide, constituting 42 percent of all mammalian species. Two-thirds of extant rodent species and the majority of rodents in Southeast Asia belong to the family *Muridae*. In this region rodents pose significant problems for crop production and human health through their capacity to act as reservoirs of important zoonotic diseases. Rodents are recognized as hosts of at least sixty zoonotic diseases, some of which represent a serious threat to human health. Southeast Asian murine rodent taxonomy, diversity of associated pathogens, and habitat preference were all topics of the CERoPath Project (Community Ecology of Rodents and Their Pathogens in a Southeast Asian Changing Environment), which aimed at better understanding of the relationships between rodent-borne diseases, rodents, and their habitats through the use of intensive fieldwork, molecular screening, and geospatial analysis (Blasdell *et al.*, 2015).

This project developed an extensive geo-referenced database containing details of more than 2,500 rodents, which were trapped at sites in Thailand, Cambodia, and Laos, with land cover layers from each site. These rodents were also screened for major rodent-borne pathogens and the data were used to estimate how these pathogens would likely be impacted by alterations in habitat structure and composition. The results confirmed the specialist and/or synanthropic status of several rodent species, although the majority of species studied demonstrated some degree of low-level habitat specialization. The prevalence of several agents of rodent-borne diseases were analyzed in relation to their rodent reservoirs and habitat, consisting of hantaviruses (the agents of hemorrhagic fever with renal syndrome (HFRS)), three groups of bacteria (*Leptopsira spp.*, the agents of leptospirosis; *Bartonella spp.*, the agents of bartonellosis; and *Orientia tsutsugamushi*, the agent of scrub typhus), and the zoonotic protozoan *Toxoplasma gondii* (the agent of toxoplasmosis). Of the bacteria, *Bartonella spp.* and *Orientia tsutsugamushi* are arthropod-borne agents, whereas *Leptospira spp.* are indirectly transmitted via contact with water or soil contaminated by the urine of infected rodents. Transmission of hantaviruses can occur either directly, via bites, or through exposure to infected aerosolized rodent excreta, whilst *T. gondii* can be transmitted vertically or via contaminated food or water sources. The richness in rodent-borne diseases was positively associated with flat agricultural land, meaning rice fields, the main agricultural landscape in Southeast Asia, suggesting that rice fields may favor microparasite transmission and should be targeted for rodent-borne disease surveillance and rodent control (Bordes *et al.*, 2013). However, the five major rodent-borne pathogens were linked to ongoing changes in habitat structure. In particular, the presence of *Bartonella spp.* and hantaviruses in rodents seemed to be favored in wooded landscapes affected by ongoing fragmentation and human encroachment (Morand *et al.*, 2015).

Recently, there have been reports of newly emerging rodent-borne arenaviruses in Southeast Asia (Li *et al.*, 2015; Blasdell *et al.*, 2016). One of these, the cardamones variant of Wēnzhōu virus, was found associated with *Rattus exulans* and *Rattus norvegicus* in Cambodia, and it has been associated with disease in humans (Blasdell *et al.*, 2016). The rodent reservoir species are synanthropic and found mostly in human-dwelling habitats.

Wildlife monitoring, surveillance, and management

Wildlife monitoring and surveillance methods

Monitoring and surveillance of wildlife are key to our understanding of EID ecology and our capacity to detect and respond in a timely fashion to an emergence. Elaborate methods for monitoring and surveillance of domestic animals have been developed in recent decades, but the ecological characteristics of wildlife species make it difficult to apply these methods to wild species (Ryser-Degiorgis, 2013). Observation of such species can be difficult depending on their habitats and behaviors, and basic information necessary to the implementation of a monitoring program, such as abundance, population structure, and migratory behavior, is often lacking. Capture and sampling are other major difficulties that can prevent optimal disease monitoring in wildlife populations. Sampling size should be estimated according to population size, which is often difficult to estimate for wild species. Sampling should also be random between individuals in a population, but in practice samples are collected from those animals that can be captured, inducing a capture bias. The diagnostic techniques also may not be easily adaptable to all wild species, and tools developed for domestic species may have to be used. Overall, then, monitoring and surveillance of wildlife is challenging, and the data collected rarely meet the requirements to adapt the analysis methods developed in human or domestic animal epidemiology. However, the monitoring and surveillance of wildlife is still informative when the data are interpreted with caution, and specific methods can be applied to wildlife studies.

When direct observation of a wildlife population is too challenging, indirect methods can be used to investigate the disease ecology. Non-invasive sampling methods can, for example, help researchers to reach the required sample size. Nipah virus was first isolated in bats through urine collected on plastic sheets deployed under the trees of a flying-fox roost; no capture or handling of any bat was necessary (Chua *et al.*, 2002). The same technique was successfully employed in Cambodia to isolate Nipah virus for the first time in that country (Reynes *et al.*, 2005). Analysis techniques based on fecal samples can help in the detection of pathogens but they can also provide information about the host when DNA can be extracted from the feces. Indirect methods can also involve investigating parasites of the host species of interest. A population genetic study showed a lack of population geographic structure of the bat fly (*Cyclopodia horsfieldi*) across species of *Pteropus* bats in Southeast Asia, suggesting frequent contacts between the different *Pteropus* species (Olival *et al.*, 2013). Such study of parasite genetics can provide information about host movements and contact between species.

Surveillance of wildlife, which involves both monitoring and dissemination of eco-epidemiological data collected from wild animals in order to implement public health policies and strategies, is particularly challenging as it requires trained observers in the field and good coordination with the national authorities responsible for public health actions. In order to improve data collection and reporting of disease events among wildlife, participatory methods can be implemented. The objective is to rely on local populations to collect information on a specific disease or to implement passive surveillance. For example, SAGIR in France is a wildlife surveillance network based on continuous collaboration between hunting federations and the French National hunting and Wildlife Agency (ONCFS) since 1955. SAGIR's surveillance relies on a locally coordinated network of observers in the field (mainly hunters, technicians from local hunting federations, and agents of the ONCFS). It focuses on dead and diseased animal observation in the field, and the data generated may then be used in passive and/or syndromic surveillance systems (Petit *et al.*, 2008).

A syndromic surveillance system can be defined as "the real-time (or near real-time) collection, analysis, interpretation, and dissemination of health-related data to enable the early identification of the impact (or absence of impact) of potential human or animal health threats that require public and/or animal health action" (Dupuy *et al.*, 2013: 221). A study based on syndrome definition from post-mortem findings recorded in the SAGIR database showed that syndromic surveillance is potentially useful for early outbreak detection because it uses the earliest available information on disease in wildlife populations (Warns-Petit *et al.*, 2010). Syndromic surveillance has proved useful in detecting emerging diseases that involved diseased animals, such as the West Nile, Ebola, Nipah, and Hendra viruses (Levinson *et al.*, 2013). However, passive surveillance (i.e., relying on diseased and dead animal detection) would not allow detection of these viruses in their reservoir – wild fruit bats – since they are not affected by the viruses. It is therefore also important to employ active surveillance (i.e., involving sampling of apparently healthy animals), especially for certain hosts, like bats and rodents, that appear to be less symptomatic than other mammals when infected by viruses (Levinson *et al.*, 2013).

Risk-based surveillance is a form of active surveillance of wildlife that can be used to target the sites, periods, species, and populations to monitor in order to increase the probability of pathogen detection. More generally, it can be defined as "a surveillance program in the design of which exposure and risk assessment methods have been applied together with traditional design approaches in order to assure appropriate and cost-effective data collection" (Stärk *et al.*, 2006: 4). A major objective of this approach is to increase the efficiency of the surveillance and to be able to detect rare events (Oidtmann *et al.*, 2013). Non-random methods would indeed be more appropriate to detect rare disease events than classical methods (Hadorn and Stärk, 2008). Different semi-quantitative or quantitative methods can be applied to identify risk factors and assess the risk of circulation of a pathogen in an area. Few risk-based surveillance programs have been designed for wildlife. However, methods combining environmental monitoring (through remote sensing, for example) and ecological monitoring of wild species (such as regular censuses of different species and populations) can allow the implementation of real-time risk mapping and the adaptation of an active surveillance program to seasonal changes (Cappelle *et al.*, 2010). In Cambodia, the monitoring of the population dynamics of *Pteropus lylei* allowed implementation of a targeted monitoring of Nipah virus circulation, and this method could be turned into a permanent and cost-effective surveillance system (Cappelle *et al.*, 2014). In addition to the risk-based approach, taking into account the cost-effectiveness of the surveillance and associated control is essential, especially where management resources are scarce, as in several countries of Southeast Asia. Modeling frameworks can therefore be used to optimize wildlife surveillance and the controls implemented to control animal or zoonotic diseases in wildlife (Gormley *et al.*, 2016).

Overall, a major requirement in the implementation of wildlife monitoring and surveillance is to establish a permanent and long-term data recording system (Ryser-Degiorgis, 2013). It is essential to catch the annual and inter-annual variability of wildlife populations and their pathogen dynamics. Seasonality can bring major changes in ecological and epidemiological patterns of EID circulation and emergence (Altizer *et al.*, 2006). Long-term trends, such as population decline, can also impact the circulation and emergence of EIDs. This kind of monitoring and surveillance program should involve research and surveillance activities, meaning that coordination is needed between wildlife researchers and national authorities responsible for animal and human public health. Reports of wildlife morbidity and mortality from the public could also be integrated into the surveillance program; indeed, using citizen science may be particularly beneficial in limited-resources settings (Lawson *et al.*, 2015). Such coordinated action is essential to ensure a timely response during an emergence event. Furthermore, such collaboration can lead to management strategies that reconcile public health and wildlife conservation.

Wildlife disease management strategies that reconcile conservation and public health

Management strategies aiming to reduce the risk of the emergence of infectious diseases from wildlife may also serve to protect wildlife from becoming victims of EIDs and thus reconcile conservation with public health. Wild species can be reservoirs of zoonotic diseases but they can also be threatened by EIDs. The spread of H5N1 from China to Europe in 2005 caused the deaths of approximately 5,000 wild birds, including approximately 3,000 bar-headed geese (*Anser indicus*), almost 10 percent of the global population of that species (Liu *et al.*, 2005). New EIDs can cause massive damage in wildlife populations, threatening several species, as exemplified by the white-nose syndrome in North American bats or chytridiomycosis in amphibian species worldwide (Kilpatrick *et al.*, 2010). Wild species may even be led to extinction because of EIDs, depending on the characteristics of the transmission. The facial tumor disease of the Tasmanian devil (*Sarcophilus harrisii*) is related to its mating behavior and appears to have a frequency-dependent transmission that may allow the disease to spread even at low host densities and therefore drive the species toward extinction (Beeton and McCallum, 2011).

Improving our understanding of transmission patterns is essential if we are to implement appropriate wildlife management strategies, and collaboration between scientists and wildlife managers should be promoted. A recent review showed that, while some epidemiological concepts have been commonly applied in wildlife management, new concepts are not always taken into account (Joseph *et al.*, 2013).

For example, the concept of density-dependence transmission is well known among wildlife managers and has been applied through host population reduction in order to reach a host density threshold under which pathogen transmission is highly reduced or even stopped (Anderson and May, 1979). Wildlife population reduction has mainly been applied through lethal methods, including shooting, trapping, and poisoning. However, this strategy is likely to fail to reduce the targeted host population over the long term, especially in large populations with high turnovers. Furthermore, it can lead to the spread of the disease when infected animals flee from the culling area to neighbouring areas, as observed with bovine tuberculosis in Eurasian badgers (*Meles meles*) in the United Kingdom (Weber *et al.*, 2013). Additionally, lethal methods are negatively perceived among the public and increasingly abandoned in favor of alternative methods. Fertility control and contraception may facilitate reductions in wildlife populations in the long term, but it has not yet been implemented in relation to disease control (Artois *et al.*, 2011). The control of rabies in France showed that red fox (*Vulpes vulpes*) vaccination was more efficient than culling (Aubert, 1994).

Wildlife vaccination is an alternative strategy to population reduction in order to control wildlife diseases. The objective of a wildlife vaccination program is to immunize a sufficient proportion of individuals to initiate a drastic reduction in transmission between hosts. The targeted immunization coverage should not allow a primary infectious case to cause more than one secondary case (i.e. R_0 < 1 in epidemiological modeling; see Artois *et al.*, 2011). Reaching this threshold of vaccinated individuals may be challenging and different methods of delivery have been developed, with the two main routes being oral (ingestion) and parenteral (injection). Oral vaccination through the distribution of bait has proven successful in controlling rabies in red foxes in Western Europe. Several other wildlife vaccination programs are being developed, such as parenteral and oral tuberculosis vaccines for badgers in Great Britain, oral plague vaccines for prairie dogs, and brucellosis vaccine for elk and bison in the USA (Cross *et al.*, 2007). Vaccination will probably be used increasingly in wildlife disease management in the future.

Reducing contacts between wildlife and domestic animals and/or humans is another strategy to mitigate the transmission of EIDs at the wild–domestic interface. Several methods can be used through environmental manipulation or increased biosecurity (Artois *et al.*, 2011). Influencing the spatial distribution of wild species to avoid high host densities close to domestic animals can be achieved by modifying attractive areas (water or food sources) or by dispersing hosts using scare devices or repellents. Preventing contact between wild and domestic animals can also be achieved by physically separating these populations through fencing or translocation, for example. However, these methods can be difficult to implement efficiently and may have ecological consequences that are difficult to anticipate (Debinski and Holt, 2000; Massei *et al.*, 2010). Increasing the bio-security of animal production systems can help to isolate domestic animals while conserving wild populations in the area. The development of zoning and compartmentalization will also help to control EIDs in areas with wild reservoirs by targeting adapted prevention and biosecurity measures to specific sub-populations that are defined geographically or based on production character-istics (Artois *et al.*, 2011). Restraining access to certain sites during high-risk periods may also help to prevent spill-overs from wildlife reservoirs, but to be efficient these measures need to be understood and accepted by the local population.

Implementing efficient and sustainable measures to control wildlife disease transmission requires collaborative management involving local populations and national services for public health, animal health, and wildlife conservation. Strong links should also be established between scientists and managers in order to apply the most recent measures. The perception of the risk related to wildlife disease among the public should be taken into account and risk communication should be adapted to local situations (Hanisch-Kirkbride *et al.*, 2013). A bottom-up approach can be implemented and community-based strategies developed to involve local communities in the management. For example, restricting access to caves during a bat colony's breeding season may be efficient only if local actors are aware of the regulation and accept it. Such a measure would reduce the risk of disease transmission, which may be higher in bat colonies during the breeding season (Drexler *et al.*, 2011), would help to protect the bat species during a critical period (Furey *et al.*, 2011), and would still allow guano collection or other human activities throughout the rest of the year.

Conclusion

Southeast Asia is a particularly interesting region with respect to the risk of infectious disease emergence from wildlife. It combines several factors that may facilitate such emergence events: considerable biodiversity, which is the source of a high richness of potentially zoonotic pathogens; high and increasing human and domestic animal population densities with lifestyles that provide opportunities for close human–animal contact; a fast-changing environment that may lead to new wild animal–domestic animal and human interfaces; and a limited health monitoring network, especially in least-developed countries such as Laos, Cambodia, and Myanmar.

This vulnerability of Southeast Asia calls for a rapid improvement in wildlife surveillance networks both nationally and regionally, which could benefit both public health and the conservation of endangered species. The Association of Southeast Asian Nations could play an important role in coordinating and implementing public health improvements (Hotez *et al.*, 2015). Increased cooperation in the region will be key to improving preparedness for a major emergence event. Cooperation with China will also be of particular importance, given the recent major emergence events in the southern provinces of that country, which border Southeast Asia, and China's increasing focus on controlling infectious diseases, including neglected tropical diseases (Yang *et al.*, 2014). New approaches, such as EcoHealth studies, which have been widely introduced in Southeast Asia over the last decade, involving interdisciplinary teams and using participatory methods to engage local communities, may help the region to face this global challenge (Nguyen Viet *et al.*, 2015).

References

Acar, J.F., G. Moulin, S.W. Page, and P. P. Pastoret. 2012. Antimicrobial resistance in animal and public health: introduction and classification of antimicrobial agents. *Revue Scientifique et Technique-Office International des Epizooties* 31 (1): 15–21.

Alders, R. and P. Spradbrow. 2001. *Controlling Newcastle Disease in Village Chickens: A Field Manual*. P. ACIAR Monograph No.82. Canberra.

Alexander, D.J., W.A. Elizabeth, and C.M. Fuller. 2012. The long view: a selective review of 40 years of Newcastle disease research. *Avian Pathology* 41(4): 329–335.

Allan, B.F., F. Keesing, and R.S. Ostfeld. 2003. Effect of forest fragmentation on Lyme disease risk. *Conservation Biology* 17(1): 267–272.

Altizer, S., A. Dobson, P. Hosseini, P. Hudson, M. Pascual, and P. Rohani. 2006. Seasonality and the dynamics of infectious diseases. *Ecology Letters* 9(4): 467–484.

Anderson, R.M. and R.M. May. 1979. Population biology of infectious-diseases. *Nature* 280: 361–367.

Artois, M., J. Blancou, O. Dupeyroux, and E. Gilot-Fromont. 2011. Sustainable control of zoonotic pathogens in wildlife: how to be fair to wild animals? *Revue Scientifique et Technique – Office International des Epizooties* 30(3): 733–743.

Aubert, M. 1994. Control of rabies in foxes: what are the appropriate measures? *Veterinary Record* 134(3): 55–59.

Beeton, N. and H. McCallum. 2011. Models predict that culling is not a feasible strategy to prevent extinction of Tasmanian devils from facial tumour disease. *Journal of Applied Ecology* 48(6): 1315–1323.

Blasdell, K., F. Bordes, K. Chaisiri, Y. Chaval, J. Claude, J.-F. Cosson, A. Latinne, J. Michaux, S. Morand, M. Pagès, and A. Tran. 2015. Progress on research on rodents and rodent-borne zoonoses in South-East Asia. *Wildlife Research* 42(2): 98–107.

Blasdell, K., J.-F. Cosson, Y. Chaval, V. Herbreteau, B. Douangboupha, S. Jittapalapong, A. Lundqvist, J.-P. Hugot, S. Morand, and P. Buchy. 2011. Rodent-borne hantaviruses in Cambodia, Lao PDR, and Thailand. *EcoHealth* 8 (4): 432–443.

Blasdell, K., V. Duong, M. Eloit, F. Chretien, S. Ly, V. Hul, V. Deubel, S. Morand, and P. Buchy. 2016. Evidence of human infection by new arenaviruses endemic to Southeast Asia. *ELife* 5: e13135.

Blasdell, K., H. Hentonnen, and P. Buchy. 2012. *Hantavirus Genetic Diversity*. In S. Morand, F. Beaudeau, and J. Cabaret (eds.), *New Frontiers of Molecular Epidemiology of Infectious Diseases*. Springer, Amsterdam: 179–216.

Bordes, F., K. Blasdell, and S. Morand. 2015. Transmission ecology of rodent-borne diseases: new frontiers. *Integrative Zoology* 10: 424–435.

Bordes, F., V. Herbreteau, S. Dupuy, Y. Chaval, A. Tran, and S. Morand. 2013. The diversity of microparasites of rodents: a comparative analysis that helps in identifying rodent-borne rich habitats in Southeast Asia. *Infection Ecology and Epidemiology* 3: 20178.

Calisher, C.H., J.E. Childs, H.E. Field, K.V. Holmes, and T. Schountz. 2006. Bats: important reservoir hosts of emerging viruses. *Clinical Microbiology Reviews* 19(3): 531–545.

Cappelle, J., V. Hul, V. Duong, A. Tarantola, and P. Buchy. 2014. Ecology of flying foxes and the risk of Nipah virus emergence in Cambodia. http://agritrop.cirad.fr/573883/1/document_573883.pdf (accessed 8 March 2017).

Cappelle, J., O. Girard, B. Fofana, N. Gaidet, and M. Gilbert. 2010. Ecological modeling of the spatial distribution of wild waterbirds to identify the main areas where avian influenza viruses are circulating in the Inner Niger Delta, Mali. *EcoHealth* 7(3): 283.

Cappelle, J., D. Zhao, M. Gilbert, M. Nelson, S.H. Newman, J.Y. Takekawa, N. Gaidet, et al. 2014. Risks of avian influenza transmission in areas of intensive free-ranging duck production with wild waterfowl. *EcoHealth* 11(1): 111–119.

Caron, A., M. de Garine-Wichatitsky, N. Gaidet, N. Chiweshe, and G.S. Cumming. 2010. Estimating dynamic risk factors for pathogen transmission using community-level bird census data at the wildlife/domestic interface. *Ecology and Society* 15(3): 25.

Caron, A., M. de Garine-Wichatitsky, M. Ndlovu, and G.S. Cumming. 2012. Linking avian communities and avian influenza ecology in southern Africa using epidemiological functional groups. *Veterinary Research* 43(1): 1–11.

Carver, S., A. Kuenzi, K.H. Bagamian, J.N. Mills, P.E. Rollin, S.N. Zanto, and R. Douglass. 2011. A temporal dilution effect: hantavirus infection in deer mice and the intermittent presence of voles in Montana. *Oecologia* 166 (3): 713–721.

Chivian, E. 2003. *Biodiversity: Its Importance to Human Health*. Harvard Medical School, Boston.

Chua, K.B. 2003. Nipah virus outbreak in Malaysia. *Journal of Clinical Virology* 26(3): 265–275.

Chua, K.B., G. Crameri, A. Hyatt, M. Yu, M.R. Tompang, J. Rosli, J. McEachern, et al. 2007. A previously unknown reovirus of bat origin is associated with an acute respiratory disease in humans. *Proceedings of the National Academy of Sciences USA* 104(27): 11424–11429.

Chua, K.B., C.L. Koh, P.S. Hooi, K.F. Wee, J.H. Khong, B.H. Chua, Y.P. Chan, M.E. Lim, and S.K. Lam. 2002. Isolation of Nipah virus from Malaysian island flying-foxes. *Microbes and Infection* 4(2): 145–151.

Chua, K.B., K. Voon, G. Crameri, H.S. Tan, J. Rosli, J.A. McEachern, S. Suluraju, M. Yu, and L.-F. Wang. 2008. Identification and characterization of a new Orthoreovirus from patients with acute respiratory infections. *PLoS ONE* 3(11). http://journals.plos.org/plosone/article?id=10.1371/journal.pone.0003803 (accessed 8 March 2017).

Chua, K.B., L.F. Wang, S.K. Lam, G. Crameri, M. Yu, T. Wise, D. Boyle, A.D. Hyatt, and B.T. Eaton. 2001. Tioman virus, a novel Paramyxovirus isolated from fruit bats in Malaysia. *Virology* 283(2): 215–229.

Coker, R.J., B.M. Hunter, J.W. Rudge, M. Liverani, and P. Hanvoravongchai. 2011. Emerging infectious diseases in Southeast Asia: regional challenges to control. *Lancet* 377(9765): 599–609.

Cross, M.L., B.M. Buddle, and F.E. Aldwell. 2007. The potential of oral vaccines for disease control in wildlife species. *Veterinary Journal* 174(3): 472–480.

Daszak, P., A.A. Cunningham, and A.D. Hyatt. 2000. Emerging Infectious diseases of wildlife-threats to biodiversity and human health. *Science* 287(5452): 443–449.

Debinski, D. and R. Holt. 2000. A survey and overview of habitat fragmentation experiments. *Conservation Biology* 14(2): 342–355.

Dowell, S.F. 2001. Seasonal variation in host susceptibility and cycles of certain infectious diseases. *Emerging Infectious Diseases* 7(3): 369–374.

Drexler, J.F., V.M. Corman, T. Wegner, A.F. Tateno, R.M. Zerbinati, F. Gloza-Rausch, A. Seebens, M.A. Mueller, and C. Drosten. 2011. Amplification of emerging viruses in a bat colony. *Emerging Infectious Diseases* 17(3): 449–456.

Dunn, R.R., T.J. Davies, N.C. Harris, and M.C. Gavin. 2010. Global drivers of human pathogen richness and prevalence. *Proceedings of the Royal Society London B*. http://rspb.royalsocietypublishing.org/content/early/2010/04/08/rspb.2010.0340 (accessed 8 March 2017).

Dupuy, C., A. Bronner, E. Watson, L. Wuyckhuise-Sjouke, M. Reist, A. Fouillet, D. Calavas, P. Hendrikx, and J.-B. Perrin. 2013. Inventory of veterinary syndromic surveillance initiatives in Europe (Triple-S Project): current situation and perspectives. *Preventive Veterinary Medicine* 111(3–4): 220–229.

Edmunds, K., S.I. Roberton, R. Few, S. Mahood, P.L. Bui, P.R. Hunter, and D.J. Bell. 2011. Investigating Vietnam's ornamental bird trade: implications for transmission of zoonoses. *EcoHealth* 8(1): 63–75.

Epstein, J.H., K.J. Olival, J.R.C. Pulliam, C. Smith, J. Westrum, T. Hughes, A.P. Dobson, et al. 2009. *Pteropus vampyrus*, a hunted migratory species with a multinational home-range and a need for regional management. *Journal of Applied Ecology* 46(5): 991–1002.

Ezenwa, V.O., M.S. Godsey, R.J. King, and S.C. Guptill. 2006. Avian diversity and West Nile virus: testing associations between biodiversity and infectious disease risk. *Proceedings of the Royal Society London B* 273(1582): 109–117.

Furey, N.M., I.J. Mackie, and P.A. Racey. 2011. Reproductive phenology of bat assemblages in Vietnamese karst and its conservation implications. *Acta Chiropterologica* 13(2): 341–354.

Gaidet, N., A. Caron, J. Cappelle, G.S. Cumming, G. Balança, S. Hammoumi, G. Cattoli, et al. 2011. Understanding the ecological drivers of avian influenza virus infection in wildfowl: a continental-scale study across Africa. *Proceedings of the Royal Society London B* 279(1731): 1131–1141.

Gao, F., E. Bailes, D.L. Robertson, Y. Chen, C.M. Rodenburg, S.F. Michael, L.B,. Cummins, *et al.* 1999. Origin of HIV-1 in the chimpanzee *Pan troglodytes. Nature* 397(6718): 436–441.

Gay, N., K.J. Olival, S. Bumrungsri, B. Siriaroonrat, M. Bourgarel, and S. Morand. 2014. Parasite and viral species richness of Southeast Asian bats: fragmentation of area distribution matters. *International Journal for Parasitology: Parasites and Wildlife* 3(2): 161–170.

Ge, X.-Y., J.-L. Li, X.-L. Yang, A.A. Chmura, G. Zhu, J.H. Epstein, J.K. Mazet, *et al.* 2013. Isolation and characterization of a bat SARS-like coronavirus that uses the ACE2 receptor. *Nature* 503: 535–538.

Gilbert, M., X. Xiao, D.U. Pfeiffer, M. Epprecht, S. Boles, C. Czarnecki, P. Chaitaweesub, *et al.* 2008. Mapping H5N1 highly pathogenic avian influenza risk in Southeast Asia. *Proceedings of the National Academy of Sciences USA* 105(12): 4769–4774.

Gormley, A.M., E.P. Holland, M.C. Barron, D.P. Anderson, and G. Nugent. 2016. A modelling framework for predicting the optimal balance between control and surveillance effort in the local eradication of tuberculosis in New Zealand wildlife. *Preventive Veterinary Medicine* 125: 10–18.

Gylfe, A., S. Bergstrom, J. Lunstrom, and B. Olsen. 2000. Epidemiology – reactivation of *Borrelia* infection in birds. *Nature* 403(6771): 724–725.

Hadorn, D.C. and K.D.C. Stärk. 2008. Evaluation and optimization of surveillance systems for rare and emerging infectious diseases. *Veterinary Research* 39: 57.

Hahn, M.B., R.E. Gangnon, C. Barcellos, G.P. Asner, and J.A. Patz. 2014. Influence of deforestation, logging, and fire on malaria in the Brazilian Amazon. *PLoS ONE* 9(1): e85725.

Hanisch-Kirkbride, S.L., S.J. Riley, and M.L. Gore. 2013. Wildlife disease and risk perception. *Journal of Wildlife Diseases* 49(4): 841–849.

Horby, P.W., D. Pfeiffer, and H. Oshitani. 2013. Prospects for emerging infections in East and Southeast Asia 10 years after severe acute respiratory syndrome. *Emerging Infectious Diseases* 19(6): 853–860.

Hotez, P.J., M.E. Bottazzi, U. Strych, L.-Y. Chang, Y.A.L. Lim, M.M. Goodenow, and S. AbuBakar. 2015. Neglected tropical diseases among the Association of Southeast Asian Nations (ASEAN): overview and update. *PLOS Neglected Tropical Diseases* 9(4): e0003575.

Impoinvil, D.E., M. Baylis, and T. Solomon. 2012. Japanese encephalitis: on the One Health agenda. *Current Topics in Microbiology and Immunology* 365: 205–247.

Jiyipong, T., S. Morand, S. Jittapalapong, D. Raoult, and J.-M. Rolain. 2013. *Bordetella hinzii* in rodents, Southeast Asia. *Emerging Infectious Diseases* 19(3): 502–503.

Jones, K.E., N.G. Patel, M.A. Levy, A. Storeygard, D. Balk, J.L. Gittleman, and P. Daszak. 2008. Global trends in emerging infectious diseases. *Nature* 451(7181): 990–993.

Joseph, M.B., J.R. Mihaljevic, A.L. Arellano, J.G. Kueneman, D.L. Preston, P.C. Cross, and P.T.J. Johnson. 2013. Taming wildlife disease: bridging the gap between science and management. *Journal of Applied Ecology* 50(3): 702–712.

Karesh, W.B., R.A. Cook, E.L. Bennett, and J. Newcomb. 2005. Wildlife trade and global disease emergence. *Emerging Infectious Diseases* 11(7): 1000–1002.

Keesing, F., L.K. Belden, P. Daszak, A. Dobson, C.D. Harvell, R..D. Holt, P. Hudson, *et al.* 2010. Impacts of biodiversity on the emergence and transmission of infectious diseases. *Nature* 468(7324): 647–652.

Keesing, F., R.D. Holt, and R.S. Ostfeld. 2006. Effects of species diversity on disease risk. *Ecology Letters* 9(4): 485–498.

Kilpatrick, A.M., C.J. Briggs, and P. Daszak. 2010. The ecology and impact of chytridio-mycosis: an emerging disease of amphibians. *Trends in Ecology and Evolution* 25(2): 109–118.

Lafferty, K.D. and C.L. Wood. 2013. It's a myth that protection against disease is a strong and general service of biodiversity conservation: response to Ostfeld and Keesing. *Trends in Ecology and Evolution* 28(9): 503–504.

Lam, T.T.-Y., J. Wang, Y. Shen, B. Zhou, L. Duan, C.-L. Cheung, C. Ma, et al. 2013. The genesis and source of the H7N9 Influenza viruses causing human infections in China. *Nature* 502 : 241–244.

Lambin, E., A. Tran, S. Vanwambeke, C. Linard, and V. Soti. 2010. Pathogenic landscapes: interactions between land, people, disease vectors, and their animal hosts. *International Journal of Health Geographics* 9(1): 54.

Lawson, B., S.O. Petrovan, and A.A. Cunningham. 2015. Citizen science and wildlife disease surveillance. *EcoHealth* 12(4): 693–702.

Le Flohic, G., V. Porphyre, P. Barbazan, and J.-P. Gonzalez. 2013. Review of climate, landscape, and viral genetics as drivers of the Japanese encephalitis virus ecology. *PLoS Neglected Tropical Diseases* 7(9). http://journals.plos.org/plosntds/article?id=10.1371/journal.pntd.0002208 (accessed 8 March 2017).

Levinson, J., T.L. Bogich, K.J. Olival, J.H. Epstein, C.K. Johnson, W. Karesh, and P. Daszak. 2013. Targeting surveillance for zoonotic virus discovery. *Emerging Infectious Diseases* 19(5): 743–747.

Li, K., X.-D. Lin, W. Wang, M. Shi, W.-P. Guo, X.-H. Zhang, J.-G. Xing, J.-R. He, K. Wang, M.-H. Li, J.-H. Cao, M.-L. Jiang, E.C. Holmes, and Y.-Z. Zhang. 2015. Isolation and characterization of a novel arenavirus harbored by rodents and shrews in Zhejiang province, China. *Virology* 476: 37–42.

Liu, J., H. Xiao, F. Lei, Q. Zhu, K. Qin, X.W. Zhang, X.L. Zhang, et al. 2005. Highly pathogenic H5N1 influenza virus infection in migratory birds. *Science* 309(5738): 1206.

Luis, A.D., D.T.S. Hayman, T.J. O'Shea, P.M. Cryan, A.T. Gilbert, J.R.C. Pulliam, J.N. Mills, et al. 2013. A comparison of bats and rodents as reservoirs of zoonotic viruses: are bats special? *Proceedings of the Royal Society London B.* http://rspb.royalsocietypublishing.org/content/280/1756/20122753 (accessed 17 March 2017).

Massei, G., R.J. Quy, J. Gurney, and D.P. Cowan. 2010. Can translocations be used to mitigate human–wildlife conflicts? *Wildlife Research* 37(5): 428–439.

McNeill, W.H. 1976. *Plagues and Peoples.* Anchor Press, New York.

Miguel, E., V. Grobois, C. Berthouly-Salazar, A. Caron, J. Cappelle, and F. Roger. 2013. A meta-analysis of observational epidemiological studies of Newcastle disease in African agro-systems, 1980–2009. *Epidemiology and Infection* 141: 1117–1133.

Miranda, M.E.G. and N.L.J. Miranda. 2011. Reston Ebola virus in humans and animals in the Philippines: a review. *Journal of Infectious Diseases* 204: S757–S760.

Morand, S., F., Bordes, K. Blasdell, S. Pilosof, J.-F. Cornu, K. Chaisiri, Y. Chaval, J.-F. Cosson, J. Claude, T. Feyfant, V. Herbreteau, S. Dupuy, and A. Tran. 2015. Assessing the distribution of disease-bearing rodents in human-modified tropical landscapes. *Journal of Applied Ecology* 52(3): 784–794.

Morand, S., S. Jittapalapong, Y. Suputtamongkol, M.T. Abdullah, and T.B. Huan. 2014. Infectious diseases and their outbreaks in Asia-Pacific: biodiversity and its regulation loss matter. *PLoS ONE* 9(2): e90032.

Morse, S.S. 1995. Factors in the emergence of infectious diseases. *Emerging Infectious Diseases* 1(1): 7–15.

Mukhtar, M.M., S.T. Rasool, D. Song, C. Zhu, Q. Hao, Y. Zhu, and J. Wu. 2007. Origin of highly pathogenic H5N1 avian influenza virus in China and genetic characterization of donor and recipient viruses. *Journal of General Virology* 88(11): 3094–3099.

Munster, V.J., C. Baas, P. Lexmond, J. Waldenström, A. Wallensten, T. Fransson, G.F. Rimmelzwaan, *et al.* 2007. Spatial, temporal, and species variation in prevalence of Influenza A viruses in wild migratory birds. *PLoS Pathogens* 3(5): e61.

Myers, N., R.A. Mittermeier, C.G. Mittermeier, G.A.B. da Fonseca, and J. Kent. 2000. Biodiversity hotspots for conservation priorities. *Nature* 403(6772): 853–858.

Nguyen-Viet, H., S. Doria, D.X. Tung, H. Mallee, B.A. Wilcox, and D. Grace. 2015. Ecohealth research in Southeast Asia: past, present and the way forward. *Infectious Diseases of Poverty* 4: 5.

Oidtmann, B., E. Peeler, T. Lyngstad, E. Brun, B.B. Jensen, and K.D.C. Stärk. 2013. Risk-based methods for fish and terrestrial animal disease surveillance. *Preventive Veterinary Medicine* 112(1–2): 13–26.

Olival, K.J., C.W. Dick, N.B. Simmons, J.C. Morales, D.J. Melnick, K. Dittmar, S.L. Perkins, P. Daszak, and R. DeSalle. 2013. Lack of population genetic structure and host specificity in the bat fly, *Cyclopodia horsfieldi*, across species of *Pteropus* bats in Southeast Asia. *Parasites and Vectors* 6: 231.

Olival, K.J., J.H. Epstein, L.-F. Wang, H.E. Field, and P. Daszak. 2012. Are bats exceptional viral reservoirs ? In A.A. Aguirre, R. Ostfeld, and P. Daszak (eds.), *New Directions in Conservation Medicine: Applied Cases of Ecological Health*. Oxford University Press, Oxford: 195–212.

Olival, K.J., C.C. Weekley, and P. Daszak. 2015. Are bats really "special" as viral reservoirs? What we know and need to know. In L.-F. Wang and C. Cowled (eds.), *Bats and Viruses: A New Frontier of Emerging Infectious Diseases*. Wiley Blackwell, Hoboken: 281–294.

Ostfeld, R.S. and F. Keesing. 2012. Effects of host diversity on infectious disease. *Annual Review of Ecology, Evolution, and Systematics* 43(1): 157–182.

Ostfeld, R.S. 2013. A candid response to panglossian accusations by Randolph and Dobson: biodiversity buffers disease. *Parasitology* 140(10): 1196–1198.

Petit, E., O. Mastain, C. Dunoyer, J. Barrat, M. Artois, and D. Calavas. 2008. Analysis of a wildlife disease monitoring network for the purpose of early disease detection. In B. Dufour (ed.), *Epidemiologie et Sante Animale*. Fondation Merieux, Lyon: 137–140.

Pritchard, L.I., K.B. Chua, D. Cummins, A. Hyatt, G. Crameri, B.T. Eaton, and L.F. Wang. 2006. Pulau virus: a new member of the Nelson Bay Orthoreovirus species isolated from fruit bats in Malaysia. *Archives of Virology* 151(2): 229–239.

Pulliam, J.R.C., J.H. Epstein, J. Dushoff, S.A. Rahman, M. Bunning, A.A. Jamaluddin, A.D. Hyatt, H.E. Field, A.P. Dobson, and P. Daszak 2012. Agricultural intensification, priming for persistence and the emergence of Nipah virus: a lethal bat-borne zoonosis. *Journal of the Royal Society Interface* 9: 89–101.

Randolph, S.E. and A.D.M. Dobson. 2012. Pangloss revisited: a critique of the dilution effect and the biodiversity–buffers–disease paradigm. *Parasitology* 139: 847–863.

Reynes, J.M., D. Counor, S. Ong, C. Faure, V. Seng, S. Molia, J. Walston, M.C. Georges-Courbot, V. Deubel, and J.L. Sarthou. 2005. Nipah virus in Lyle's flying foxes, Cambodia. *Emerging Infectious Diseases* 11(7): 1042–1047.

Roche, B. and J.-F. Guégan. 2011. Ecosystem dynamics, biological diversity and emerging infectious diseases. *Comptes Rendus Biologies* 334(5–6): 385–392.

Ryser-Degiorgis, M.-P. 2013. Wildlife health investigations: needs, challenges and recommendations. *BMC Veterinary Research* 9(1): 223.

Salkeld, D.J., K.A. Padgett, and J.H. Jones. 2013. A meta-analysis suggesting that the relationship between biodiversity and risk of zoonotic pathogen transmission is idiosyncratic. *Ecology Letters* 16(5): 679–686.

Schmid, B.V., U. Büntgen, W.R. Easterday, C. Ginzler, L. Walløe, B. Bramanti, and N.C. Stenseth. 2015. Climate-driven introduction of the black death and successive plague reintroductions into Europe. *Proceedings of the National Academy of Sciences USA* 112(10): 3020–3025.

Searle, C.L., L.M. Biga, J.W. Spatafora, and A.R. Blaustein. 2011. A dilution effect in the emerging amphibian pathogen *Batrachochytrium Dendrobatidis*. *Proceedings of the National Academy of Sciences* 108(39): 16322–16326.

Simmons, N.B. 2005. Chiroptera. In K.D. Rose and J.D. Archibald (eds.), *The Rise of Placental Mammals: Origins and Relationships of the Major Extant Clades*. Johns Hopkins University Press, Baltimore: 159–174.

Sinclair, A., S.A. Mduma, J.G.C. Hopcraft, J.M. Fryxell, R. Hilborn, and S. Thirgood. 2007. Long-term ecosystem dynamics in the Serengeti: lessons for conservation. *Conservation Biology* 21(3): 580–590.

Sodhi, N.S., L.P. Koh, B.W. Brook, and P.K.L. Ng. 2004. Southeast Asian biodiversity: an impending disaster. *Trends in Ecology and Evolution* 19(12): 654–660.

Sohayati, A.R., L. Hassan, S.H. Sharifah, K. Lazarus, C.M. Zaini, J.H. Epstein, N.S. Naim, *et al.* 2011. Evidence for Nipah virus recrudescence and serological patterns of captive *Pteropus vampyrus*. *Epidemiology and Infection* 139(10): 1570–1579.

Stärk, K.D.C., G. Regula, J. Hernandez, L, Knopf, K. Fuchs, R.S. Morris, and P. Davies. 2006. Concepts for risk-based surveillance in the field of veterinary medicine and veterinary public health: review of current approaches. *BMC Health Services Research* 6(1): 1–8.

Swaddle, J.P. and S.E. Calos. 2008. Increased avian diversity is associated with lower incidence of human West Nile infection: observation of the dilution effect. *PLoS ONE* 3(6): e2488.

UNAIDS. 2013. *Report on the Global AIDS Epidemic: 2013*. www.unaids.org/en/resources/campaigns/globalreport2013/factsheet/ (accessed 8 March 2017).

Victoriano, A., L. Smythe, N. Gloriani-Barzaga, L. Cavinta, T. Kasai, K. Limpakarnjanarat, B. Ong, *et al.* 2009. Leptospirosis in the Asia Pacific region. *BMC Infectious Diseases* 9(1): 147.

Warns-Petit, E., E. Morignat, M. Artois, and D. Calavas. 2010. Unsupervised clustering of wildlife necropsy data for syndromic surveillance. *BMC Veterinary Research* 6: 56.

Weber, N., S. Bearhop, S.R.X. Dall, R.J. Delahay, R.A. McDonald, and S.P. Carter. 2013. Denning behaviour of the European badger (*Meles meles*) correlates with bovine tuberculosis infection status. *Behavioral Ecology and Sociobiology* 67(3): 471–479.

WHO. 2013. *Global Health Observatory Data Repository*. http://apps.who.int/gho/data/node.country.country-KHM (accessed 8 March 2017).

WHO. 2016. *Ebola Situation Reports*. http://apps.who.int/ebola/ebola-situation-reports (accessed 8 March 2017).

Wibbelt, G., M.S. Moore, T. Schountz, and C.C. Voigt. 2010. Emerging diseases in Chiroptera: why bats? *Biology Letters*. http://rsbl.royalsocietypublishing.org/content/early/2010/04/26/rsbl.2010.0267 (accessed 8 March 2017).

Wilcove, D.S., X. Giam, D.P. Edwards, B. Fisher, and L.P. Koh. 2013. Navjot's nightmare revisited: logging, agriculture, and biodiversity in Southeast Asia. *Trends in Ecology and Evolution* 28(9): 531–540.

Wolfe, N.D., C.P. Dunavan, and J. Diamond. 2007. Origins of major human infectious diseases. *Nature* 447: 279–283.

Wood, C.L., K.D. Lafferty, G. DeLeo, H.S. Young, P.J. Hudson, and A.M. Kuris. 2016. Does biodiversity protect humans against infectious disease? Reply. *Ecology* 97(2): 543–546. .

Woolhouse, M.E.J. and S. Gowtage-Sequeria. 2005. Host range and emerging and reemerging pathogens. *Emerging Infectious Diseases* 11: 1842–1847.

Worobey, M., G.-Z. Han, and A. Rambaut. 2014. A synchronized global sweep of the internal genes of modern Avian Influenza virus. *Nature* 508: 254–257.

Yang, G.-J., L. Liu, H.-R. Zhu, S.M. Griffiths, M. Tanner, R. Bergquist, J. Utzinger, and X.-N. Zhou. 2014. China's sustained drive to eliminate neglected tropical diseases. *Lancet Infectious Diseases* 14(9): 881–892.

Zhang, G., C. Cowled, Z. Shi, Z. Huang, K.A. Bishop-Lilly, X. Fang, J.W. Wynne, *et al.* 2013. Comparative analysis of bat genomes provides insight into the evolution of flight and immunity. *Science* 339(6118): 456–460.

Zhou, S.Z. 2009. A seasonal influenza theory and mathematical model incorporating meteorological and socio-behavioral factors. *Journal of Tropical Meteorology* 15(1): 1–12.

Part III

Ecosystem services and biodiversity

7 Diversity of weed and agricultural management

Rachanee Nam-Matra

Introduction: agriculture and its change in Thailand

Agriculture in Thailand was the most important sector in the economy before the manufacturing sector started to play an increasing role in the late 1970s. Currently the agricultural sector provides employment for half of the population (Siamwalla *et al.*, 1991). There are rapid changes occurring in Thai agriculture that are likely to have effects on both the negative impact of weeds (infestation) and their contribution to biodiversity across the Thai agricultural landscape. More than half of the total land area in Thailand is used for agricultural production (Chinawong and Suwanketnikom, 2001). Rice is the most economically important crop. In the 1990s, Thai agriculture was still traditional subsistence farming and the main output was for household consumption or for sale in local markets (Panyakul, 2001). However, more recently (2001–2006), with a rising population, food consumption has been continually increasing. Farm outputs are now rising, with the effect of increasing farmers' incomes and building a major national agricultural export trade. This increase in farm outputs has been achieved by marked changes in production practices of paddy rice and some other major crops, including soybean and sugarcane. New technology has been reaching farmers – for instance, production has become more industrial with the use of crop protection chemicals, modern farm machinery, precision farming and multi-cropping.

These changes in cultivation practice have been accompanied by changes in weed management; indeed, in some cases weed infestations have been the major cause of crop reductions. For instance, conversion from high-yielding transplanted rice to direct seeding has led to yield reductions as a result of increased competition with greater weed incidence (Tomita *et al.*, 2003).

The linkage between changed agricultural practices and weed infestation is highlighted by the fact that one component of these changes is the direct management of the weeds in the crop. Traditional weed controls, such as hand weeding, plowing and mulching have been replaced by a greater reliance on herbicides. Moreover, herbicides are misused, to the extent that there are problems of herbicide residues in food products, including cereals, fruits and vegetables, which can be determined by chromatographic methods (Tadeo *et al.*, 2000). Clearly, these direct weed control techniques may impact on the diversity of weeds that establish and survive.

There are various reasons why changes in agricultural practice are likely to have effects on weed diversity. Firstly, farmers have switched from transplanted to direct-seeded flooded rice cultivation (De Datta and Finn, 1986). Rice cultivation in Thailand covers about 10 million hectares. During the 1990s, farmers used transplanted rice throughout the rain-fed northern and north-eastern areas, and also in the irrigated area in the central plain and the rain-fed surrounding areas in the central part of the country. By the end of the 1990s, farmers had switched from transplanted to dry direct-seeded rice in the rain-fed area in the north-east, and to wet-seeding in the central plain. However, transplanted rice is still used in the rain-fed area of the north (Azmi et al., 2004). In the irrigated area in the central plain, most farmers use multi-cropping, for instance growing rice three to four times a year instead of once a year. However, a major problem of the direct-seeded method of rice production is that weeds emerge at the same time as germination of rice seed. Because of this, effective weed control is essential in order to achieve desired crop yields.

Secondly, there has been a decrease in farm size despite the demand for increased agricultural production. Reduction in farm size occurs because of: partitioning of land between family members; and rising land prices, which make it difficult to purchase a large area. The intensification of the agricultural sector has facilitated an increase in exports. However, further enlargement of scale of production through increased farm size is no longer possible. Therefore, farm management at a small scale must be intensive to ensure sufficient production for export.

Thirdly, chemical weed control was introduced to reduce weed competition, crop losses and labor costs, which collectively may contribute to the decline of both weed density and diversity (Masako and Keiya, 2001).

Thus, at least in rice, there are noticeable examples of how the widespread introduction of direct seeding, the repeated use of herbicides and limited irrigation supplies may all be factors that affect shifts in weed flora composition in rice ecosystems (Azmi et al., 2004).

Irrigation in Thailand

Panyakul (2001) stated that irrigation is one of the important factors for the success of rice cultivation in central Thailand. Irrigation technologies are helping supply water to farms and control water levels. However, they may cause problems as a consequence of weed dissemination in irrigation water from one place to another.

LePoer (1987) stated that Thai farmers produced rice by traditional cultivation which relied on natural resources, such as rain and flood water. However, the natural water supply is insufficient for rice production. The construction of canals to carry flood waters from the Chao Phraya River begin in the mid-1800s. These canals provide a distribution network but do not form a controlled irrigation system. Additional water for agriculture depends on the level of the rivers. However, in 1930, data showed that water from the rivers was insufficient for a third of the year, with subsequent losses in crop yields. In 1902, a Dutch expert helped the Thai government to develop a controlled irrigation plan. However, it failed and droughts continued until 1910. In 1911, a British irrigation specialist was hired to improve this work. The first irrigation project was finally completed in 1922.

About 440,000 hectares had been irrigated by 1938. During World War II, further projects were held up by supply problems. Nevertheless, nearly 650,000 hectares were irrigated by 1950. Moreover, the first of a series of loans from the World Bank enabled the construction of the Chainant Diversion Dam on the Chao Phraya and the completion of a number of major canals in 1950. Irrigation supplied over 105 million hectares in central and northern Thailand by 1960.

The World Bank also assisted and financed the continuation of the First Economic Development Plan (1961–1966). Therefore, the irrigation system was developed in the north, including the Bhumibol Dam on the Ping River and Sirikit Dam on the Nan River, both of which have associated hydroelectric power-generating facilities. By the end of the 1970s, the controlled water flow could cover nearly 1.3 million hectares in the rainy season and approximately 450,000 hectares in the dry season.

On the other hand, in the north-east, the topography and water systems are not suited to large-scale irrigation. Only 10 percent of the 3.5 million hectares of paddy in this area are under controlled irrigation. In the 1960s, irrigation work started in the Mae Kong River Basin, which contained nearly 400,000 irrigable hectares of paddy. A multiple dam was completed in the late 1970s, and in the 1980s distribution systems were constructed to provide sufficient water for double cropping of more than 250,000 hectares. Additionally, in the 1960s, small irrigation projects were started in the south which could supply wet-season water to about 75,000 hectares; and in the 1980s it was possible to irrigate about 52,000 hectares in the Pattani River Basin. However, water supplies were still insufficient for this area, especially on the east coast, where more than 600,000 hectares of paddy are located (LePoer, 1987).

Agricultural statistics for Thailand

Agricultural statistics show that since the major agricultural changes have occurred, the greatest cropped area and production of rice are in the north-eastern region. For example, in 2005, there were approximately 5.6 million hectares of rice production in the north-east, around 2.08 million hectares in the north, 1.36 million hectares in central Thailand and 0.26 million hectares in the south. Because of the large area under rice cultivation, roughly 10 million tonnes are produced in the north-east, double the amount produced in the north and central regions, and more than ten times that produced in the south. However, the central plain is the most productive region, with a yield of 3.75 tons per hectare, while in the north the figure is about 2.13 tons per hectare. In addition, a large part of the central plain area is planted for a second (and often a third) rice harvest, and it is highly productive for these additional crops grown out of season. Not surprisingly, this intense production is accompanied by increased herbicide application (Uraikul, 2006).

Farm structure

Change of farm structure: past and present

In the mid-eighteenth century, the principal feature of the Thai rural sector was small-scale survival or subsistence farming which was directed at meeting the basic needs of the family, with additional produce for barter exchange or for 'tribute' to the local elite (Panyakul, 2001). Export of Thai rice was opened up to other colonial states, particularly Singapore and Malaysia, following the British colonial power's Bowring Treaty (1855) and several other treaties which followed. Subsequently, there was a noticeable change in the Thai agricultural sector, with a steady integration into the international trade system. Subsistence production shifted towards commercially adjusted farm production and commercial farming occurred in areas that were accessible by transport, such as the central plain and near sea coasts. Production could be expanded simply by bringing more land under cultivation with more-or-less unchanged farming technology, because land was still abundant and family labor was under-utilized. Rice was the predominant commercial crop during this period. It was exported mainly to feed migrant labor in other colonial countries in Southeast Asia, especially Singapore and Malaysia.

The rapid expansion of world trade in agricultural raw materials was observed after World War II. Several cash crops were introduced into Thailand to utilize the available upland areas that were not suitable for rice farming. A massive expansion of farmland occurred in this period, together with the growth of export crops, such as maize, kenaf (Java jute), sugarcane and cassava. New farm inputs were introduced to boost farm productivity under the banner of the 'Green Revolution' (Panyakul, 2001).

De Datta (1981) stated that the most important options for increasing world rice supplies are an expansion in the area where rice is grown and an increase in the productivity of existing rice lands. In Southeast Asia, irrigation development and the application of science-based technologies were implemented to maintain production increases. Half to three-quarters of small rice farmers still rely on rain and continue to struggle with old rice technology, producing a subsistence crop. The Green Revolution has bypassed those farmers who lack improved varieties or access to new technology. However, research for developing and adopting the technology needed to produce more rice for the developing world has been encouraged to solve farmers' problems in the field.

Synthetic fertilizers were introduced and applied to increase yields of the high-yielding rice varieties. The change in farm management took place from the 1960s to the early 1980s due to agricultural growth, especially in the central plain, where double or triple rice cropping is widespread, with heavy reliance on fertilizer and pesticide applications. Furthermore, farm mechanization has predominated since the mid-1980s with the introduction of tractors and combine-harvesters (Panyakul, 2001).

Crops and farming

In the north, most farmers grow glutinous rice intercropped with cash crops such as tobacco, soybean, garlic, peanuts and vegetables. Alternatively, soybean, peanuts, maize, sugarcane and cassava are grown as upland crops, with agro forestry in the highlands including shifting cultivation and clear-cut monoculture of high-value export crops. In the north-east, lowland farming includes glutinous rice and upland crops such as cassava, kenaf (Java jute), maize and sugarcane. Intensive rice farming has developed in the central plain, known as the Green Revolution area. Upland areas cultivate maize, sorghum, soybean, cotton, cassava and sugarcane. In the south, there are four farming categories: rice farming; rubber and palm oil plantations; fruits and perennial crops, such as pineapple, coconut and pepper; and aquaculture, especially shrimp farming and fisheries (Vijarmsorn and Eswaran, 2001).

Weeds and agriculture

Weed diversity

Weeds are always associated with crop production, with some species ubiquitous. Despite thousands of years of controlling them, they are still problematic. This is because of the wide array of available weed species (Dekker, 1997). For example, if one species is controlled, another will occupy the niche it has vacated. In addition, weeds can adapt and occupy an alternative cropping system (Dekker, 1997). For this reason, some populations within weed species can take advantage of resource availability and become adapted to changes in cropping practices and the environment.

Definition of weeds as components of diversity

A weed is simply a plant that grows where it is not desired (King, 1966). Mustafa *et al.* (2007) mention that a weed can be thought of as any plant growing in the wrong place at the wrong time and doing more harm than good. Several studies have revealed that weeds are plants that adversely affect human activity, such as toxicity to livestock (poisonous plants), and may harbor pests and diseases of useful plants and interfere with conventional crop production methods (Akobundu, 1987; Auld *et al.*, 1987). Sullivan (2003) stated that 'weeds are the natural result of defiance of nature's preference for high species diversity and covered ground'. Zimdahl (2004) highlighted that weed–crop competition affects crops both directly and indirectly. Furthermore, weeds compete with the crop for water, light, nutrients and space, and therefore reduce crop yields and affect the efficient use of machinery (Mustafa *et al.*, 2007). Consequently, farmers have to spend large amounts of money and labor to control them.

However, although weeds seem to have many disadvantages, there are some that can be beneficial (Akobundu, 1987). Weeds can enhance soil structure and reduce soil erosion by rain or wind. In addition, they can return nutrients and organic matter to the soil when they die and decay. Furthermore, many plants referred to as weeds in one crop can be considered nutritious or have economic value in another, such as in pasture for grazing (Auld *et al.*, 1987). Some are even sold in markets as vegetables. Humans also use weeds as medicinal plants. Moreover, weeds can be used for feeding animals and as a source of shelter for them (Sullivan, 2003). Some wild species of crop plants can be used as important sources of genetic material for crop improvement, such as in crop breeding for resistance to insects and plant diseases (Zimdahl, 2004). Some useful insects use weeds as hosts or alternative food sources; for instance, bees use them as major sources of pollen and nectar. In addition, many weed species possess colorful flowers that can be used in gardens for ornamental and aesthetic purposes.

Species diversity and weed diversity

Some estimates suggest there are about 400,000 known species of seed plants in the world (Govaerts, 2001). However, others have contested that this figure is too high. Thorne (2002) argued that the estimate of accepted species of angiosperms is around 260,000. However, the total number of seed plants, including undescribed species, might approximate Govaerts' estimate of published accepted species (Thorne, 2002).

Holm *et al.* (1979) claimed that there are 50,000 species of plants that may behave as weeds. However, fewer than 8,000 of these are designated as weeds in agriculture and only about 250 of these have become problematic as major weeds around the world. Of twelve important world food crops – barley, maize, wheat, millet, oats, rice, sorghum, sugarcane, potato, sweet potato, cassava and soybean – eight are of the family Poaceae, which also contains the majority of weed species. Interestingly, approximately 70 percent of weeds in agriculture are concentrated in just twelve families, with more than 40 percent of these in Poaceae (Gramineae) and Asteraceae (Compositae). Respectively twelve and eight 'worst' species in Cyperaceae and Polygonaceae are not associated with any particular crop. There are seven or fewer 'worst' weed species in families such as Amaranthaceae, Brassicaceae, Fabaceae, Convolvulaceae, Euphorbiaceae, Chenopodiaceae, Malvaceae and Solanaceae occurring in one or more major crops. Additionally, forty-seven other families each have three species or fewer. Some weeds are infested in rice cultivation, whereas others are distributed in several of the world's other main food crops.

A review of the weed literature provides evidence of a changing weed flora over time, which may be linked with changes in weed management practices. Some species may decline in abundance and become rare, whereas others may remain static or even increase in importance. The first weed review consulted was by Holm *et al.* (1977), who presented forty species as 'serious' in their evaluation of the 'world's worst weeds'. Swarbrick and Mercado (1987) recorded 138 weed species when studying the 'distribution of the world's worst weeds in different plant families'. Moody (1989) enumerated 1,800 weed species in South and Southeast Asia, 472 of which were weeds of rice. In addition, the first study of weeds in Thailand was published more than sixty years ago: Suwatabandhu (1950) listed thirty-seven species when discussing 'weeds in paddy fields in Thailand', illustrating that the weed problem for deep-water rice is more serious than that for transplanted rice, and that the problem species are grasses, sedges, broadleaf and aquatics. Noda *et al.* (1984) listed a total of 116 major weeds in Thailand (a few of which are considered serious), which are classified into four groups: grass weeds (23); cyperaceous weeds (17); broad-leaved weeds (65); miscellaneous weeds (11). Smitinand (1986) documented weeds of shifting cultivation in Thailand, including 145 species in annual shifting cultivation and 396 species in rotary shifting cultivation. Harada *et al.* (1987) recorded 92 weeds in the highlands of northern Thailand. The following decade, Machacheep (1995) published a list of 148 weeds in Thailand. Radanachaless and Maxwell (1997) stated that there were 1,710 weeds of arable fields, including both wet- and dry-seeded rice fields such as paddy and fields with soft fruits. The numbers of species associated with specific crops were: 195 in paddy fields; 115 in deep-water rice; 116 in upland rice; 203 in soybean; 43 in corn; 49 in cotton; 57 in sugarcane; 35 in cassava; 70 in fruit orchards; and 27 in vegetable fields. According to Leangarpapong (1997), about 100 species of weeds are classified as 'major weeds' in Thailand. Most of these have become distributed in many areas and also among various crops. Kawmeechai (2005) noted 212 weeds in arable crops, while Chantarasamee *et al.* (2002) referred to 102 species and Suwankul and Suwanakatnikom (2001) cited 233.

Clearly, then, the number and frequency of weed species vary significantly among these published reports. It is likely that the occurrence and relative importance of individual weed species in Thailand has changed over the past sixty years. However, this needs to be confirmed and requires comprehensive investigation. This, therefore, is the focus of the remainder of this chapter.

Weed severity

A system for categorizing weed severity is important for research. The system used here is that of Holm *et al.* (1977). According to Holm *et al.* (1977), weeds in agricultural areas can be categorized as 'serious', 'principal' or 'common'. Serious weeds are defined as being those two or three weed species in one farm that cause most of the problems in the fields; principal weeds are the second most important species, while common weeds are very widespread in many or all crops but do not seriously threaten crop yields. However, many weed species are associated with particular crops. For instance, grassy weeds are mostly found and are most severe in rice production, while broad leaf weeds occur in vegetable fields.

Specific weeds in agricultural areas under different agricultural practices are likely to be associated with specific crops. Some weeds can be serious problems in some areas but not in others; hence, they might be classified only as common weeds in specific situations, such as cultural systems or crop vegetation.

As mentioned above, Holm *et al.* (1977) argued that there are fewer than 250 important weeds worldwide. The weed species of the major crops are all well known on a global scale, especially their distribution and severity as well as their biology. However, more research is needed into the dynamics of weeds in Thai agriculture, as well as the reasons for the decline in diversity and implications for their conservation. As a consequence of changes in agricultural practice, the future implications for weed biodiversity should be considered.

Serious weeds in Thailand

In Thailand, a weed is called '*wat-cha-peud*', although Vongsaroj (1997) mentioned that this word was quite new in Thailand in the early 1980s. In the past, a weed had been called '*ya*', which was used for both grass weeds and broad leaf weeds, although strictly '*ya*' means only the former. This was because the most serious weeds in fields belong to the Poaceae or grass family.

Suwatabandhu (1950) stated that the most serious weeds of rice farms are *Echinochloa crus-galli*, *Cyperus diffomis*, *Sphenoclea zeylanica*, *Marsilea crenata*. Vongsaroj (1997) stated that weeds can adapt to difficult or bad conditions: for instance, *Xyris indica* and *Ericaulon sp.* were found in acid and salty soils; *Monochoria vaginalis* was found in soil with a high nitrogen content; *Aeschynomene spp.* were found in soils with low nitrogen; and *Fimbristylis miliacea* was observed in soils with a high phosphorus content.

Rao and Moody (1990) pointed out that some weeds were contaminants in rice seed production. For example, *Aeschynomene indica*, *Monochoria vaginalis*, *Echinochloa colona*, *Ischaemum rugosum*, *Fimbristylis miliacea*, *Oryza rufipogon*, *Scirpus grossus* and *Scirpus juncoides* were all spread as contaminated crop seed. However, weed dispersal has also occurred by water flow or irrigation: for instance, *Aeschynomene spp.*, *Cyperus iria*, *Cyperus procerus*, *Echinochloa colona*, *Echinochloa crus-galli*, *Eclipta prostata*, *Limnocharis flana* and *Monochoria vaginalis* (Rao and Moody 1990).

Finally, some weeds are more serious because of their dispersal ability, such as *Echinochloa colona*, *Cyperus iria*, *Monochoria vaginalis*, *Echinochloa crus-galli*, *Fimbristylis miliacea* and *Cyperus diffformis*, which are serious problems not only in Thailand but also in Bangladesh, Cambodia, India, Indonesia, Lao PDR, Malaysia, Pakistan, Philippines, Sri Lanka and Vietnam (Vongsaroj, 1997).

Diversity of weeds and rice production management

Changes in weed communities

Several attempts have been made to identify the associations between weeds in arable land and soil type and crops. Some weed species are often found restricted to certain soils, although many others occur in many different conditions (Marshall *et al.*, 2003). A survey of weed incidence in central southern England found that certain species were dominant in the east, such as *Alopecurus myosuroides* Huds., whereas others, such as *Fumaria officinalis* L., were common in the west (Froud-Williams and Chancellor, 1982; Chancellor and Froud-Williams, 1984). Sutcliffe and Kay (2000) surveyed similar areas in central southern England in the 1960s, 1977 and 1997 to assess change in species occurrence caused by the intensification of agricultural practices. Some species have become rarer, after being common in the past, others have become more common, and still others have remained stable. For example, *Anthemis arvensis*, *Galeopsis angustifolia* and *Silence noctiflora* have all declined significantly, while *Scandix pectin-veneris*, *Sherardia arvensis* and *Petroselinum segetum* were frequently found in the field resurveys (Sutcliffe and Kay, 2000).

There are several factors affecting changes in weed communities, including weed seedbank communities, weeds invasion, late-season weed infestation and agricultural landscapes.

Weed seedbank communities

The effects of management systems on weed seedbanks have been studied in the northern Great Plains region of the USA. This reasearch has found that the decrease in diversity is highest in conventionally managed fields compared with organic fields. Moreover, fluctuations in environmental factors each year have significant impacts on weed seedbanks (Harbuck *et al.*, 2009).

Weed invasion

This is a serious biotic stress on fallow land. However, sowing high-diversity seed mixtures containing non-weedy disturbance-tolerant and generalist members of the flora can suppress weeds. For this reason, the composition of multi-species seed mixtures in major climatic and soil conditions can reduce weeds (Csontos *et al.*, 2009).

Late-season weed infestation

This is an important factor leading to changes in weed communities. Although these weeds do not affect yields, if allowed to mature and contribute seeds to the soil seedbank, they may lead to the future establishment of competitive weed complexes. Therefore, effective long-term weed management strategies must incorporate practices to reduce late-season weed seed production by weed complexes (Walker and Oliver, 2008).

Agricultural landscapes

These are important sources of biodiversity, such as a mosaic of farmers' fields, semi-natural habitats, human infrastructures (e.g. roads) and occasional natural habitats. The edges of agricultural fields, field margin habitats, serve as refuges for biodiversity, particularly in north-western Europe. In contrast, North America and northern Europe have wilderness areas and dedicated sites for nature conservation which can be integrated into the managed farmed landscape. For this reason, margins can have agricultural, environmental and conservation roles as they contain a range of plant communities in a variety of structures that maintain weed biodiversity. Thus, some margin floras may spread into crops and become field weeds (Marshall, 2005).

Changes in farming systems

Changes in arable management may have affected the weed flora (Marshall *et al.*, 2003). Herbicide usage and intensive arable cropping have both resulted in significant changes over the past forty years (Marshall, 2001).

First, after the introduction of national subsidies in 1989, organic farming has increased considerably. For example, in Germany, there have been positive effects on the environment and nature, such as diversity and abundance of wild animal and plant species. Even though the landscape structure and farm size vary, the intensity of farm management, mechanical weed control, availability of landscape elements and abandonment of cultivation on infertile land all affect species conservation (Neumann *et al.*, 2008). Furthermore, weed control with herbicides is impossible in perennial organic agricultural systems. Alternatively, in these systems, weeds can be minimized mechanically, thermally or by mulching with plastic film, but this is usually insufficient to exterminate them. Thermal weed control requires knowledge of the plants' thermal sensitivity. The most common weeds growing between strawberry rows in Lithuania are Shepherd's purse (*Capsella bursa-pastoris*), common groundsel (*Senecio vulgaris*) and common chick-weed (*Stellaria media*) (Sniauka and Pocius, 2008).

Second, as reported by Koocheki *et al.* (2009), the manipulation of cropping systems to improve weed management has led to changing dynamics of weed populations in arable fields. However, these systems require significant understanding of the spatial and temporal dynamics of weeds, seed losses and seed production. Changes in weed seedbank density and species composition often occur when crop management practices and crop rotations are changed. For example, continuous winter wheat fields have more annual grass weeds, but broad leaf weeds are more abundant in sugar beet–winter wheat rotations (Koocheki *et al.*, 2009).

Third, ecological weed management practices, such as cultural weed control (e.g. hoeing, plowing), are starting to replace heavy reliance on chemical weed control. There are three main advantages of this approach: reduced recruitment of weed seedlings from the soil seedbank; alteration of crop–weed competitive relations to the benefit of the crop; and gradual reduction of the size of the weed seedbank. However, further adoption of cultural weed management strategies depends on the needs and skills of individual farmers (Bastiaans *et al.*, 2008).

Fourth, plant allelopathy has been used in weed control. For instance, potent allelochemicals from rice tissues and root exudates have been studied worldwide and substantial progress has been made recently. Rice allelopathy has been known since the mid-1980s and numerous compounds have been identified. Furthermore, allelopathic rice not only releases allelochemicals to suppress the growth of neighboring plant species, but also detects the presence of interspecific neighbors and responds through increased release of certain allelochemicals, and may modify soil microorganisms to its advantage by release of allelochemicals. Allelopathic rice cultivars combined with integrated cultural management practices may facilitate a reduction in the use of herbicides (Kong, 2008).

Finally, tillage exerts one of the greatest effects on weed diversity and density, as is evident in a study of four farms (thirty-six fields) in Ontario, Canada, over six years. No tillage promoted the highest weed species diversity; chisel plow was intermediate; and moldboard plow resulted in the lowest species diversity. Practically, reduced tillage in combination with good crop rotation may reduce weed density and expenditure on weed management (Murphy *et al.*, 2006).

Current rice production in Thailand

Rice is the main crop of Thailand (Chinawong and Suwanketnikom, 2001), but Vongsaroj (1993) has pointed out that weeds are a major problem in this sector. There are good reasons why the level of weed infestation varies, such as the method of cultivation (including transplanting or direct seeding with dry or wet seed) and the locality. Rice cultivation in Thailand is divided into five types – upland rice, dry-seeded rice, deep-water rice, transplanted rice and wet-seeded rice (i.e. pre-germinated, direct-seeded rice) (Vongsaroj, 1993) – and weeds are associated with all five types. Additionally, weeds are one of the limiting factors in direct-seeded rice, whereas yield losses from competition with weeds are observed in all field types. The average yield of direct-seeded rice is significantly lower than that of transplanted rice, especially in resource-medium and resource-poor conditions. However, the low yield of direct-seeded rice in such conditions is not primarily caused by competition with weeds. Nonetheless, from an economic viewpoint, weeding has been applied to only limited parts of the north-east region with stable and relatively high rice yields (Tomita *et al.*, 2003). As a result, 2–9 percent of potential rice yield is lost because of weeds (Tomita *et al.*, 2003).

The infestation of weeds in Thai agriculture

As already mentioned, Vongsaroj (1997) has stated that rice cultivation in Thailand has been developed in different ways depending on topography, soil type, economics and the culture of the community.

Upland rice is grown in dry soil in highland areas, such as in northern Thailand, by broadcasting rice seeds directly onto the soil, as with other field crops (e.g. corn). Farmers practice this method only in small areas to serve their own families. It is mostly used in newly opened forest areas. The rice-growing area will be moved from one place to another if a weed problem develops. The most serious weeds in this area, according to Vongsaroj (1997), are *Ageratum conyzoides, Richardia brasiliensis, Spilanthes paniculata, Bidens pilosa, Mimosa invisa, Mitracapus villosus, Dactylocnenium ciliaris, Eleusine indica, Cyperus rotundus* and *Commelina benghalensis*.

Dry direct-seeded rice is sown into dry soil before water is channeled onto the land. This form of cultivation is popular in the lowlands, especially in the north-east region, and in the rain-fed areas with deep water in the central region, such as Ratchaburee, Pichit, Ayutaya, Prachinburee, Nakornayok and Nakonswan provinces, where there is irrigation. Farmers sow the rice seed directly into the soil, as in upland rice. With this method, weeds germinate at the same time as the rice, but they are less problematic than in upland rice cultivation because of flooding during the rice-growing period. Therefore, irrigation is very important for this cultivation method. The weeds that are most associated with this method are *Oryza rufipogon, Ipomoea aquatica, Aeschynomene spp., Pentapetes phoenicea, Melochia corchorifolia, Echinochloa colona, Ischaemum rugosum, Setaria geniculata, Panicum cambogiense* and *Eleocharis dulcis* (Vongsaroj, 1997).

In this cultivation method, farmers previously started growing their rice at the beginning of the rainy season by sowing seed into the soil after tillage. Weeds were less problematic because plowing by animal power can be done in wet soil after rain. The advantage of this method is that weed seeds can germinate before the second plowing. Weed seedlings are buried and then composted in the soil. Some weed seeds in the seedbank can grow, but these are then destroyed by the second and third tillage operations. As a result, weeds can be efficiently controlled. Nowadays, though, tractors are used and therefore tillage must be done before the onset of rain. Hence, weed seeds can be buried and accumulate in the weed seedbank, which can result in weed competition with the crop. Although this cultivation method has fewer problematic weeds than upland rice, it is not greatly practiced in Thailand.

Transplanted rice was developed to minimize weed problems. Harrows are used to remove the rough finish left by plowing. Harrowing not only breaks up clods of soil to provide a fine tilth but also contributes to weed removal. Subsequently, the fields are flooded with water to a depth of 5–10 centimeters, while the rice seedlings are growing. Weeds can be well controlled under these conditions – particularly grass weeds, which cannot grow. Although other weeds can grow in these conditions, the rice is able to compete successfully with them. Consequently, weeds are less of a problem with this method. However, labor costs are an issue. Some weeds which are found in this type of cultivation are *Sphenoclea zeylanica, Mimulus orbiculalis, Monochoria vaginalis, Cyperus difformis, Fimbristylis miliacea, Cyperus pulcherrimus, Cyperus iria, Leptochloa chinensis, Echinochloa crus-galli, Chara zeylanica* and *Marsilea crenata* (Vongsaroj, 1997).

Red rice has now become a serious weed in Thailand's rice-producing areas (Maneechote and Jamjod, 1999) as well as various other regions around the world (Burgos et al., 2006; Olsen et al., 2007). Prathepha (2009) points out that weedy rice (*Oryza sativa f. spontanea*) is problematic in rice production areas in north-east Thailand, particularly in Thung Kula Ronghai. Most weedy rice has a red pericarp to the seed; hence, it is known as 'red rice' (Arrieta-Espinoza *et al.*, 2005). It competes for production inputs, increases weed control costs, and reduces yield and grain quality (Maneechote and Jamjod, 1999; Burgos *et al.*, 2006). Maneechote and Jamjod (1999) point out that weedy rices are classified into three groups: Khao Harng (long awn and shattering rachis); Khao Deed (no awn and shattering); and Khao Dang (awnless and not shattering). The first two categories are serious weeds due to the high degree of seed shattering before the rice harvest.

Chemical weed control

Changes in cultural practices have magnified the weed problem. Although hand weeding is the most common method of weed control worldwide, herbicides have become increasingly important, particularly in Asian countries (Kim, 2001). In Malaysia, uncontrolled weed competition with rice may cause 10–35 percent grain yield reduction (Karim *et al.*, 2004). Recently a number of sulfonylurea herbicides (i.e. bensulfuron, metsulfuron, pyrozosulfuron and cinosulfuron) have been introduced as alternatives to the older herbicides based on phenoxy-alkanoic acids, such as 2, 4-D, in order to control grass weeds, which became dominant after continuous use of the latter (Karim *et al.*, 2004). It was then that farmers started using graminicides, especially molinate, which is applied seven days before seeding (DBS) or after seeding (DAS). Sulfonylurea herbicides are often used for broad leaf weed control, while chlorimuron is used in areas heavily infested by sedges (Morris and Therivel, 1995).

Azmi (2002) pointed out that some particular weed species have become dominant because of the serial application of certain herbicides. For instance, *Echinochloa crus-galli* was found dominant in 2, 4-D treated plots; *Scirpus grossus* dominated under bensulfuron; and *Monochoria vaginalis* was dominant under repeated use of molinate/propanil, thiobencarb/propanil, pretilachlor, quinclorac, propanil and fenoxaprop-ethyl. For this reason, a combination of graminicides with one of the herbicides for sedges and broad leaf weed control was used for broad-spectrum weed control in rice. For example, quinclorac + bensulfuron, molinate + 2, 4-D isobutyl ester (IBE), molinate + bensulfuron and thibencarb + pyrazosulfuron are the suggested herbicide mixtures for direct-seeded rice. Farmers cultivating transplanted rice use cinosulfuron at four to seven days after transplanting (DAT). Bensulfuron-methyl can control *Marsilea crenata* and *Sagittaria guyanensis*, which are tolerant to 2, 4-D. A mixture of cinosulfuron and 2, 4-D IBE is suitable for broad-spectrum weed control, including *Ludwigia spp.* and *Fimbristylis miliacea*. Moreover, the standard of land preparation and the proper cultural practices have important impacts on the success of chemical weed control (Azmi, 2002).

Integrated weed management

Weed management, such as herbicide usage, might cause unforeseen problems or have unintended destructive environmental and/or economic effects. Although humans often depend on single solutions for weed management, single weed-control strategies enable weed adaptation to management. Nevertheless, herbicides normally control weeds effectively. Herbicide misuse can lead to problems, however, such as groundwater contamination, residual carry-over, cropping restriction and the development of genetically based herbicide resistance (Murphy et al., 2006). Weed control with herbicides reduces the vigor of weeds, but ultimately reduces the yield and profitability of crops, probably because of their phytotoxicity to the crop. Therefore, this approach may lead to instability in weed management because, once one weed species is removed, another will occupy the niche created (Murphy et al., 2006).

On the other hand, an ecologically based approach to weed management may lead to a more balanced and diverse weed community. A diverse weed community should be managed for the benefit of the weed community dynamics. Diversity in weed communities may be considered to support and enhance bio-diversity management and natural ecosystems (Booth et al., 2003). Additionally, in a three-crop rotation, no-tillage systems increased weed species diversity above ground and within the seed bank (Booth et al., 2003). Although it is necessary to increase production to meet food demands, cultural weed control and weed biodiversity should still be supported for sustainable production methods.

The shift in weed diversity in Thailand

Smitinand (1986) stated that some weeds occur both in the lowlands and in the highlands of Thailand, such as: *Imperata cylindrica*, *Cyperus rotundus* and *Chromolaena odorata*. Annual shifting cultivation is practiced in the northern highlands, where the inhabitants grow their crops annually. Serious weeds found on the open ridges and slopes, such as *Ageratina adenophora*, *Artemisia dubia* and *Artemisia roxburghiana*, whereas *Tithonia diversifolia* is observed along the rather humid valleys. If there are annual fires in these areas, there is widespread growth of grasses, such as: *Arundinaria setosa*, *Capillipedium assimile*, *Capillipedium parriflorum*, *Cymbopogon flexuosus*, *Heteropogon triticeus*, *Imperata cylindrica*, *Saccharum arundinaceum*, *Thermeda arundinacea*, *Thermeda triandra* and *Thysanolaena maxima*. Ferns are also found, such as *Dicranopteris linearis* and *Pteridium aquilinum*.

Furthermore, rotary shifting cultivation is practiced by hill tribes who inhabit elevations above 600 meters. Farmers in these areas grow wet rice with additional shifting cultivation to supplement their income. However, the land is left fallow between the rotations and it is not cleaned. As a result, associated annual weeds include *Ageratum conyzoides*, *Cassocephalum crepidioides* and *Physalis minima*, as well as other perennials, such as *Arundinella hispida*, *Hyparrhenia rufa*, *Imperata cylindrica*, *Saccharum arundinaceum* and *Thysanolaena maxima* (Smitinand, 1986).

In north-east Thailand, direct dry seeding is more popular than direct wet seeding. Tomita et al. (2003) pointed out that transplanting has been replaced by direct dry seeding to increase the frequency of successful planting and save labor. Additionally, the fields have more species-rich vegetation and greater species diversity due to this cultivation change. The dominant weed species observed in the study areas (179 paddy fields) were: *Fimbristylis miliacea, Ludwigia hyssofolia, Ludwigia adscendens, Cynodon dactylon* and *Panicum repens*. The most common weeds found in both direct dry seeded fields (DSF) and transplanted fields (TF) were *Fimbristylis miliacea* and *Ludwigia hyssofolia*. In other words, these two species grew in both DSF and TF, indicating that the change in cultivation method did not affect them differentially. The most frequently observed weeds in DSF were *Cynodon dactylon* and *Panicum repens*, whereas *Ludwigia adscendens* was most common in TF. These weed species are abundant not only in rain-fed paddy areas in north-east Thailand but also in other areas of Southeast Asia, as well as other tropical and subtropical regions (Holm *et al.*, 1977; Noda *et al.*, 1984).

Comparison with weed research in other Southeast Asian countries

The shift in cropping systems was motivated by the shortage of labor for hand weeding (Kim, 2001). De Datta and Flinn (1986) pointed out that many farmers in Thailand, Malaysia and the Philippines have switched from transplanted to direct-seeded flooded rice because of increases in labor costs, poor access to irrigated areas, the development of modern early-maturing varieties and improved weed management techniques. Direct seeding has also been widely adopted in Vietnam and Sri Lanka (Moody, 1989).

The change in rice cultivation has led to a change in weed management. In Malaysia, farmers usually controlled broad leaf weeds with the phenoxy compound 2, 4-D (Saharan cited in Karim *et al.*, 2004). Consequently, grassy weeds became dominant in rice fields. Then the farmers started using graminicides, especially molinate, applying seven days before seeding (DBS) or after seeding (DAS). Sulfonylurea herbicides, such as bensulfuron, metsulfuron, pyrozosulfuron, and cinosulfuron, were often used for broad leaf weed control, while chlorimuron was used in areas heavily infested by sedges (Morris and Therivel, 1995). Ho (1995) mentioned that the most dominant weeds in the Muda rice-growing area of Malaysia are *Monochoria vaginalis*, followed by *Fimbristylis miliacea, Leersia hexandra, Jussiaea repens, Scirpus grossus, Isachne globosa* and *Scirpus juncoides*, together with *Cyperus haspen* and *Marsilia crenata*. Recently, an increase in the use of machinery in Muda has led to weed seed contamination and widespread emergence of *Echinochloa crus-galli, Echinochloa glabrescens* and *Leptochloa chinensis*.

Direct seeding of rice in tropical Asia includes broadcasting and drilling seed onto puddle wet (wet seeding) or dry soil (dry seeding). De Datta (cited in De Datta and Flinn, 1986) stated that there were more sedges and broad leaf weeds and grasses growing in shallow (less than 2.5 centimeters) water with continuous flooding, while grasses and sedges were almost completely suppressed in 15 centimeters of standing water.

Vongsaroj (1993) stated that the infestation of weeds in broadcast-seeded flooded rice areas is more severe with pre-geminated rice. The major weeds in this form of rice cultivation are: *Echinochloa crus-galli, Leptochloa chinensis, Echinochloa colona, Sphenoclea zeylanica, Mimurus orbicularis, Monochoria vaginalis, Cyperus difformis, Fimbristylis miliacea, Eleocharis dulcis, Marsilia crenata* and *Chara zeylanica*.

In the Philippines, the principal weed species are *Echinochloa glabrescens*, *Echinochloa crus-galli*, *Cyperus difformis*, *Monochoria vaginalis* and *Spenoclea zeylanica*.

Karim *et al.* (2004) stated that some weeds observed in the rice fields of Malaysia were: *Oryza sativa* (weedy rice), *Echinochloa spp.*, *Leptochloa chinensis*, *Fimbristylis miliacea* and *Limmocharis flava*. The most serious weed which caused serious yield loss in that country was *Echinochloa crus-galli* (Azmi and Mashhor, 1995), followed by broad leaf weeds and sedges. However, the weed species found in the rice fields of Malaysia has changed due to changes in cultural practices and the use of agro-chemicals (Karim *et al.*, 2004). Species abundance and dominance in rice fields changed markedly between 1989 (100 percent transplanting method), 1991 (50.4 percent direct seeding) and 1993 (79.6 percent direct seeding) in the Kamubu area (Azmi and Mashhor, 1995). In 1989, five weeds were dominant in Malaysia – *Monochoria vaginalis*, *Sagittaria guyanensis*, *Fimbristylis miliacea*, *Microcarpaea minima* and *Limnocharis flava* – whereas by 1993 *Echinochloa crus-galli*, *Fimbristylis miliacea*, *Monochoria vaginalis*, *Echinochloa colona* and *Ludwigia hyssopifolia* had achieved dominance. Furthermore, following continuous cultivation of direct-seeded rice, perennial weeds such as *Echinochloa stagina*, *Paspalum distichum* and *Cyperus babakan* occurred increasingly (Morris and Therivel, 1995). The most dominant weed species in the Muda area was *Echinochloa crus-galli*, followed by *Leptochloa chinensis*, *Oryza sativa* (weedy rice), *Ludwigia hyssopifolia*, *Fimbristylis miliacea*, *Sphenoclea zeylanica* and *Scirpus grossus*, whereas the dominant weeds in the Besut area were *Echinochloa crus-galli*, *Ludwigia hyssopifolia*, *Fimbristylis miliacea*, *Ischaemum rugosum*, *Leptochloa chinensis*, *Oryza sativa*, *Cyperus babakan* and *Cyperus iria*. Consequently, *Echinochloa cruss-galli* and *Leptochloa chinensis* remain dominant in direct-seeded rice fields (Karim *et al.*, 2004).

Inamura *et al.* (2003) stated that there was competition between weeds and wet-season transplanted paddy rice for nitrogen, growth and yield in the central and northern regions of Laos. Thirteen weeds species were found, although few major weeds were abundant at the survey sites. Weed growth was poor in infertile soils under rain-fed conditions. Rice yield and the amount of nitrogen in rice were suppressed by competition with weeds. The major weed species at the survey sites were broad leaf, such as *Ludwigia octovalvis*, followed by *Sagittaria trifolia*, *Ageratum conyzoides*, *Ludwigia hyssopifolia*, *Xyris indica* and *Marsilea crenata*; Cyperaceous weeds, such as *Fimbristylis littoralis*, *Scirpus juncoides* and *Cyperus difformis*; and grass weeds, such as *Ischaemun rugosum*, *Echinochloa colonum*, *Paspalum distichum* and *Axonopus compressus* (Tomita *et al.*, 2003).

Although weeds are plants that farmers do not want on their land, humans have long used them in various ways, for instance as food, medicine and animal feed (Akobundu, 1987; Auld *et al.*, 1987). This has prompted several interesting weed research projects in Southeast Asia and Asia-Pacific.

Towards a conservation of weeds

Red List weeds/threatened weeds

The International Union for Conservation of Nature (IUCN)'s Red List of Threatened Species was used increasingly during the 1980s to assess the conservation status of species for policy and planning purposes. This stimulated the development of a new set of quantitative criteria for listing species in the categories of threat: 'critically endangered', 'endangered' and 'vulnerable'. Since then, the system and the criteria have been widely used by conservation practitioners and scientists, and they now underpin one indicator that is used to assess the Convention on Biological Diversity's 2010 biodiversity target (Mace et al., 2008).

Recent evidence suggests that the weed flora in the UK has changed over the past century because of changes in agricultural practices. Although many species of plants can survive within an arable system, some of those that are now threatened with extinction were once problematic agricultural weeds (Sutcliffe and Kay, 2000). There have been significant decreases in some species due to improvements in agricultural efficiency (Marshall et al., 2003). For instance, both the cornflower (*Centaurea cyanus*) and the corncockle (*Agrostemma githago*) were once major weeds in agricultural areas, particularly in wheat fields, but they are now virtually extinct in Britain as wild flowers (Marren, 1999). Furthermore, thorow-wax (*Bupleurum rotundifolium*) is probably already extinct in Britain (Sutcliffe and Kay, 2000). The last Red List for the UK (Cheffings and Farrell, 2005) classified 9 species as extinct, 4 species as extinct in the wild, 35 species as critically endangered, 90 species as endangered, 220 species as vulnerable, 39 species as data deficient, 98 species as near threatened and 1,261 species as of least concern. The rare or threatened species in the UK include twenty-two in arable habitats, of which 14 percent are classed as probably extinct and 50 percent as endangered, compared with overall percentages of only 6 percent extinct and 16 percent endangered among rare species from the entire range of habitats (Perring and Farrell, 1983).

Concerns unique to weed species diversity and weed conservation

Evidence that the weed flora of Thailand has changed over recent decades, with some species declining whereas others have increased, supports the thesis that changing agricultural practice may affect weed diversity. The significant and surprising fall in weed species diversity can be attributed to weed management systems and the dynamics of crop protection.

According to research by Mortensen et al. (2000), weed management systems have been constrained by a number of factors, including weed species diversity. Bastiaans et al. (2007), on the other hand, maintain that weed management systems with less chemical control are globally essential. This is because the nature of weed management strategies is determined by the diversity of weed communities (Derksen et al., 1993). For example, weed problems can be tackled by: reducing the weed seedbank in the soil; reducing recruitment of weed seeds from the soil seedbank; and strengthening the relative competitive ability of crops (Bastiaans et al., 2007). Furthermore, environmental and soil characteristics as well as cropping system and management practices can influence the dynamics of weed populations in arable fields (Koocheki et al., 2009). As a result, improved weed management approaches may provide the basis for increased weed diversity.

First, herbicide-resistant crops have been introduced over the past few years while transgenic crops have been introduced through management programs to fit with the use of singular chemical tactics. However, this 'silver-bullet' approach has failed as a result of fundamental ecological relationships. The reduction of herbicides has been replaced by management practices to maintain weed populations at low-equilibrium densities and reduce the relative fitness of weeds (Mortensen et al., 2000). For this reason, in many countries, pesticide policies have called for significant reductions in use, together with the promotion of biodiversity in agro-ecosystems.

Second, high-level pest control has high priority in competitive agriculture to stimulate competition and enhance efficiency. Within pest control, weed management has the highest priority in terms of agronomy and economics. Weed management technologies are introduced to increase yield and reduce the costs of crop production. The great advantage of precision farming technology is that it reduces environmental damage and preserves soil productivity (Lencses and Takacs-Gyorgy, 2008). Therefore, the environmental impact of herbicide utilization has stimulated research into new methods of weed control, such as selective herbicide application in highly infested crop areas (Naeem et al., 2007).

Finally, researchers have shown increased interest in bioherbicides (Hoagland et al., 2007). Many microorganisms have been shown to possess bioherbicidal activity and several phytopathogenic fungi and bacteria have been patented as weed-control agents. Furthermore, although the phytotoxic components of most agents have not been elucidated, some phytotoxins and other secondary compounds produced by such microbes may be toxic to mammalian systems. However, there is a lack of definitive research on the overall toxicological risk of bioherbicidal microorganisms to the degree achieved or required for synthetic herbicides (Hoagland et al., 2007).

Thus, more herbicide usage has led to declines in the populations of farmland birds, invertebrates and plants in Europe (Marshall, 2001). Marshall (2001) also stated that although herbicides are popular and useful in agriculture, they have an effect on non-target plants which are outside the target area.

Future challenges for weed diversity and weed conservation

Several studies have reported successes and failures in agriculture over recent years (Fernandez-Quintanilla et al., 2008). Efficient, safe technologies to manage weeds can lead to lower food prices and perhaps even global food security. However, these technologies must be implemented with proper advice and options for the end-users based on broad scientific knowledge, which will aid the development of effective practices for new crops, new production systems and new weeds. Moreover, increased concern about the conservation of biodiversity and food safety must be addressed, and new clients in non-agricultural sectors should be offered proven expertise and know-how (Fernandez-Quintanilla *et al.*, 2008).

Most countries currently have action plans for the conservation of biological diversity because changes in agricultural management have reduced weed flora diversity (Marshall *et al.*, 2003). The concept of conservation of species diversity in farmlands can be widened in many ways. First, through the introduction of complex farming systems, including research and development, traditional farming systems and conservation. Such systems have been developed and inherited from traditional farmers through a series of traditional resource-conserving practices, variety conservation, weed, pest, nutrient and water management practices to deal with socio-environmental changes. In addition, scientists involved in agricultural research and development must incorporate farmers' practices and traditional farming systems which are rapidly disappearing in the face of major social, economic and political changes occurring in developing societies (Singh and Sureja, 2008).

Second, many of the challenges faced by weed ecologists can be met by predicting the responses of weed populations to changes in their environment and/or management. However, several limitations can lead to the failure of predictive population modeling. Without the further development of models for weed population dynamics, the ability to predict long-term dynamics will be restricted (Freckleton and Stephens, 2009).

In contrast, there are advanced technologies for more efficient weed control that may contribute to a decline in weed diversity, such as: aerial crop imaging; automated recognition of crops and weeds; intelligent real-time automatic weed control systems; and novel herbicidal compounds of natural origin.

Aerial recognition using digital cameras has been used to identify weed infestations within field crops. Photographs were taken from a Cessna airplane at an altitude of 300 meters above a maize field. Applying this method of identifying mapped areas with high densities of weeds allows a formulated plan to pinpoint the application and dosage of herbicides to the field crop (Pudelko *et al.*, 2008).

Zhu *et al.* (2008) have pointed out that the automated recognition of crops and weeds through the use of Vis/NIR spectral imaging is one of 'hottest' research branches of agricultural engineering. The reflectance rate of green plant leaves can be used to identify crop varieties. As the colors and surface textures of crops and weeds change in different living phases, these changes may exert great influence on the reflectance spectra of plant leaves. Hence, the NIR spectra can be used to identify the crop from weeds with no need to worry about the respective living stages of these plants (Zhu *et al.*, 2008).

As pointed out by Fennimore and Doohan (2008), major crops, such as maize, cotton, rice, soybean and wheat, are widely distributed and treated with a wide variety of herbicides, whereas fewer herbicides are developed for minor-crop usage. Although minor crops are less important, they still require practical solutions to weed control, besides the application of herbicides (Mustafa *et al.*, 2007). Fennimore and Doohan (2008) argued that the heavy reliance on chemicals raises many environmental and economic concerns, causing many farmers to seek alternatives for weed control in order to reduce their use of chemicals. An intelligent real-time automated weed control system using image processing to identify and discriminate between weed types, such as narrow and broad leaf, may provide the solution. One system consists of a mechanical structure which includes a sprayer, a digital camera, a 12-volt motor coupled with a pump system and a small CPU (Fennimore and Doohan, 2008).

Finally, phytotoxins produced by fungi or natural amino-acids may be useful in parasite weed management strategies as they could interfere with early growth stages of the parasites (Vurro *et al.*, 2009).

For the reasons mentioned above, such developments could lead to a significant decline in global weed diversity.

The conservation and sustainable use of biological diversity

Conservation of biological diversity is compatible with agricultural sustainability or low-input agriculture that attempts to minimize the use of non-renewable inputs. Sustainable agriculture is also the activity of growing food and fiber in a productive and economically efficient manner. Such practices maintain or enhance the quality of the local and surrounding environment, including the soil, water, air and all living things. Thus, *sustainable agriculture* helps to maintain or improve the health and quality of life of individual farmers, their families and their communities; *organic agriculture* eliminates all synthetic chemical inputs of fertilizers and pesticides; *ecological agriculture* emphasizes the relationship with the surrounding environment and takes measures to maintain its integrity; and *alternative agriculture* incorporates some of the practices from all three. However, sustainable agriculture practices must include consideration of social and economic issues as well as those related to the biophysical environment (Loon *et al.*, 2005).

Weed management strategies based on weed biology and ecology enable a more diverse agro-ecosystem. For instance, Stone *et al.* (2008) described the widespread adoption of perennial plants in Australian agriculture to produce more sustainable and biodiverse systems. This strategy incorporates pre-trial screening, weed risk assessment, experimental site hygiene practices and species management guidelines.

In the UK, agri-environment schemes aim to arrest declines in arable biodiversity through, for example, cereal field margin management (Walker *et al.*, 2007). This includes uncropped cultivated margins (UCMs), spring fallow (SF) and cropped conservation headlands with fertilizer inputs (CH) or without fertilizer inputs (CHNF), designed to sustain plant species diversity and endangered arable plants. For instance, species diversity, including rare species, was highest at UCMs, followed by SF and CHNF margins. Additionally, diversity was generally lower on cropped margins due to competition from the crop. Species diversity was greatest at the edge of all except UCMs and there was a strong latitudinal decline in overall diversity and rare species. These results confirm that agri-environment schemes are effective in conserving arable plants, including rare species, across a variety of landscape types.

Conversion of native land to agriculture represents a major threat to bio-diversity and loss of habitat for wildlife. However, increased productivity on existing agricultural land can be achieved through technological innovation, such as irrigation, mechanization, improved crop vigor, crop protection and nutrition. These strategies play a role in preventing further land conversion to agriculture. Therefore, such technologies must be integrated and engaged appro-priately and locally within land management strategies to improve agricultural production, increase rural livelihoods and achieve biodiversity conservation goals (Dollacker and Rhodes, 2007).

Biodiversity and ecosystem services

Agricultural practice

According to Matson *et al.* (1997), agricultural practices have had significant effects on natural ecosystems over recent decades. However, the world's popu-lation is expected to grow by more than a billion people (or about 20 percent) over the next decade, which will require more use of agrochemicals for efficient agriculture (Casida and Quistad, 1994).

Changes in the intensification of agriculture and the utilization of land have caused broad changes in farming practices (Lee and Lee, 2003). Additionally, these intensive practices have often resulted in a loss of biological diversity. Salisbury (1961) stated that there has been a significant change in weed evolution. Common weeds have the capacity to become serious weeds in the future as a result of changes in cultural practice (Shiau *et al.*, 2005). In addition, the evolution of herbicide-resistant weeds is of increasing concern, having now been identified in developing regions as well as intensive agriculture (Sangakkara *et al.*, 2002). Thus, the continued use of herbicides may result in selection of resistant biotypes which will present new challenges for farmers.

Effects of changing agricultural practices on weed diversity

Regarding the success of weed control, strategies of weed management are required, such as scouting, prevention, mechanical practices, cultural practices, biological control and chemical control.

The growth in demand for food products may affect agricultural practices as conventional agriculture may be unable to meet customers' needs. Consequently, production methods will have to be developed to increase agricultural efficiency. In the past, conventional agriculture has sometimes used inappropriate techno-logies, such as the heavy use of off-farm inputs to produce high-yield mono-cropping. This can lead to a decline in soil fertility through the overuse of chemical fertilizers.

Furthermore, intensification of agricultural activities can lead to a decline in weed communities. For example, rice productivity in paddy fields in the Tama Hills, central Japan, has been increased by employing drainage after the harvest to reduce the incidence of competitive species. Under these conditions, many paddy weed species have become threatened (Yamada *et al.*, 2007). The recovery of indigenous plants or arable flora becomes important to restore these species by cultural ecosystem practices, such as traditional rice cultivation or low-intensity farming. Typical paddy weed communities can be rapidly restored after restarting agricultural practices in abandoned paddy fields (Yamada *et al.*, 2007).

Additionally, there are alternatives to chemical weed control based on ecological agricultural practices which may affect weed diversity: for instance, the utilization of plant pathogens in biological weed control; allelochemicals and novel phytotoxins; in-field surveys and automated detection systems; and cultural approaches, such as weed-suppressive crop cultivars (Haefele *et al.*, 2004).

A considerable number of plant pathogens have been studied for their possible use in weed control. Some have proven virulent enough to control weed species and compete commercially with chemical herbicides. These approaches may well lead to sustainable systems of biological control of parasitic weeds (Sands and Pilgeram, 2009). Moreover, smut fungi may have potential as classical biological control agents for use against *Imperata cylindrica* and *Rottboellia cochinchinensis* by reducing these weeds' seed dispersal (Wood *et al.*, 2009).

Allelochemicals may be used in place of synthetic herbicides for reduced-risk weed control. For example, a mixture of allelochemicals was highly effective in suppressing *Echinochoa crus-galli* in paddy rice (He *et al.*, 2009).

Over the last two decades, the demand for organic products has grown rapidly in the world due to increased concern about the side-effects of pesticides on the environment and human health. Hence, crops or crop residues such as *Vicia villosa* and *Sorghum bicolor* have been used to suppress weeds, for instance in early-season organic lettuce production (Isik *et al.*, 2009).

Automatic detection of weeds is necessary for site-specific application of herbicides or precise physical weed control. Leaf reflectance is mainly determined by photosynthetic pigments, leaf structural properties and water content. In other words, spectral reflectance characteristics can be used for weed discrimination (Chen *et al.*, 2009).

In-field surveys, which directly estimate weed population densities, typically utilize either random or non-random field-selection methods. This method develops a database for tracking weed shifts, control failures and the presence of other herbicide-resistant biotypes over time (Davis *et al.*, 2008).

Weed-suppressive cultivars, intercropping and rotational cover cropping all have potential to contribute significantly to weed management in agro-ecosystems. The weed-suppressive effect is largely determined by the combined effects of genotype (or species) and management. The major strategies depend on a sufficient level of weed suppression while maintaining yield potential (Bastiaans *et al.*, 2007).

Implications of weed conservation

Ecological, physiological and molecular methods must be integrated in future agricultural research in order to understand agricultural crops in situ and their interaction with the environment. Additionally, concern about organisms impacting on long-term human health and crop productivity has led to the introduction of so-called 'agricultural eco-genomics'. This will enable an ecologically based approach that is necessary for successful weed control management. Crop-breeding technologies for appropriate agricultural systems have been introduced to meet the demands of food consumers. In addition, Weih *et al.* (2008) have highlighted two possible avenues of agricultural research: suppressing weeds through crop allelopathic activity rather than chemical control; and developing environmentally friendly and sustainable production of perennial energy crops on agricultural land. It is often difficult to define 'sustainability' in agriculture, but it normally implies the increased utilization of ecological processes.

References

Akobundu, I.O. 1987. *Weed Science in the Tropics*. Norwich, Page Bros. Ltd.

Arrieta-Espinoza, G., E. Sanchez, S. Vargas, J. Lobo, T. Quesada, and A.M. Espinoza. 2005. The weedy rice complex in Costa Rica. I. Morphological study of relationships between commercial rice varieties, wild Oryza relatives and weedy types. *Genetic Resources and Crop Evolution* 52: 575–587.

Auld, B.A., K.M. Menz and C.A. Tisdell. 1987. *Weed Control Economics*. Academic Press, London.

Azmi, M. 2002. Weed succession as affected by repeated application of the same herbicide in direct-seeded rice. *Journal of Tropical Agriculture and Food Science* 30: 151–161.

Azmi, M., D.V. Chin, P. Vongsaroj, and D.E. Johnson. 2004. Emerging issues in weed managment of direct-seeded rice in Malaysia, Vietnam, and Thailand. In K. Toriyama, K.L. Heong, and B. Hardy (eds.), *Rice is Life: Scientific Perpectives for the 21st Century*. Tsukuba, Japan.

Azmi, M. and M. Mashhor. 1995. Weed succession from transplanting to direct-seeding method in Kemubu rice area. *Malaysian Journal of Bioscience* 6: 143–154.

Bastiaans, L., R. Paolini, and D.T. Baumann. 2008. Focus on ecological weed management: what is hindering adoption? *Weed Research* 48(6): 481–491.

Bastiaans, L., D.L. Zhao, N.G. den Hollander, D.T. Baumann, H.M. Kruidhof, and M.J. Kropff. 2007. Exploiting diversity to manage weeds in agro-ecosystems. *Scale and Complexity in Plant Systems Research: Gene–Plant–Crop Relations* 21: 267–284.

Booth, B.D., S.D. Murphy, and C.J. Swanton. 2003. *Weed Ecology in Natural and Agricultural Systems*. CABI Publishing, Cambridge.

Burgos, N.R., R.J. Norman, D.R. Gealy, and H. Black. 2006. Competitive N uptake between rice and weedy rice. *Field Crops Research* 99: 96–105.

Casida, J.E. and G.B. Quistad. 1994. *Safer and More Effective Insecticides for the Future. Modern Agriculture and the Environment*. Kluwer Academic Publishers, Rehovot.

Chancellor, R.J. and R.J. Froud-Williams. 1984. A second survey of cereal weeds in central southern England. *Weed Research* 24(1): 29–36.

Chantarasamee, W., C. Premathatien, T. Sangtong, C. Prokongwong, and C.Supatharung. 2002. *Common Weeds of Central Thailand*. Funnypublishing, Bangkok.

Cheffings, C.M. and L. Farrell. 2005. *The Vascular Plant Red Data List for Great Britain*. Joint Nature Conservation Committee, Peterborough.

Chen, S.R., Y.X. Li, H.P. Mao, B.G. Shen, Y.Z. Zhang, and B. Chen. 2009. Research on distinguishing weed from crop using spectrum analysis technology. *Spectroscopy and Spectral Analysis* 29(2): 463–466.

Chinawong, S. and R. Suwanketnikom 2001. Trends and expectations for research and technology in weed science in Thailand. *Weed Biology and Management* 1: 25–27.

Csontos, P., J.Tamas, O. Szecsy, C. Szinetar, and M. Kiehn. 2009. Old-field succession on abandoned soils, and opportunity to reduce biotic stress by skipping early weedy stages. *Cereal Research Communications* 37: 69–72.

Davis, V.M., K.D. Gibson, and W. G. Johnson. 2008. A field survey to determine distribution and frequency of glyphosate-resistant horseweed (*Conyza canadensis*) in Indiana. *Weed Technology* 22(2): 331–338.

De Datta, S.K. 1981. *Principles and Practices of Rice Production*. International Rice Research Institute, Manila.

De Datta, S.K. and J.C. Flinn. 1986. Technology and economics of weed control in broadcast-seeded flooded tropical rice. Symposium of the 10th Conference, Asian-Pacific Weed Science Society and Japan International Cooperation Agency, Chiang Mai, Thailand.

Dekker, J. 1997. Weed diversity and weed management. *Weed Science* 45(3): 357–363.

Derksen, D.A., G.P. Lafond, A.G. Thomas, H.A. Loeppky, and C.J. Swanton. 1993. Impact of agronomic practices on weed communities – tillage systems. *Weed Science* 41(3): 409–417.

Dollacker, A. and C. Rhodes. 2007. Integrating crop productivity and biodiversity conservation pilot initiatives developed by Bayer CropScience. *Crop Protection* 26(3): 408–416.

Fennimore, S.A. and D.J. Doohan. 2008. The challenges of specialty crop weed control, future directions. *Weed Technology* 22(2): 364–372.

Fernandez-Quintanilla, C., M. Quadranti, P. Kudsk, and P. Barberi. 2008. Which future for weed science? *Weed Research* 48(4): 297–301.

Freckleton, R.P. and P.A. Stephens. 2009. Predictive models of weed population dynamics. *Weed Research* 49(3): 225–232.

Froud-Williams, R.J. and R.J. Chancellor. 1982. A survey of grass weeds in cereals in central southern England. *Weed Research* 22(3): 163–171.

Gill, G.J. and D. Carney. 1999. *Competitive Agricultural Technology Funds in Developing Countries*. Overseas Development Institute, London.

Govaerts, R. 2001. How many species of seed plants are there? *Taxon* 50: 1085–1090.

Haefele, S.M., D.E. Johnson, D. M'Bodj, M.C.S. Wopereis, and K.M. Miezan. 2004. Field screening of diverse rice genotypes for weed competitiveness in irrigated lowland ecosystems. *Field Crops Research* 88(1): 39–56.

Harada, J., Y. Paisookandtivatana, and S. Zungsontiporn. 1987. *Weeds in the Highlands of Northern Thailand*. Mass & Media Co., Bangkok.

Harbuck, K.S.B., F.D. Menalled, and F.W. Pollnac. 2009. Impact of cropping systems on the weed seed banks in the northern Great Plains, USA. *Weed Biology and Management* 9(2): 160–168.

He, H.B., H.B. Wang, C.X. Fang, Y.Y. Lin, C.M. Zeng, L.Z. Wu, W.C. Guo, and W.X. Lin. 2009. Herbicidal effect of a combination of oxygenic terpenoids on *Echinochloa crus-galli*. *Weed Research* 49(2): 183–192.

Ho, N.K. 1995. Management innovations and technology transfer in wet-seeded rice: a case study of the Muda Irrigation Scheme, Malaysia. In K. Moody (ed.), *Constraints,*

Opportunities and Innovations for Wet-Seeded Rice. IRRI Discussion Paper No. 10. International Rice Research Institute, Los Banos.

Hoagland, R.E., C.D. Boyette, M.A. Weaver, and H.K. Abbas. 2007. Bioherbicides: research and risks. *Toxin Reviews* 26(4): 313–342.

Holm, L.G, J.V. Pancho, J.P. Herberger and D. L. Plucknett. 1979. *A Geographical Atlas of World Weeds*. John Wiley & Sons, New York.

Holm, L.G., D.L. Plucknett, J.V. Pancho, and J.P. Herberger. 1977. *The World's Worst Weeds: Distribution and Biology*. University Press of Hawaii, Honolulu.

Inamura, T., S. Miyagawa, O. Singvilay, N. Sipaseauth, and Y. Kono. 2003. Competition between weeds and wet season transplanted paddy rice for nitrogen use, growth and yield in the central and northern regions of Laos. *Weed Biology and Management* 3(4): 213–221.

Isik, D., E. Kaya, M. Ngouajio, and H. Mennan. 2009. Summer cover crops for weed management and yield improvement in organic lettuce (*Lactuca sativa*) production. *Phytoparasitica* 37(2): 193–203.

Karim, R.S.M., A.B. Man, and I.B. Sahid. 2004. Weed problems and their management in rice fields of Malaysia: an overview. *Weed Biology and Management* 4(4): 177–186.

Kawmeechai, S. 2005. Weed in arable crops. In MoA, Thailand (ed.), *Plant Protection*. Co-operative Society, Bangkok.

Kim, K.U. 2001. Trends and expectations for research and technology in the Asia-Pacific region. *Weed Biology and Management* 1: 20–24.

King, J.J. 1966. *Weeds of the World: Biology and Control*. Leonard Hill Books, New York.

Kong, C.H. 2008. Rice allelopathy. *Allelopathy Journal* 22(2): 261–273.

Koocheki, A., M. Nassiri, L. Alimoradi, and R. Ghorbani. 2009. Effect of cropping systems and crop rotations on weeds. *Agronomy for Sustainable Development* 29(2): 401–408.

Leangarpapong P. 1997. *Weed Science*. Lincorn Printing, Bangkok.

Lee, Y.-I. and N. Lee. 2003. Plant regeneration from protocorm-derived callus of *Cypripedium formosanum*. *In Vitro Cellular and Developmental Biology – Plant* 39(5): 475–479.

Lencses, E. and K. Takacs-Gyorgy. 2008. Economic aspects of different weed management systems in corn production. *Cereal Research Communications* 36: 707–710.

LePoer, B.L. 1987. Thailand: a country study. In B.L. LePoer (ed.), *Area Handbook Series*. Library of Congress Federal Research Division, Washington, D.C.

Loon, G.W.V., S.G. Patil, and L.B. Hugar. 2005. *Agricultural Sustainability: Strategies for Assessment*. Sage India, New Delhi.

Mace, G.M., N.J. Collar, K.J. Gaston, H.C. Taylor, H.R. Akcakaya, N.M. Williams, E.J. Gulland, and S.N. Stuart. 2008. Quantification of extinction risk: IUCN's system for classifying threatened species. *Conservation Biology* 22(6): 1424–1442.

Machacheap, S. 1995. *Weed in Thailand*. Phaepithaya Printing, Bangkok.

Maneechote, C. and S. Jamjod. 1999. *Weedy Rice Infestation in Thailand*. Plant Protection Research and Developement Office, Department of Agriculture, Thailand.

Marren, P. 1999. *Britain's Rare Flowers*. Poyser with Plantlife and English Nature, London.

Marshall, E.J.P. 2001. Biodiversity, herbicides and non-target plants. BCPC *Symposium Proceedings: Pesticide Behaviour in Soils and Water*: 419–426.

Marshall, E.J.P. 2005. Field margins in northern Europe: integrating agricultural, environmental and biodiversity functions. *Proceedings of Field Boundary Habitats: Implications for Weed, Insect and Disease Management*: 39–67.

Marshall, E.J.P., V.K. Brown, N.D. Boatman, P.J.W. Lutman, G.R. Squire, and L.K. Ward. 2003. The role of weeds in supporting biological diversity within crop fields. *Weed Research* 43(2): 77–89.

Masako, U. and I. Keiya. 2001. Rice paddy field herbicides and their effects on the environment and ecosystems. *Weed Biology and Management* 1(1): 71–79.

Matson, P.A., W.J. Parton, A.G. Power, and M.J. Sweift. 1997. Agricultural intensification and ecosystem properties. *Science* 277: 504–509.

Moody, K. 1989. *Weeds Reported in Rice in South and Southeast Asia.* International Rice Research Institute, Manila.

Morris, P. and R. Therivel. 1995. *Methods of Environmental Impact Assessment.* University of British Columbia Press, Vancouver.

Mortensen, D.A, L. Bastiaans, and M. Sattin. 2000. The role of ecology in the development of weed management systems: an outlook. *Weed Research* 40(1): 49–62.

Murphy, S.D., D.R. Clements, S. Belaoussoff, P.G. Kevan, and C. J. Swanton.2006. Promotion of weed species diversity and reduction of weed seedbanks with conservation tillage and crop rotation. *Weed Science* 54(1): 69–77.

Mustafa, M., A. Hussain, K.H. Ghazali, and S. Riyadi. 2007. Implementation of image processing technique in real time vision system for automatic weeding strategy. *Ieee International Symposium on Signal Processing and Information Technology:* 1108–1111.

Naeem, A.M., I. Ahmad, M. Islam, and S. Nawaz. 2007. Weed classification using angular cross sectional intensities for real-time selective herbicide applications. *Proceedings of the International Conference on Computing 2007: Theory and Applications:* 731–735.

Neumann, H., R. Loges, H. Roweck, and F. Taube. 2008. Nature conservation and organic agriculture – framework, stand of the art of science and concept of the research project 'Hof Ritzerau'. *Faunistisch-Oekologische Mitteilungen* Supplement 35: 7–19.

Noda, K., M. Teerawatsakul, C. Prakongvongs, and L. Chaiwirtnukul. 1984. *Major Weeds in Thailand.* Department of Agriculture, Thailand.

Olsen, K.M., A.L. Caicedo, and Y.L. Jia. 2007. Evolutionary genomics of weedy rice in the USA. *Journal of Integrative Plant Biology* 49: 811–816.

Panyakul, V.R. 2001. *Organic Agriculture in Thailand: A National Report Prepared for ESCAP Exploring the Potential of Organic Farming for Rural Employment and Income Generation in Asia.* Earth Net Foundation, Bangkok.

Perring, F.H. and L. Farrell. 1983. *British Red Data Books*, Vol: 1: *Vascular Plants.* Royal Society for Nature Conservation, Lincoln.

Prathepha, P. 2009. Seed morphological traits and genotypic diversity of weedy rice (*Oryza sativa* f. *spontanea*) populations found in the Thai Hom Mali rice fields of north-eastern Thailand. *Weed Biology and Management* 9: 1–9.

Pudelko, R., J. Kozyra, and P. Nierobca. 2008. Identification of the intensity of weeds in maize plantations based on aerial photographs. *Zemdirbyste-Agriculture* 95(3): 130–134.

Radanachaless, T. and J.F. Maxwell. 1997. Lists of weeds reported in Thailand. *Thai Studies in Biodiversity* 1:1–286.

Rao, A.N. and K. Moody. 1990. Weed seed contamination in rice seed. *Seed Science and Technology* 18(1): 139–146.

Salisbury, E.J. 1961. *Weeds and Aliens.* Collins, London.

Sands, D.C. and A.L. Pilgeram. 2009. Methods for selecting hypervirulent biocontrol agents of weeds: why and how. *Pest Management Science* 65(5): 581–587.

Sangakkara, U.R., K. Hurle, and B. Rubin. 2002. Herbicide resistance and sustaining food security in south Asian rice culture – a case study. Paper delivered at the international symposium 'Sustaining Food Security and Managing Natural Resources in Southeast Asia – Challenges for the 21st Century', Chiang Mai. www.uni-hohenheim.de/file admin/einrichtungen/sfb564/events/uplands2002/Poster24_Sangakkara.pdf (accessed 9 March 2017).

Shiau, Y.J., S.M. Nalawade, C.N. Hsia, V. Mulabagal, and H.S. Tsay. 2005. In vitro propagation of the Chinese medicinal plant *Dendrobium candidum Wall. Ex Lindl.*, from axenic nodal segments. *In Vitro Cellular and Developmental Biology – Plant* 41(5): 666–670.

Siamwalla, A., S. Setboonsarng, and P. Werakarnjanapongs. 1991. *Changing Comparative Advantage in Thai Agriculture.* OECD Development Centre Working Paper.

Singh, R.K. and A.K. Sureja 2008. Indigenous knowledge and sustainable agricultural resources management under rainfed agro-ecosystem. *Indian Journal of Traditional Knowledge* 7(4): 642–654.

Smitinand, T. 1986. *Weed in shifting cultivation in Thailand.* Asian-Pacific Weed Science Society, Japan International Cooperation Agency, Tokyo.

Sniauka, P. and A. Pocius 2008. Thermal weed control in strawberry. *Agronomy Research* 6: 359–366.

Stone, L.M., M. Byrne, and J.G. Virtue. 2008. Identifying and managing environmental weed risk in perennial pasture research. *Plant Protection Quarterly* 23(2): 73–76.

Sullivan, P. 2003. *Principles of Sustainable Weed Management for Croplands.* https://attra. ncat.org/attra-pub/summaries/summary.php?pub=109 (accessed 9 March 2017).

Sutcliffe, O.L. and Q.O.N. Kay. 2000. Changes in the arable flora of central southern England since the 1960s. *Biological Conservation* 93(1): 1–8.

Suwankul, D. and R. Suwanketnikom. 2001. *Weeds in Thailand.* Kasetsart University, Bangkok.

Suwatabandhu, K. 1950. *Weeds in Paddy Field in Thailand.* Department of Agriculture, Bangkok.

Swarbrick, J.T. and B.L. Mercado. 1987. *Weed Science and Weed Control in Southeast Asia.* Food and Agricultural Organization of the United Nations, Rome.

Tadeo, J.L., C. Sanchez-Brunete, R.A. Perez, and M.D. Fernandez. 2000. Analysis of herbicide residues in cereals, fruits and vegetables. *Journal of Chromatography A* 882 (1–2): 175–191.

Thorne, R.F. 2002. How many species of seed plants are there? *Taxon* 51: 511–522.

Tomita, S., S. Miyagawa, Y. Kono, C. Noichana, T. Inamura, Y. Nagata, A. Sributta, and E. Nawata. 2003. Rice yield losses by competition with weeds in rainfed paddy fields in north-east Thailand. *Weed Biology and Management* 3(3): 162–171.

Turnhout E., B. Bloomfield, and M.B. Hulme. 2012. Conservation policy: listen to the voices of experience. *Nature* 488: 454–455

United Nations. 1992. *Report of the United Nations Conference on Environment and Development.* A/CONF.151/26 (Vol. 1). New York, United Nations.

Unnikrishnan, P.M. and B.N. Prakash. 2007. Traditional herbal medicines for malaria prevention. *Endogenous Development Magazine* 1: 16–18.

Unnikrishnan, P.M. and M.S. Suneetha. 2012. *Biodiversity, Traditional Knowledge and Community Health: Strengthening Linkages.* UNU–IAS, Yokohama.

Uraikul, A. 2006. *Agricultural Statistics of Thailand.* Center for Agriculture Information, Ministry of Agriculture and Cooperatives, Bangkok.

Van den Hove, S. 2007. A rationale for science–policy interfaces. *Futures* 39(7): 807–826.

Ved, D.K. and G.S. Goraya. 2008. *Demand and Supply of Medicinal Plants in India.* Bishen Singh and Mahendra Pal Singh, Dehradun.

Vijarmsorn, P. and H. Eswaran. 2001. *The Soil Resources of Thailand.* Ministry of Agriculture, Bangkok.

Von Hertzen, L., L. Hanski, and T. Haahtela. 2011. Natural immunity: biodiversity loss and inflammatory diseases are two global megatrends that might be related. *EMBO* 12(11):1089–1093.

Von Schirnding, Y. 2002. Health and sustainable development: can we rise to the challenge? *Lancet* 360(9333): 632–637.

Vongsaroj, P. 1993. *Integrated Management of Paddy Weeds in Thailand*. Food and Fertirizer Technology Center, Bangkok.

Vongsaroj, P. 1997. *Weed Management in Rice*. Ministry of Agriculture, Bangkok.

Vurro, M., A. Boari, A. Evidente, A. Andolfi, and N. Zermane. 2009. Natural metabolites for parasitic weed management. *Pest Management Science* 65(5): 566–571.

Walker, B., S. Barrett, S. Polasky, V. Galaz, C.Folke, G. Engström, F. Ackerman, K. Arrow, S. Carpenter, K. Chopra, G. Daily, P. Ehrlich. T. Hughes, N. Kautsky, S. Levin, K. Göran Maler, J. Shogren, J. Vincent, T. Xepapadeas, and A. de Zeeuw 2009. Looming global-scale failures and missing institutions. *Science* 325(5946): 1345–1346.

Walker, E.R. and L.R. Oliver. 2008. Weed seed production as influenced by glyphosate applications at flowering across a weed complex. *Weed Technology* 22(2): 318–325.

Walker, K.J., C.N.R. Critchley, A.J. Sherwood, R. Large, P. Nuttall, S. Hulmes, R. Rose, and J.O. Mountford. 2007. The conservation of arable plants on cereal field margins: an assessment of new agri-environment scheme options in England, UK. *Biological Conservation* 136(2): 260–270.

Weih, M., U.M.E. Didon, A.C. Ronnberg-Wastljung, and C. Bjorkman. 2008. Integrated agricultural research and crop breeding: allelopathic weed control in cereals and long-term productivity in perennial bion ass crops. *Agricultural Systems* 97(3): 99–107.

Wilby, A., C. Mitchell, D. Blumenthal, P. Daszak, C.S. Friedman, P. Jutro, A. Mazumder, A. Prieur-Richard, M. Desprez-Loustau, M. Sharma, and M.B. Thomas. 2009. Biodiversity, food provision, and human health. In O.E. Sala, L.A. Meyerson and C. Parmesan (eds.), *Biodiversity Change and Human Health: From Ecosystem Services to Spread of Disease*. Island Press, Washington, D.C.

Wilcove, D.S., X. Giam, D.P. Edwards, B. Fisher, and L.P. Koh. 2013. Navjot's nightmare revisited: logging, agriculture and biodiversity in Southeast Asia. *Trends in Ecology and Evolution* 28: 531–540.

Wood, A.R., A. den Breeyen, and F. Beed. 2009. First report of smut on *Imperata cylindrica* caused by *Sporisorium schweinfurthianum* in South Africa. *Plant Disease* 93(3): 322.

Yamada, S., S. Okubo, Y. Kitagawa, and K. Takeuchi. 2007. Restoration of weed communities in abandoned rice paddy fields in the Tama Hills, central Japan. *Agriculture Ecosystems and Environment* 119: 88–102.

Zhu, D.S., J.Z. Pan, and Y. He. 2008. Identification methods of crop and weeds based on Vis/NIR spectroscopy and RBF-NN model. *Spectroscopy and Spectral Analysis* 28(5): 1102–1106.

Zimdahl, R.L. 2004. *Weed–Crop Competition*. Blackwell, Iowa.

8 From landscape patterns to ecosystem functions

Watershed services based on different land use and climate change scenarios

Yongyut Trisurat

Introduction

It is widely recognized that the integrity and ecological processes of tropical ecosystems are increasingly threatened by land use/land cover (LU/LC) change due to rapid developments and uncertainties of commodity prices. In addition, those changes are anticipated to increase by the combined drivers of land use and climate change (Corlett, 2012).

To address some of their causes, the Natural Capital Project has developed the Integrated Valuation of Ecosystem Services and Tradeoff (InVEST) tool for quantifying ecosystem services and their values. This software tool is a spatially explicit model (Tillis *et al.*, 2011) that has been used worldwide at the local scale (e.g., Willamette Basin, Oregon – see Nelson *et al.*, 2009; Kushiro watershed in northern Japan – see Shoyama and Yamagata, 2014) and at the regional scale (e.g., Amazon Basin – see Tallis *et al.*, 2009; and West Africa – see Leh *et al.*, 2013). The results of recent studies were used to inform decision-makers and planners to manage natural resources effectively and efficiently. This initiative is relevant to the conceptual framework of The Intergovernmental Science-Policy Platform on Biodiversity and Ecosystem Services (IPBES), established to strengthen the implementation of the Millennium Ecosystem Assessment (Dias *et al.*, 2015). However, few projects have been conducted in Asia, particularly in Thailand.

This chapter provides an overview of forest situation and watershed management in Thailand. It also focuses on the consequences of climate and LU/LC changes on two ecosystem service provisions, water yield and mitigation of soil erosion, because they are highly relevant to ecosystem processes and human well-being in the Thadee watershed. The approaches and results used in this chapter may also lead to it becoming as a key regional case study that can be applied in other countries of Southeast Asia.

Recent trends in forest cover

The Royal Forest Department (RFD) has employed aerial photographs and satellite images to monitor forest cover in Thailand since 1961. Most of the

Table 8.1 Trend in forest area in Thailand, 1961–2008

Year	Area (ha)	% of country
1961	27,362,850	53.33
1973	22,172,500	43.21
1976	19,841,700	38.67
1978	17,522,400	34.15
1982	15,660,000	30.52
1985	15,086,616	29.40
1988	14,380,349	28.03
1989	14,341,700	27.95
1991	13,669,805	26.64
1993	13,355,355	26.03
1995	13,148,506	25.62
1998	12,972,229	25.28
2000*	17,011,080	33.15
2008*	17,158,565	33.40

Source: Royal Forest Department (2013).

Note: * Satellite imagery at scale 1:50,000.

interpretations have been done visually using black-and-white or false-color composite images at the scale of 1:250,000. The results of image interpretation have revealed that Thailand has experienced a long-term decline in forest area (Table 8.1). In 1961 forest still covered 53 percent of the country, but surveys in 1989 and 1998 revealed declines to 28 percent and 25 percent of national land area, respectively (Charuphat, 2000). Thus, over thirty-seven years (1961–1998), the average annual deforestation rate was about 388,935 hetares or 2.0 percent, and the forest had been reduced by about a half by 1991. Although the commercial logging concession was banned in 1989, the rate of deforestation remained high, mainly due to agricultural encroachment.

The latest surveys, carried out in 2004 and 2008 using high image resolution (scale 1:50,000) and a new classification system, appeared to show a sharp and sustained rise in forest area – to 33.1 percent of the country in 2000 and 33.4 percent in 2008 (Royal Forest Department, 2013; Table 8.1).

There are many reasons of deforestation in the tropics, but underlying these causes is that humans place a relatively low value on ecosystems compared with the benefits that potentially degrade them because some services (e.g. regulating and supporting functions) are difficult to quantify and the consequences of deterioration are long term (Daily, 1997). In addition, people assume ecosystem components are public property rather than privately owned. The causes of deforestation are different across Thailand. In the south, for example, there has been strong pressure to clear forest for rubber and oil palm plantations (Trisurat et al., 2016).

Watershed services and watershed management in Thailand

Watershed services

The Millennium Assessment (2005) defined ecosystem services as the conditions and the processes through which natural ecosystems and the species that comprise them sustain and fulfill human life. Tropical ecosystems are recognized as essential natural capitals and unique economic assets. The services provided by tropical forests include: food, energy and timber for people; regulation of water yield; prevention of soil erosion; and provision of habitats for plants and animals (Sukhdev, 2010). Using different economic valuation methods, the average value of ecosystem services derived from tropical forests ranges from US$6,120 – 16,362 ha/year (TEEB, 2009).

Watershed (or catchment or river basin) is defined as a basic environmental unit where water drains downhill into a common stream (Gordon *et al.*, 2004). Landscape features within a watershed include forests, grasslands, cultivated areas, riparian areas, wetlands, etc. Important services provided by a watershed include provision of water resource, maintenance of water quality, and erosion, sedimentation and flood prevention. Smith *et al.* (2006) indicated that watershed services are usually controlled by the forms of land and water use within them.

Watershed management concepts and strategies have been gradually developed and adopted. The main concept has focused on the whole watershed system, integrating the human and social development involved in environmental management, integrated watershed management, integrated river basin management and integrated water resources management (Abell *et al.*, 2002). Recently, an eco-hydrology concept has been proposed, which aims to examine the consequences of landscape patterns on hydrological processes and their effects on the distribution, structure, and function of watershed ecosystems (Zalewski *et al.*, 1997). Research topics in eco-hydrology comprise the effects of land use/land cover on stream flow and function, the relationship between ecological processes and the water cycle, and adaptation of biotic elements to their aquatic ecosystem.

As a watershed system is very complex and involves different stakeholders, watershed management projects do not benefit all stakeholders equally. In most developing countries the upstream inhabitants in a watershed are mainly small farmers, whereas downstream people are urban-dwellers. Government investments usually focus on protecting downstream infrastructures, such as reservoirs, irrigation installation, water supply schemes, electricity generation stations and minimizing flood damage. Upstream inhabitants may not benefit from these projects (Gonzales Inca, 2009). However, the conflict over water use between upland and downstream inhabitants in northern Thailand is quite different. It has been observed by the Electricity Generation Authority of Thailand and the Royal Irrigation Department that the inflow in most reservoirs has recently declined due to upstream development schemes, such as small-storage reservoirs and water diversion, developed by government and local farmers (Tangtham, 1996). This has resulted in insufficient water for domestic use in dry periods as well as additional problems, such as sedimentation, toxic chemical runoff and poisoned drinking water.

Currently, watershed management employs a payment for ecosystem service (PES) mechanism through scoping, recognition and valuation of watershed services (Mayrand and Paquin, 2004). This is due to the fact that economic valuation can provide an optimum management approach and set priorities for programs and strategic actions that protect and maintain ecosystems and their services in a sustainable manner (King and Mazzotta, 2000). This approach has been implemented successfully in the Fondo para La Conservacion del Agua, Quito, Ecuador, to maintain a clean water supply for the downstream population (Goldman-Benner *et al.*, 2012).

Watershed classification

Thailand has a tropical monsoon climate but precipitation varies from region to region due to topographic variations and latitudinal gradients. For instance, peninsular Thailand and the east are the wettest parts of the country, receiving annual rainfall of 2,000 millimeters or more, which means they are classified as "permanently humid." Annual rainfall in the north and north-eastern parts of the country varies from 600–1,000 millimeters in lowland areas to over 1,000 millimeters in mountainous areas (Trisurat and Grainger, 2012). Although water resources in Thailand are sufficient for the country's needs, the quality and flow regime are becoming more critical in all regions due to inadequate management from the government and private companies. In addition to water shortages, several tributaries are now being polluted by fertilizers and other chemical substances draining from non-conservation farming areas and irresponsible industrial sectors. Water use conflicts between farmers and industrial groups are also becoming more serious (Tangtham, 1996).

The water resource problems mentioned above, together with the forest degradation in the upper watershed, are driving the critical need for proper land use planning of forest areas and land allocation of various uses (Tangtham, 1996). In 1979, a National Watershed Classification Committee was appointed by the Thai government, and Kasetsart University was assigned to conduct a watershed classification project. This project defined watershed classification (WSC) as the identification of the inherent capacity of landscape units to be properly managed for natural resources and the environment (Tangtham and Chunkao, 1990). The overall goal is to make the human use of land as compatible as possible with the features of the environment and to mitigate the onsite and offsite effects of such usage.

The specific objectives of watershed classification are as follows:

1. Develop a method for classifying watershed class.
2. Delineate watershed areas for protected forests, commercial forests, agronomic use and other uses.
3. Prepare watershed classification maps of all river basins in Thailand.
4. Suggest land use practices to be promulgated as a cabinet resolution.

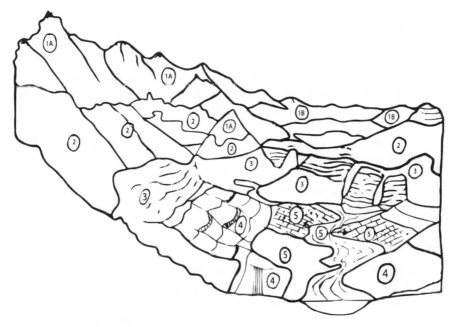

Figure 8.1 Pictorial model of watershed classes in Thailand

Source: Tangtham (1996).

After consultation among government agencies and technical experts, the watershed areas were grouped into five classes (Figure 8.1) on the basis of slope, elevation, landform, soil, geology and forest. The definitions and characteristics of each watershed class are as follows:

- WSC 1: Protected or conservation forest and headwater source. This class is divided into two subclasses:
  - WSC 1A: Protected forest areas, including the headwaters of rivers. These areas are usually at high elevations and have very steep slopes and should remain under permanent forest cover.
  - WSC 1B: Areas with similar physical and environmental features to WSC 1A, but portions of these areas have already been cleared for agriculture. Where possible, they should be rehabilitated to forest or maintain permanent agro-forestry.
- WSC 2: Commercial forest. These areas are designed for protection and/or commercial forests where mining and logging will be allowed within legal boundaries, usually at high elevations with steep to very steep slopes. Landforms are usually less prone to erosion than WSC 1A or WSC 1B.
- WSC 3: Fruit-tree plantation. These areas cover uplands with steep slopes and less erosive landforms. This class is suitable for commercial forestry,

grazing, fruit trees, or other agricultural practices with appropriate soil conservation measures.

- WSC 4: Upland farming. This class describes those areas of gentle sloping land that are suitable for row crops, fruit trees, and grazing with a moderate need for some soil conservation measures.
- WSC 5: Lowland farming. Gentle slopes or flat terrains are dominant in this class and the areas are suitable for paddy fields or other agricultural practices with limited soil conservation measures.

The watershed classification and its land use practices resolution have been implemented since 1985 for the northern watershed areas, and for the whole country since 1991. However, implementation on the ground has been ineffective due to poor law enforcement and lack of involvement from the relevant government agencies.

Soil erosion

A number of human activities disturb the land surface of the earth, altering natural conditions, causing soil degradation, and transporting pollutants in watercourses which consequently impact the aquatic ecosystem. Erosion occurs when soil is left bare and exposed to erosive agents (Foster *et al.*, 2003). For most of soil types, a soil erosion of 10 ton/ha/year is typically used as a tolerance threshold. In Thailand, 12.5 ton/ha/year is proposed by the Land Development Department (2000).

Four major factors affect soil erosion: rainfall, soil, topography, and land use. Rainfall drives erosion according to its intensity and amount (USDA, 2008). Soil types differ in their erodibility, so soil texture and organic matter content provide indications of erodibility. Slope length and steepness are the topographic characteristics that most affect erosion and deposition. Among these factors, rainfall and land use are the important factors in soil erosion (Foster *et al.*, 2003). The greater the rainfall intensity, the higher the soil loss; in contrast, the greater the vegetation cover, the lower the soil loss. Land cover is strongly altered by human activities, but land cover management practices (e.g., vegetative type, cropping system) and conservation measures (e.g., contouring, multilayer cropping) can reduce erosion, mainly by decreasing surface runoff and accelerating deposition (USDA, 2008).

Several models have been developed to estimate soil erosion. The Universal Soil Loss Equation (USLE; Wieschmeier and Smith, 1978) is the most popular model and used worldwide because it is simple and requires quite simple data (Ongsomwang and Thinley, 2009). When used in conjunction with raster-based GIS, the USLE model can isolate locations of erosion on a cell-by-cell basis, and identify the spatial patterns of soil loss within a watershed (Millward and Mersey, 1999). Other models include the Revised Universal Soil Loss Equation (RUSLE; Renard *et al.*, 1997) and the Soil and Water Assessment Tool (SWAT; Neitsch *et al.*, 2005).

Table 8.2 Average soil erosion for each land use type in Thailand (ton/ha/year)

Major land use type	North-east	North	Central	South
Forest	42.69	16.00	47.94	5.19
Paddy	1.19	0.63	0.69	1.06
Cash crop	132.19	125.44	35.56	224.63
Fruit orchard	84.38	80.06	48.13	42.06
Horticulture	14.13	7.75	8.06	24.06
Grassland	5.63	5.31	6.31	9.56
Abandoned area	140.69	133.44	157.88	238.94
Settlement	42.69	16.00	47.94	5.19

Source: Land Development Department (2000)

It should be noted that all of these models provide only rough estimates of the possible amount of soil erosion; their results should not be taken as precise figures. Determining the inputs for soil erosion is difficult. Input parameters vary in time and space, and some are difficult to measure, which leads to considerable uncertainty. For example, there have been reports of substantial variations in soil erosion between similar sites that experienced the same rainfall (Quinton, 2004). Hence, increasing the number of processes and inputs in soil erosion models in the hope of achieving a more complete picture of the erosion process will not typically lead to better predictions of erosion and sedimentation than the simple models (Quinton, 2004). Therefore, it might be worthwhile to develop a simple way to simulate erosion and sedimentation with reliable data and consistent models. Average soil erosion in Thailand calculated using the USLE model is shown in Table 8.2.

Sediment yield calculation at the watershed scale

A large amount of sediment reaching downstream areas from a watershed usually indicates watershed degradation (Lane *et al.*, 2001; Tangtham, 1996). The input of sediment by erosion processes into reservoirs shortens the lifespan of dam operations, affects aquatic ecosystem functioning by modifying hydraulic morphology and water quality (Kinnell, 2008), and increases the risk of flooding in the wet season (Rompaey and Dostal, 2007).

The USLE model is used to estimate soil erosion for the entire study area or watershed (onsite erosion), and a large portion of the eroded soil is deposited while traveling to the watershed outlet (delivered sediment). Thus, it is essential to develop the Sediment Delivery Ratio (SDR) for a given watershed in order to estimate the total sediment transported to the watershed outlet (Lim *et al.*, 2005). The SDR can range from 0 to 1. Small watersheds have high SDRs, meaning that most of the eroded soil moves to downstream areas without significant deposition. The SDR decreases as the size of the watershed increases (Lim *et al.*, 2005). In fact, onsite erosion is derived from two main sources: gully and channel erosion, and rill and interrill erosion (Lim *et al.*, 2005). As was mentioned above, gully and channel erosion could be responsible for more than 50 percent of sedi-ment production in any catchment (Blong, 1985; Quinton, 2004), so the USLE result obtained from the calculation of the sediment yield in a basin should be treated with caution. Lim *et al.* (2005) indicated that the estimation is reliable if there is no substantial erosion occurring from the gully and channel processes.

Several methods have been developed and can be classified into three general categories (Lu *et al.*, 2004). The first group – sediment rating curve-flow duration – is suitable where sufficient sediment yield and stream flow data are available from all catchments (Gregory and Walling, 1973). However, this approach is inappropriate for a large basin because measurement results are normally not available for all sub-watersheds.

The second group uses empirical relationships or the so-called blackbox approach, which relates SDR to the most important morphological characteristics of a catchment, such as its area (Lu *et al.*, 2004). This is a simple method, but it has some limitations because it ignores the mechanisms that cause sediment trans-port (e.g., soil texture and rainfall) and catchment conditions (e.g., vegetation and topography).

The third group attempts to build models based on fundamental hydrologic and hydraulic processes (Flanagan and Nearing, 1995; Lu *et al.*, 2004). However, this method is rarely used because it requires a lot of data.

Potential impacts of land use and climate change on Thadee watershed services in Thailand

Thadee watershed

The Thadee watershed is located in Nakhon Srithammarat province, Thailand, between 8°24' and 8°29'N and 99°42' and 99°48'E. It covers approximately 112 km^2 and is subdivided into thirteen sub-watersheds (Figure 8.2). Its altitude ranges from 60 to 1,835 meters, and approximately 63 percent of the watershed is classified as WSC 1 (Trisurat *et al.*, 2016). In addition, 54 percent of the watershed is located in Khao Luang National Park, which is nationally recognized as one of Thailand's biodiversity hotspots (Trisurat *et al.*, 2013). Mean annual temperature during 2000–2010 was 26.6°C, and average annual rainfall during the same period was 2,825 millimeters. According to the Centro Internacional de Agricultura Tropical–Global Climate Model (CIAT–GCM, www.ccafs-climate. org/), mean temperature is projected to rise to 27.6°C by 2020, while annual rainfall is projected to decline to 1,980 millimeters. The average evaporation is 981 mm/year and mean humidity is 91 percent (Bansopit *et al.*, 2004). Extreme rainfall of 3,838 millimeters was recorded in 2000. The average annual water yield at the outlet recorded by the Nakhon Srithammarat Provincial Irrigation Office (unpublished data) was approximately 224 million m^3.

LU/LC in the watershed is classified into eight categories:

- intact evergreen forest;
- degraded forest (from landslide in 1988);

- multilayer cropping;
- fruit orchard;
- rubber plantation;
- settlement;
- water body; and
- miscellaneous land use.

The percentages of each class in 2009 were 46.03, 14.72, 12.54, 8.43, 11.12, 0.88, 3.96 and 2.32 percent of the watershed area, respectively (Land Development Department, 2009). Intact forest was mainly distributed in the WSC 1 watershed and the national park. Durian (*Durio kutejensis*) is a dominant fruit tree in multilayer cropping, while mangosteen (*Garcinia mangostana*) and rambutan (*Nephelium lappaceum*) are important species in monocultural plantations.

The Thadee watershed has been selected as a pilot site of the Enhancing Economic and Financial Tools for Biodiversity and Ecosystem Services in Southeast Asia (ECO-BEST) project. The overall objective of this project is to reduce terrestrial biodiversity loss for the benefit of local communities by enhancing the provision of ecosystem services and increasing the capture of ecosystem service values. Based on stakeholder consultation workshops, three land use scenarios for 2020 were defined: trend or business as usual; development; and watershed conservation. The percentages of forest cover will change from 46 percent of the watershed in 2009 to 56, 43 and 58 percent, respectively

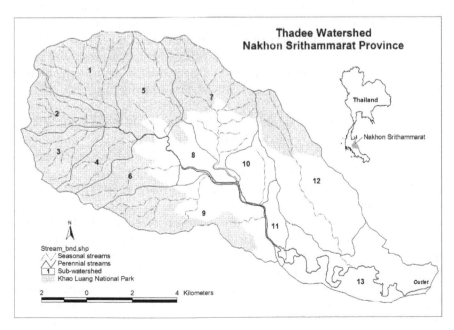

Figure 8.2 Location of Thadee watershed in Nakhon Srithammarat province, Thailand

(Trisurat and Kalliola, 2013; Verburg and Veldkamp, 2004). The results are shown in Figure 8.3.

Predicted water yield

The InVEST (Integrated Valuation of Ecosystem Services and Tradeoff) reservoir hydropower production model was used to estimate the amount of water runoff from grid cells of the Thadee watershed and how it might change as a result of predicted future LU/LC and rainfall options (Tillis *et al.*, 2011). The water yield model is based on the Budyko curve developed by Zhang *et al.* (2001) and calculated using annual precipitation, soil depth, annual reference evapotranspiration (ETo), plant available water content (PAWC), an LU/LC map, watershed, sub-watershed, maximum root depth and evapotranspiration coefficient (Kc).

The ETo was calculated from the "modified Hargreaves" equation (Subburayan *et al.*, 2011). The raster grid of PAWC was derived from soil texture and organic matter using the equation developed by Saxton and Rawls (2006). It should be noted that field surveys and laboratory analysis were conducted to obtain information on root restricting layer depth and soil characteristics in slope complex (>30 percent), where these data were not available in the existing soil map (Land Development Department, 2000). Evapotranspiration coefficient (Kc) of each land use type was gathered from previous studies in Thailand (Pukngam, 2001; Tanaka *et al.*, 2008; Guardiola-Claramonte *et al.*, 2010). In addition to the three possible LU/LCs, three possible climate conditions in 2020 that would affect the amount of water yield were considered: average rainfall during 2000–2010 (2,825 mm); less rainfall, as predicted by the CIAT–GCM (1,980 mm); and extreme rainfall, as in 2000 (3,838 mm – worst case). Finally, the estimated water yields were validated against actual measurement data at the watershed outlet (Nakhon Srithammarat Provincial Irrigation Office, unpublished data).

The predicted annual water yields in 2020 derived from combinations of the three LU/LC scenarios and three rainfall options are shown in Table 8.3. The climate change scenario provides lowest water yield (138–141 million m³), followed by average rainfall during 2000–2010 (223–225 million m³) and extreme rainfall (253–257 million m³). If rainfall were treated as constant, the conservation scenario shows slightly lower annual water yields than other LU/LC scenarios due to high evapotranspiration from forest cover. In contrast, water yield is expected to increase significantly with expansion of farmland (development scenario). In general, the amount of water yield from the entire watershed is quite similar for all LU/LC scenarios under the same rainfall. These results are consistent with findings of Hamilton and King (1983) and Shoyama and Yamagata (2014), which indicated that water flow was not significantly different between forested and non-forested areas in downstream or flat areas. However, the predicted water yields are significantly different for upstream (1 and 2) and middle stream (4, 5, 6 and 12) catchments (see Figure 8.3).

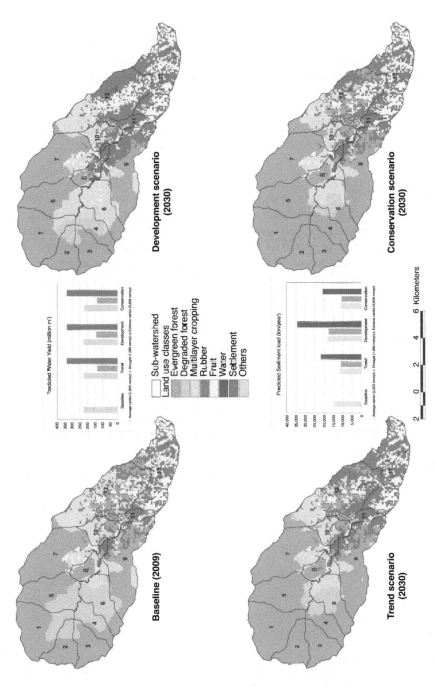

Figure 8.3 Current and future land use/land cover and consequences for sediment in the Thadee watershed

Source: Trisurat and Kalliola, 2013.

Table 8.3 Predicted water yields in 2020 derived from different scenarios

Land use scenarios	Predicted water yields (million m³)		
	Average rainfall (2,528 mm/year)[1]	Extreme rainfall (3,838 mm/year)[2]	Climate change (1,980 mm/year)[3]
LU/LC 2009	225.06	337.34	140.2
Trend	222.04	335.57	138.82
Development	225.04	338.07	141.33
Conservation	221.84	335.35	138.64

Notes: [1] average rainfall during 2000–2010; [2] rainfall in 2000; [3] predicted rainfall in 2020.

Onsite soil erosion

The results of soil loss calculated using the USLE model are presented in Table 8.4 and Figure 8.2. The estimated sediment derived from the conservation land use scenario was predicted as the lowest, at approximately 60 percent of the total sediment yield generated from the development land use scenario. This pattern is also observed for the extreme rainfall and predicted climate in 2020.

Table 8.4 Percentage of the Thadee watershed belonging to each soil loss class under different rainfall and LU/LC scenarios

Erosion classes	Average rainfall (2,528 mm/year)[1]				Extreme rainfall (3,838 mm/year)[2]			Climate change (1,980 mm/year)[3]		
	2009	Trend[4]	Devl.[5]	Cons.[6]	Trend	Devl.	Cons.	Trend	Devl.	Cons.
Very slight (0–10 ton/ha/year)	69.78	76.86	68.11	78.34	73.38	64.93	75.01	77.05	68.41	78.62
Slight (10–25 ton/ha/year)	7.16	5.57	5.99	5.24	5.51	5.39	5.39	5.63	6.08	5.25
Moderate (25–50 ton/ha/year)	7.13	4.99	6.57	4.53	3.93	4.52	3.59	5.10	6.71	4.61
High (50–100 ton/ha/year)	7.61	5.65	7.88	5.18	5.12	6.74	4.63	5.64	7.89	5.15
Very high (100–200 ton/ha/year)	4.97	4.09	6.58	3.94	5.63	7.83	5.14	3.90	6.33	3.76
Extreme (>200 ton/ha/year)	3.35	2.84	4.88	2.78	6.42	10.59	6.23	2.68	4.58	2.61
Total	100.00	100.00	100.00	100.00	100.00	100.00	100.00	100.00	100.00	100.00

Notes: [1] average rainfall during 2000–2010; [2] rainfall in 2000; [3] predicted rainfall in 2020; [4] land use trend scenario; [5] land use development scenario; [6] land use conservation scenario.

Table 8.5 Estimated onsite and delivered sediment

Land use scenarios	Onsite soil erosion and delivered sediment (ton/year)		
	Average rainfall (2,528 mm/year)[1]	Extreme rainfall (3,838 mm/year)[2]	Climate change (1,980 mm/year)[3]
Trend	275,249[4]	514,447	262,643
	11,588[5]	21,658	11,057
Development	435,664	813,294	416,106
	18,341	34,240	17,518
Conservation	262,894	491,436	250,836
	11,068	20,690	10,561

Notes: [1] average rainfall during 2000–2010; [2] rainfall in 2000; [3] predicted rainfall in 2020; [4] onsite soil erosion; [5] delivered sediment.

High soil erosion is observed mainly in the middle (sub-watersheds 4 and 6) and lower (sub-watersheds 10 and 12) sectors of the watershed (Figure 8.2), where degraded forest and rubber plantations, respectively, are dominant. In contrast, lower soil loss is predicted in sub-watersheds 1–3 and 13 due to high percentages of forest cover and flat terrain, respectively. More than 70 percent of the watershed area has very slight soil loss (0–10 ton/ha/year), which means that the erosion problem is not severe for the trend and conservation scenarios due to approximately 56 and 58 percent of the watershed being under forest cover. However, nearly 20 percent of the watershed, especially in the development land use/extreme rainfall scenario, will have high to extreme erosion and generate gross soil loss of approximately 813,294 ton/year (Table 8.5). This is because rainfall (runoff erosivity factor) has been recognized as the most important factor in the generation of soil erosion in the Thadee watershed (Trisurat *et al.*, 2016).

Delivered sediment

The average annual sediment load during 2000–2004 measured at the outlet of the Thadee watershed was 64.2 mg/L (Bansopit *et al.*, 2004). According to the Nakhon Srithammarat Provincial Irrigation Office (unpublished data), the average annual water yield measured at the same location in the same period was 224.75 million m^3. Therefore, the total suspended sediment load was 14,429 ton/ year. Using the empirical relationship, the SDR for the Thadee watershed is 0.042 and the estimated sediment yields transported to the watershed outlet are predicted to be 11,588, 18,341 and 11,068 ton/year in 2020 for the trend, development and conservation scenarios, respectively, given average annual rainfall of 2,528 mm/year (Table 8.5). Slightly lower amounts of sediment loads are predicted for the climate change scenario (1,980 mm/year). In contrast, very high values of sediment yield (> 20,000 ton/year) are predicted downstream for the combination of extreme rainfalls and land use development scenario.

Conclusions

Deforestation and climate change are becoming serious threats to watershed services because they have significant effects on hydrological processes and related services. This chapter has focused on understanding and predicting the impacts of the combination of transformative land use and climate change on water yield and water quality. The results clearly suggest that intensifying land use change through rapid expansion of rubber plantations, coupled with extreme rainfall, will generate huge sediment loads and overland flow due to the force of the rainfall and less evapotranspiration from vegetation. In contrast, the land use conservation scenario in combination with average or lower annual rainfall will generate approximately half of the sediment load of previous conditions. The results of this research advance the scientific understanding of the relationship between ecosystem structure and composition, and ecological processes and the practice of accounting for the provision of ecosystem services in the watershed decisions of local multi-stakeholders and responsible government agencies.

To maximize the resilience of watershed ecosystems, the ECO-BEST project must persuade the municipality and people downstream to participate in the PES mechanism for maintaining ecosystem processes in the Thadee watershed. If the mechanism and approach are adopted, this will greatly increase the resilience of species at moderate to high risk from climate change.

Acknowledgements

I would like to thank the Kasetsart University Research and Development Institute and the Enhancing Economic and Financial Tools for Biodiversity and Ecosystem Services in Southeast Asia (ECO-BEST) project for funding this project. In addition, I am grateful to the Austrian Agency for International Cooperation in Education and Research (OeAD-GmbH) and Klaus Katzensteiner for providing fellowship and for hosting me at the University of Natural Resources and Life Sciences in Austria, where much of this chapter was written. Thanks also to the Royal Forest Department, Department of National Parks, Wildlife and Plant Conservation and Land Development Department for providing data.

References

Abell, R., M. Thieme, E. Dinerstein, and D. Olson. 2002. *A Sourcebook for Conducting Biological Assessments and Developing Biodiversity Visions for Ecoregion Conservation.* Volume II: *Freshwater Ecoregions.* Conservation Science Program, WWF-US, Washington, D.C.

Bansopit, P., P. Thamrongwang, and N. Tanmanee. 2004. *Abiotic and Chemical Quality of Water at Thadee Watershed, Nakhon Srithammarat Province.* Watershed Conservation Office, Department of National Parks, Wildlife and Plant Conservation, Bangkok.

Blong, R.J. 1985. Gully sidewall development in New South Wales, Australia. In S.A. El-Swaify, W.C. Moldehauer, and A. Lo (eds.), *Soil Erosion and Conservation.* Soil Conservation Society of America, Ankeny, IA: 575–584.

Charuphat, T. 2000. Remote sensing and GIS for tropical forest management. *Proceedings of the Ninth Regional Seminar on Earth Observation for Tropical Ecosystem Management, Khao Yai, Thailand, November 2000*: 42–49.

Corlett, R.T. 2012. Climate change in the tropics: the end of the world as we know it? *Biological Conservation* 151: 22–25.

Daily, G.C. 1997. *Nature's Services: Societal Dependence on Natural Ecosystems*. Island Press, Washington, D.C.

Dias, S., S. Demissew, C. Joly, W.M. Lonsdale, and A. Larigauderie. 2015. A rosetta stone for nature's benefits to people. *PLoS Biology* 13(1): e100204.

Flanagan, D.C. and M.A. Nearing. 1995. *Water Erosion Prediction Project – Hill-Slope Profile and Watershed Model Documentation*. USDA, Agricultural Research Service, National Soil Erosion Research Laboratory, Washington, D.C.

Foster, G.R., D.C. Yoder, G.A. Weesies, D.K. Mccool, K.C. Mcgregor, and R.L. Bingner. 2003. *User's Guide Revised Universal Soil Loss equation*. Version 2. USDA, Agricultural Research Service, Washington, D.C.

Goldman-Benner, R.L., S. Benitez, T. Boucher, A. Calvache, G. Daily, P. Kareiva, T. Kroger, and A. Ramos. 2012. Water funds and payments for ecosystem services: practice learns from theory and theory can learn from practice. *Oryx* 46 (1): 55–63.

Gonzales Inca, C.A. 2009. Assessing the land cover and land use change and its impact on watershed services in a Tropical Andean watershed in Peru. MSc. University of Jyräskylä, Finland.

Gordon, N.D., T.A. McMahon, B.L. Finlayson, C.J. Gippel, and R.J. Nathan. 2004. *Stream Ecology: An Introduction for Ecologists*. Second edition. John Wiley & Sons, New York.

Gregory, K.J. and D.E. Walling. 1973. *Drainage Basin Form and Processes: A Geomorphological Approach*. John Wiley & Sons, New York.

Guardiola, C.M., J. Fox, T.W. Giambelluca, and P.A. Troch. 2006. Changing land use in the Golden Triangle: where the rubber meets the road. In J. Roumasset, K.M. Burnett, and A. Molina Balisacan (eds.), *Sustainability Science for Watershed Landscapes*. ISEAS Publishing, Singapore: 235–250.

Guardiola-Claramonte, M., P.A. Troch, A.D. Ziegler, T.W. Giambelluca, M. Durcik, J.B. Vogler, and M.A. Nullet. 2010. Hydrologic effects of the expansion of rubber (*Hevea brasiliensis*) in a tropical catchment. *Ecohydrology* 3: 306 :314.

Hamilton, L.S. and P.N. King. 1983. *Tropical Forested Watersheds: Hydrological and Soil Response to Major Uses or Conversions*. Westview Press, Boulder.

King, D.M. and M.J. Mazzotta. 2000. *Ecosystem Valuation*. US Department of Agriculture, Natural Resources Conservation Service and National Oceanographic and Atmospheric Administration, Washington, D.C. www.ecosystemvaluation.org/index.html. (accessed 10 March 2017).

Kinnell, P.I.A. 2008. Sediment delivery from hill slopes and the Universal Soil Loss Equation: some perceptions and misconceptions. *Hydrological Processes* 22: 3168–3175.

Land Development Department. 2000. *Soil Loss Map of Thailand*. Ministry of Agriculture and Cooperatives, Bangkok.

Land Development Department. 2009. *Land Use Map 2009*. Ministry of Agriculture and Cooperatives, Bangkok.

Lane, L.J., M.H. Nichols, L.R. Levick, and M.R. Kidwell. 2001. A simulation model for erosion and sediment yield at the hill slope scale. In R.S Harmon, and W.W. Doe (eds.), *Landscape Erosion and Evolution Modeling*. Kluwer Academic/Plenum, New York: 201–237.

Leh, M.D.K., M.D. Matlock, E.C. Cummings, and L.L. Nalley. 2013. Quantifying and mapping multiple ecosystem services change in West Africa. *Agricultural Ecosystem Environment* 165: 6–18.

Lim, K.J., M. Sagong, B.A. Engel, Z. Tang, J. Choi, and K. Kim. 2005. GIS-based sediment assessment tool. *Catena* 64: 61–80.

Lu, H., C. Moran, I. Prosser, and M. Sivapalan. 2004. Modeling sediment delivery ratio based on physical principles. Paper presented at the International Conference on Environmental Modeling and Software, Osnabruck, Germany.

Mayrand, K. and M. Paquin. 2004. *Payment for Environmental Services: A Survey and Assessment of Current Schemes*. Unisfera International Center. www.cec.org/files/ PDF/ ECONOMY/PES-Unisfera_en.pdf. (accessed 25 April 2014).

Millennium Assessment. 2005. *Ecosystem and Human Well-being: Synthesis*. World Resources Institute, Washington, D.C.

Millward, A.A. and J.E. Mersey. 1999. Adapting the RUSLE to model soil erosion potential in a mountainous tropical watershed. www.sciencedirect.com/science/article/pii/ S0341816299000673 (accessed 10 March 2014).

Neitsch, S.L, J.G. Arnold, J.R. Kiniry, and J.R. Williams. 2005. *Soil and Water Assessment Tool (SWTA): Tutorial Documentation*. USDA, Agricultural Research Service, Grassland and Soil Land Water Research Laboratory, Washington, D.C.

Nelson, E., G. Mendoza, J. Regetz, S. Polasky, H. Tallis, D. Cameron, K. Chan, G. Dailey, J. Goldstein, P. Kareiva, E. Lonsdorf, R. Naidoo, T.H. Ricketts, and R. Shaw. 2009. Modeling multiple ecosystem services, biodiversity conservation, commodity production and tradeoffs at landscape scales. *Frontiers in Ecology and the Environment* 7(1) : 4–11.

Ongsomwang, S. and U. Thinley. 2009. Spatial modeling for soil erosion assessment in Upper Lam Phra Phloeng watershed, Nakhon Ratchasima, Thailand. *Suranaree Journal of Science and Technology* 16(3): 253–262.

Pukngam, S., 2001. The comparative studies on evapotranspiration of paddy field and different forest types in the northern Thailand. Ph.D. dissertation. Kasetsart University, Bangkok.

Quinton, J.N. 2004. Erosion and sediment transport. In J. Wainwright and M. Mulligan (eds.), *Environmental Modeling: Finding Simplicity in Complexity*. John Wiley & Sons, Chichester: 187–194.

Renard, K.G., G.R. Foster, G.A. Weesies, D.K. McCool, and D.C. Yoder. 1997. *Predicting Soil Erosion by Water: A Guide for Conservation Planning with the Revised Universal Soil Loss Equation (RUSLE)*. Agricultural Handbook Number 703.USDA, Agricultural Research Service, Washington, D.C.

Rompaey, V.J. and K.T. Dostal 2007. Modelling the impact of land cover changes in the Czech Republic on sediment delivery. *Land Use Policy* 24: 576–583.

Royal Forest Department. 2010. *Statistical Data 2009*. Office of the Secretary, Royal Forest Department, Bangkok.

Royal Forest Department. 2012. *Statistical Data 2012*. Ministry of Natural Resources and Environment, Bangkok.

Saxton, K.E. and W.J. Rawls. 2006. Soil water characteristic estimates by texture and organic matter for hydrologic solutions. *Soil Science Society of American Journal* 70:1569–1578.

Shoyama, K. and Y. Yamagata. 2014. Predicting land-use change for biodiversity conservation and climate-change mitigation and its effect on ecosystem services in a watershed in Japan. *Ecosystem Services* 8 : 25–34.

Smith, M., D. De Groof, D. Perrot-Maître, and G. Berkamp. 2006. Establishing payment for watershed services. International Union for Conservation of Nature and Natural

Resources. http://data.iucn.org/dbtw-wpd/edocs/2006-054.pdf (accessed 8 December 2008).

Subburayan, S., A. Murugappan, and S. Mohan. 2011. Modified Hargreaves equation for estimation of ETO in a hot and humid location in Tamilnadu State, India. *International Journal of Engineering Science and Technology* 3(1): 592–600.

Sukhdev, P. 2010. The economics of biodiversity and ecosystem services in tropical forests. *ITTO Forest Update Bulletin* 20(1): 8–10.

Tallis, H. and S. Polasky. 2009. Mapping and valuing ecosystem services as an approach for conservation and natural-resource management. *Annual New York Academic Science* 1162: 265–283.

Tanaka, N., T. Kume, N. Yoshifuji, K. Tanaka, H. Takizawa, K. Shiraki, C. Tantasirin, N. Tangtham, and M. Suzuki. 2008. A review of evapotranspiration estimates from tropical forests in Thailand and adjacent regions. *Agricultural and Forest Meteorology* 148: 807–819.

Tangtham, N. 1996. Watershed classification: the macro land-use planning for the sustainable development of water resources. Paper presented at the International Seminar Workshop on Advances in Water Resources Management and Wastewater Treatment Technologies, Nakhon Ratchasima Province, Thailand.

Tangtham, N. and W. Chunkao. 1990. Methodology and application of watershed classification in Thailand. Paper presented at the Workshop on Watershed Management Planning, MRC, Bangkok.

TEEB. 2009. *Climate Issues Update.* The Economics of Ecosystems and Biodiversity, Bonn, Germany.

Tillis, H.T., T. Ricketts, A.D. Guerry, E. Nelson, D. Ennaanay, S. Wolny, N. Olwero, K. Vigerstol, D. Pennington, G. Mendoza, J. Aukema, A. Foster, J. Forrest, D. Cameron, E. Lonsdorf, C. Kennedy, G. Verutes, C. K. Kim, G. Guannel, M. Papenfus, J. Toft, M. Marsik, J. Bernhardt, S. Wood, and R. Sharp. 2011. *InVEST 2.1 Beta User's Guide: Integrated Valuation of Ecosystem Services and Tradeoff.* The Nature Conservancy. www.naturalcapitalproject.org/pubs/InVEST_2.0beta_Users_Guide.pdf (accessed 10 March 2017).

Trisurat, Y., Chimchome, V. Pattanavibool, A. Jinamoy, S. Thongaree, S. Kanchanasaka, B. Simcharoen, S. S. Sribuarod, N. Mahannop, and P. Poonswad. 2013. Assessing distribution and status of hornbill species in Thailand. *Oryx* 47(3): 441–450.

Trisurat, Y., P. Eiwpanich, and R. Kalliola. 2016. Integrating land use and climate change scenarios and models into assessment of forested watershed services in Southern Thailand. *Environmental Research* 147: 611–612.

Trisurat, Y. and A. Grainger. 2012. Assessing ecosystem services at Sakaerat Environmental Research Station, Thailand. Paper presented at the 9th International Long-Term Ecological Research Annual Meeting, Lisbon, Portugal.

Trisurat, Y. and R. Kalliola. 2013. Ecohydrology consequences of transformative land use and climate change at Thadee watershed, Nakhon Sri Thammarat Province. *Proceedings of the International Workshop on Ecological Knowledge for Adaptation on Climate Change:* 38–41.

USDA. 2008. *Revised Universal Soil Loss Equation, RUSLE2.* USDA Agricultural Research Services, Washington, D.C.

Verburg, P.H. and A. Veldkamp 2004. Projecting land use transitions at forest fringes in the Philippines at two spatial scales. *Landscape Ecology* 19: 77–98.

Wieschmeier, W.H. and D.D. Smith. 1978. *Predicting Rainfall Erosion Losses: A Guide For Conservation Planning.* Agriculture Handbook Number 537. USDA, Washington, D.C.

Zalewski, M., G.A. Janauer, and G. Jolankai. 1997. *Ecohydrology: A New Paradigm for the Sustainable Use of Aquatic Resources*. International Hydrological Program, UNESCO, Paris.

Zhang, L., W.R. Dawes, and G.R. Walker. 2001. Response of mean annual evapotranspiration to vegetation changes at catchment scale. *Water Resources Research* 37: 701–708.

9 Which ecosystem services for disease regulation and health improvement, taking into account biodiversity and health interactions?

Aurélie Binot and Serge Morand

Introduction

We have entered the Anthropocene, a new geological period in which human-kind has a direct impact on major geochemical cycles, climate and biosphere. The epidemiological environment is thereby also affected by the emergence or reemergence of new infectious diseases affecting humans, their crops, their pets and wildlife. Climate change, biodiversity erosion, land use changes and increased use of natural resources are thus considered as ecosystem degradation factors (60 percent of ecosystems are considered more or less degraded; see Millennium Ecosystem Assessment, 2005). International organizations are increasingly linking ecosystem degradation to poverty, insecurity and political conflict. The degradation of ecosystem functions and services, as well as biodiversity loss, is also regularly mentioned as aggravating health risk factors, including the emergence of zoonotic infectious diseases. A link between ecosystem health and human and animal health is clearly expressed and increasingly assumed.

However, from the ecological point of view, one should not only take into account negative health impacts and risks associated with parasitism and infectious diseases. Would a "parasite-free" world be more "healthy"? Many empirical and theoretical studies seem to indicate the opposite. Parasites are key elements of ecosystems' functioning due to their roles in structuring and regulating animal (and plant) communities. Diversity and abundance of pests improve food chains' functions (predator and prey relationships) and the resistance of communities to invasive species. However, parasitism can also cause extinction. The introduction of new vectors or pathogens has been the cause of high mortality or even extinction, as in the case of avian malaria and endemic birds in Hawai'i.

On this basis, in this chapter we will discuss the links between an ecosystem's functions and associated ecosystem services in terms of infectious disease regulation, agricultural pest protection, resistance to biological invasion, and impact on animal and plant communities' structures. We will address these issues at the ecological and social interface, considering the usefulness of nature.

The emergence of the ecosystem services approach

A new understanding of the notion of "biodiversity" appeared with the concept of "ecosystem services," which emerged from the "ecosystem approach," as defined by the Convention on Biological Diversity:

> The ecosystem approach is a strategy for integrated management of land, water and natural resources that promotes conservation and sustainable use in an equitable way. Thus, the application of an ecosystem approach will help to achieve a balance between the three objectives of the Convention: conservation, sustainable use and the fair and equitable sharing of benefits arising from the utilization of genetic resources. An ecosystem approach is based on the application of appropriate scientific methodologies focused on levels of biological organization, which includes the processes, functions and interactions between organisms and their environment. It acknowledges that humans, with their cultural diversity, are an integral component of many ecosystems.
>
> (Secretariat of the Convention on Biological Diversity, 2004: 9)

A more precise definition of ecosystem services has been provided in the framework of the Millennium Ecosystem Assessment (2005) as the benefits that individuals and societies obtain from ecosystems, including the provision of services (water, food, etc.), the regulation of services (flooding, climate, disease, etc.), cultural services (spirituality, recreation, etc.) and supporting services (biogeochemical cycles).

Any change in the status of biodiversity is likely to influence the contribution of ecosystem services to human well-being. The conservation of biodiversity is designed, according to this paradigm, as a key element for human development and poverty reduction.

Among the benefits (or amenities) of ecosystem services, we may mention:

- the regulation of the emergence of diseases in plants, animals and humans as well as any changes that may impact on human pathogens and their vectors;
- biological control and all changes impacting agricultural pests and plant diseases; and
- the mitigation of biological invasions, particularly those caused by international trade.

The ecosystem approach has led to improved coordination among various international forums and organizations. The UN Food and Agriculture Organization (FAO), the World Organization for Animal Health (OIE) and the World Health Organization (WHO) have all shown an interest in matters relating to the interactions between biodiversity and health. In parallel, non-governmental organizations for nature conservation, such as the World Conservation Society (WCS), have focused on biodiversity issues and zoonotic infectious diseases as a new conceptual framework in conservation biology. For example, the WCS's international initiative "One World, One Health" connects wildlife conservation, human health and animal health in a single charter. From an institutional perspective, cross-sectoral collaborations are generally framed within the "One Health" approach, in which FAO, WHO and OIE work together for earlier detection and better management of pathogen emergence at the animal/man/ environment interface.

Almost ten years after the concept of ecosystem services was formulated, the International Science-Policy Platform on Biodiversity and Ecosystem Services (IPBES) emerged as an global institution to coordinate scientific expertise about biodiversity management, in much the same way as the Intergovernmental Panel on Climate Change (IPCC) covers climate change. COP10 of the CBD, held in Nagoya in 2012, requested the creation of this platform, and it has since been endorsed by the United Nations General Assembly. The objective is to improve the supply of information to decision-makers and policy-makers in contexts where international stakeholders advocate an ecosystem approach to health and biodiversity management, involving ecosystem services as a key notion for the implementation of actions (Morand, 2011).

What services for biodiversity management and wildlife health?

The ecology of evolution shows us that no "free" biological organisms can escape parasitism, as parasites are levers for evolution (including human evolution). It also shows us that some hosts provide shelter for many parasites while others host few, and that there are "hotspots" of pathogen diversity. Infectious diseases are thus more numerous and diverse in high-biodiversity areas, such as the tropics (compared to temperate zones). The diversity of human pathogens appears to be linked to the diversity in birds and mammals. Thus, a country with high biodiversity will host a high diversity of human pathogens (Dunn et al., 2010), as has been observed across Europe (Morand and Waret-Szkuta, 2012) and in Asia-Pacific (Morand et al., 2014a).

There seems to be a contradiction here: biodiversity is a source of pathogens for both humans and animals, but the Millennium Ecosystem Assessment presents it as helping in the regulation of health risk with respect to the emergence of infectious zoonotic diseases. However, looking a step beyond, it seems that the variety of infectious diseases does not matter as much as their actual impacts: epidemics, emergencies. Indeed, several studies have shown that high species diversity is associated with decreased prevalence of many zoonotic diseases in wildlife (Civitello et al., 2015). Analysis at the national level and focusing on the number of outbreaks rather than the number of infectious diseases reveals a link with diversity loss (assessed by the number of endangered mammals and birds according the IUCN's Red List; see Morand et al., 2014b).

However, preservation of biodiversity can lead to higher health risks: for instance, stopping deforestation seems to have induced greater risk of malaria in Brazil (Valle and Clark, 2013). In this framework, Lafferty and Wood (2013) stress that seeing biodiversity as a protection against health risks associated with wildlife is a myth which can conflict with conservation objectives.

What are disease regulation services (including zoonotic issues)?

The literature on ecosystem services is considerable, and rapidly evolving. Nevertheless, studies specifically devoted to disease control and regulation services are rare and remain mostly in the form of conceptual papers. Thus far, there has been insufficient development of implementable and validated indicators for these regulation services (Walpole *et al.*, 2009).

Studies focusing on the impacts of biodiversity on disease regulation provide conflicting results (Cardinale *et al.*, 2012). This is largely explained by their disparate methodologies, their focus on either human or plant diseases, and especially their small number in comparison with other work on regulating services (climate, water).

Another problem that could hinder the definition and validation of health regulation ecosystem services and functions relates to the method for their economic evaluation, mainly based on the costs linked to the lack of regulation: morbidity increase (DALYs), increased utility costs of human healthcare or agriculture, etc. Methodology to assess the effective impact of disease regulation services is missing, and there should be improved dialogue between the ecological and social sciences. Such methodology should also incorporate ecology and evolution, which are currently absent, as ecology is often referring only to ecological epidemiology (or eco-epidemiology).

Moreover, the concept of ecosystem services inherently has a highly utilitarian dimension as it is the result of social and institutional dynamics in the framework of environmental management. Therefore, the development of efficient methodological tools for analysis and assessment should take into account constructivist underlying processes involving scientists and civil society.

How to design disease regulation ecosystem services?

We should consider three key elements: the identification of the specific ecosystem service; the description of the ecological function; and the representation (and perception) of the service (Pedrono *et al.*, 2015). These three aspects require dialogue between scientists, experts, policy-makers and citizens. Beyond this dialogue, sharing knowledge is fundamental, involving a deep understanding of different points of view and perceptions (academic, political, socio-cultural) as well as local parameters (ecological and seasonal local drivers). Co-designing management schemes is a powerful tool for collective decision-making which leans on participatory methods, taking into account the consequences of individual and collective decisions on communities' future social and natural environments.

Linking the identification of services and human values

Wallace (2007) proposes a methodology to establish a conceptual framework on the basis of the identification of human values (Table 9.1). Regarding human values, we will address protection against disease, obtaining adequate resources (power) and socio-cultural satisfaction (preservation of biological and cultural diversity). Ecosystem services experienced at the individual and collective levels can then be defined together with the mechanisms (ecological and socio-economic), capital (social, economic and political) and assets (ecological and socio-cultural) that must be managed in order to produce these services.

The value of such a framework is that it immediately calls for multidisciplinary scientific skills (ecologist, agronomist, anthropologist, political scientist, etc.) and diverse points of view (scientific, political, and secular).

Describing socio-ecological mechanisms and assets

Wallace (2007) then proposes a functionalist approach to the services, starting from the understanding of the abiotic (climate, temperature, precipitation) and geophysics determinants of biological production (plant and animal biomass), focusing on energy as the key driver of the system. Land tenure and ownership modalities are shaping landscapes' and communities' structure (animal and vegetal, domestic and wild), predators and parasites (see Figure 9.1). Biological regulation (host–parasite interactions, pollination, predation, competition) contributes to the production of ecosystem services focusing on food, disease regulation and conservation.

Regulation services representation

The service can be represented at different spatial and temporal scales. For operational reasons, mapping ecosystem services can help understanding of land use modalities (conservation, water management, carbon sequestration) that are

Table 9.1 Human values determining ecosystem services, mechanisms, capital and assets to be managed

Human values	Ecosystem services perceived at individual level	Mechanisms and capital/assets to be managed in order to produce ecosystem services
Protection against parasites/diseases	Regulation/protection against diseases/parasites	Biological regulation
Adequate resources	Production (food)	Pollination
Socio-cultural	Opportunity values Capacity of cultural and biological evolution	Socio-cultural interactions

Source: Adapted from Wallace, 2007.

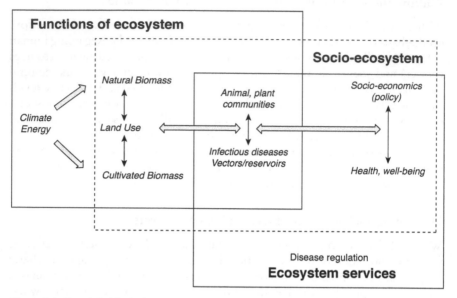

Figure 9.1 Simplified links between ecological processes (functions of ecosystems) and
three ecosystem services (including disease regulation) embedded in the
socio-ecosystem

Source: Adapted from Wallace, 2007.

directly related to those services, involving, for example, innovative participatory
approaches.

However, representing disease regulation services seems to be very tricky to
achieve. For example, if this regulation is based on an index linked to habitat
fragmentation, mapping the service will reflect only a recommendation about the
acceptable level of forest area fragmentation. The representation also involves
the danger of "freezing" the landscape, excluding the possibility of adaptive land
management (to climate change, for example, or socio-economic or political
changes that may affect land use choice).

Conclusion: co-designing public action at health/environment interface by involving scientists and social actors

Taking health into account in development policy results in the integration
of health issues in the development of ecological engineering, or even in the
development of public health engineering schemes in the framework of an ecosystem approach. This allows us to understand the impacts of natural habitat
fragmentation on health risks for humans and animals (and plants), but also the
effects of afforestation or ecological corridors established in terms of health risk
management.

Dialogue between conservation and health is reviewed through a new integrative discipline named "health ecology" (implemented through ecosystem approaches, such as "EcoHealth"), because the increase in the emergence of zoonotic diseases raises questions about the relationship between "wilderness," production systems and human well-being. This dialogue should facilitate co-design of the definition of ecosystem services and evaluation based on shared scientific approaches that will be socially accepted and acknowledged.

References

Cardinale, B.J, J.E. Duffy, A. Gonzalez, D.U. Hooper, C. Perrings, P. Venail, A. Narwani, G.M. Mace, D. Tilman, D.A. Wardle, A.P. Kinzig, G.C. Daily, M. Loreau, J.B. Grace, A. Larigauderie, D.S. Srivastava, and S. Naeem. 2012. Biodiversity loss and its impact on humanity. *Nature* 486: 59–67.

Civitello, D.J., J. Cohen, H. Fatima, N.T. Halstead, J. Liriano, T.A. McMahon, C.N. Ortega, E.L. Sauer, T. Sehgal, S. Young, and J.R. Jason. 2015. Biodiversity inhibits parasites: broad evidence for the dilution effect. *Proceedings of the National Academy of Sciences USA* 112(28): 8667–8671.

Dunn, R.R., T.J. Davies, N.C. Harris, and M.C. Gavin. 2010. Global drivers of human pathogen richness and prevalence. *Proceedings of the Royal Society London B* 277(1694): 2587–2595.

Lafferty, K.D. and C.L. Wood. 2013. It's a myth that protection against disease is a strong and general service of biodiversity conservation: response to Ostfeld and Keesing. *Trends in Ecology and Evolution*: 503–504.

Millennium Ecosystem Assessment. 2005. *Ecosystems and Human Well-being: Biodiversity Synthesis.* Island Press, Washigton, D.C.

Morand. S., 2011. La biodiversité comme dimension du changement global. *Recherches Internationales* 89: 213–230.

Morand, S. and Binot, A. 2014. Quels services rendus par les écosystèmes ? In S. Morand, F. Moutou, and C. Richomme (eds.), *Faune Sauvage, Biodiversité et Santé, Quels Défis?* Editions Quae, Versailles:147–155.

Morand, S., S. Jittapalapong, Y. Supputamongkol, M.T. Abdullah, and T.B. Huan. 2014a. Infectious diseases and their outbreaks in Asia-Pacific: biodiversity and its regulation loss matter. *PLoS One* 9(2): e90032.

Morand, S., K. Owers, and F. Bordes. 2014b. Biodiversity and emerging zoonoses. In A. Yamada, L.H. Kahn, B. Kaplan, T.P. Monath, J.J. Woodall, and L.A. Conti (eds.), *Confronting Emerging Zoonoses: The One Health Paradigm.* Springer Japan, Tokyo: 27–41.

Morand, S. and A. Waret-Szkuta. 2012. Determinants of human infectious diseases in Europe: biodiversity and climate variability influences. *Bulletin Epidémiologique Hebdomadaire* 12–13:156–159.

Pedrono, M., B. Locatelli, D. Ezzine-de-Blas, D. Pesche, S. Morand, and B. Binot. 2015. Impact of climate change on ecosystem services. In E. Torquebiau (ed.), *Climate Change and Agriculture Worldwide.* Quae & Springer, Versailles: 251–261.

Secretariat of the Convention on Biological Diversity. 2004. *The Ecosystem Approach* (CBD guidelines). Secretariat of the Convention on Biological Diversity, Montreal.

Valle, D. and J. Clark. 2013. Conservation efforts may increase malaria burden in the Brazilian Amazon. *PLoS ONE* 8: e57519 10.

Wallace, K.J. 2007. Classification of ecosystem services: problems and solutions. *Conservation Biology* 139: 235–246.

Walpole, M., R.E.A. Almond, C. Besançon, S.H.M. Butchart, D. Campbell-Lendrum, G.M. Carr, B. Collen, L. Collette, N.C. Davidson, E. Dulloo, A.M. Fazel, J.N. Galloway, M. Gill, T. Goverse, M. Hockings, D.J. Leaman, D.H.W. Morgan, C. Revenga, C.J. Rickwood, F. Schutyser, S. Simons, A.J. Stattersfield, T.D. Tyrrell, J.-C. Vié, M. Zimsky. 2009. Tracking progress toward the 2010 Biodiversity target and beyond. *Science* 325: 1503–1504.

Part IV

Managing biodiversity and living resources

10 Natural systems and climate change resilience in the Lower Mekong Basin

Future directions for biodiversity, agriculture and livelihoods in a rapidly changing environment

Jeremy Carew-Reid and Luke Taylor

Introduction

The Lower Mekong Basin (LMB) is experiencing a total transformation of its social, economic and natural environment. Exceptional advances have been made in reducing poverty, in expanding infrastructure networks, in trade and commerce and in improving the quality of life in many of the region's diverse communities and cultures. That progress has come at a substantial cost with long-term implications for its sustainability. Much development is proceeding without adequate knowledge of its impacts. The information available to decision-makers has been insufficient to avoid and mitigate unplanned and unwanted social and ecological effects.

Some 80 percent of families in the region are small-scale farmers dependent on healthy natural systems for their livelihoods and subsistence. Throughout the LMB average farm size is 2.8ha, with about 70 percent ranging in area from 1.5 to 3ha.[1] Small-scale farmers depend on wild plants and animals as much as they do on domesticated breeds. Their farming activities are interwoven with surrounding natural systems and are shaped by the seasons and by variable and changing climate. Agricultural crops are supplemented by wild fish, forest products and other natural resources (ICEM, 2014c). Natural systems contribute to the food security and well-being of all people of the region by supporting agricultural production and in the provision of ecosystem services, and to people worldwide through legal trade in wild products, tourism, science and intrinsic values. Also, they provide resilience and options for adaptation to climate change. In a degraded state, resilience of Mekong communities is reduced, and vulnerability[2] to climate change increases.

In this chapter, the concept of farming ecosystems recognizes that farms and their surrounding areas are integrated systems in which all habitats, species and their genetics interact with one another and the physical environment and contribute to farm productivity. In the Mekong region, farming ecosystems consist of human-modified and natural environments which both provide services and products essential to farming livelihoods and subsistence. Understanding and carefully managing that relationship is essential to the sustainability of Mekong farms, to continued poverty reduction and to building adaptive capacity to climate change.

Natural systems refer to areas that are in a relatively natural state. They may be impacted or degraded but still maintain a high level of integrity and functionality in supporting biodiversity and ecosystem services. The focus here is on protected areas (PAs) and on biodiversity beyond them, including wetlands, capture fisheries, non-timber forest products (NTFPs) and crop wild relatives (CWRs), all critical to the maintenance and productivity of farming ecosystems.

Natural systems in the LMB are degrading and changing at a rate and scale never before experienced. For example, over the past fifteen years, for the first time since human habitation in the region, the hydrodynamics and geomorphology of the Mekong River and its tributaries are being fundamentally and permanently altered by development – in this case mainly by full channel dams for hydropower (ICEM, 2010). The Mekong forest landscape has been transformed for agriculture and other developments. In the last thirty-five years, close to one-third of the forest area has been lost and at current rates little more than 10–20 percent of the original cover will remain by 2030 (WWF, 2013). Large connected areas of "core" forest – defined as areas of at least 3.2km² of uninterrupted forest – have declined from over 70 percent in 1973 to about 20 percent in 2009, with negative implications for the species they can sustain (WWF, 2013). Deforestation and linked agricultural expansion are the main causes of land degradation in the region, affecting between 10 and 40 percent of the land in each country (UNEP and TEI, 2007).

The continued degradation of natural systems demonstrates that the links between healthy natural ecosystems, economic development and human well-being are not understood or appropriately valued then reflected in adequate investment in conservation actions. The costs of ecosystem changes could be high and changes irreversible – in situations where the benefits and costs are not entirely known, a precautionary approach is needed (MEA, 2005).

Altered natural systems can recover or be rehabilitated. They have an inherent capacity for renewal. Yet, as an added threat in the region, climate change could push natural systems, many of which are already seriously degraded, beyond their capacity to adapt. Changes in precipitation and temperature and sea level rise, with associated water availability, flooding, drought and storm surge impacts, are likely. Natural systems, which have adapted to climatic changes in the past, need to be given the best possible chance of maintaining their overall productivity and functionality in the face of climate changes of increased severity and frequency in extremes. Ecosystem services, including disaster reduction, water, food and public health benefits to local communities, will become even more important as the incidence of extreme events associated with climate change increases (Dudley *et al.*, 2010). Climate change impacts on agricultural and other economic sectors may require adaptation options found within or provided by natural systems.

Reducing the existing threats to natural systems, restoring their integrity and implementing any additional climate change adaptation measures are important for the benefits they provide now and for storing and maintaining resources and services which will be needed as the region's climatic regime alters.

Natural systems and resilience

There is no hard distinction between natural and human-modified systems. Rather, there is a continuum of degradation and alteration of system characteristics and function (Figure 10.1). Typically, with increasing modification to natural systems, *resilience*[3] decreases and there is *simplification* in terms of biodiversity[4] and interactions between components and functions. When natural or traditional rural ecosystems are modified through development, species composition and the structure of ecosystems tend to become simplified, contributing to reduced adaptability to future change (Hashimoto, 2001). Ecosystems retaining their original complexity are more resilient, with species and habitat diversity offering more flexibility and buffers to change (Turner *et al.*, 2013). For example, removing just one of many species of fish from a river can worsen freshwater quality (Taylor *et al.*, 2006). Other relationships are more complex – for example, eliminating biodiversity from landscapes contributes to Lyme disease and hantavirus pulmonary syndrome can become epidemic, opening pathways for human infection on a large scale (Keesing *et al.*, 2010).

Also, ecosystem simplification leads to reduced cycling of inputs and outputs (of nutrients, for example) from the more cyclical operation of less altered systems. An industrial agricultural system (e.g. monoculture cropping) is a highly altered system, much simplified (i.e. less biodiversity and fewer ecosystem functions) and more linear (large external inputs required to maintain outputs) than a subsistence or agro-ecological based system having higher biodiversity and

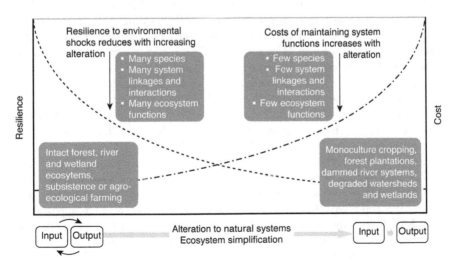

Figure 10. 1 Relationships between alteration of natural systems, resilience and maintenance cost

more functions with greater cycling of nutrients and internal self-regulation. In general, as natural systems become increasingly altered, their vulnerability to climate-related or other environmental shocks increases as resilience decreases and significant human intervention and investments are required to maintain or rehabilitate functionality. Conversely, systems with low alteration tend to be more resilient and require less intervention and investment to maintain functionality.

Overall, most responses of highly altered systems to climate change will rely on technology or systemic interventions and will involve increasing costs as changes become more extreme. The benefits or options provided by natural systems may be free or at least considerably more cost-effective.

In some cases the productivity of a particular output or service from a natural system is increased with alteration – for example, when species competition is reduced, allowing one to thrive. However, usually, this is unsustainable or detrimental for other important functions or for the system as a whole, potentially leading to collapse.[5] There is a need to reach a balance that maintains broader productivity and benefits as well as resilience over the long term.

Climate change projections and impacts on ecosystems and species

Overview

At the global scale, approximately 20 to 30 percent of plant and animal species are at risk of extinction due to climate change-associated temperature increases (Fischlin *et al.*, 2007). Although many ecosystems are expected to have some ability to adapt naturally to changes in climate (Gitay *et al.*, 2001), limited information is available regarding critical thresholds from which ecosystems will be driven into novel states that are poorly understood (Fischlin *et al.*, 2007). Loss of species can have knock-on effects for ecosystem functioning and the maintenance of ecosystem services (Hooper *et al.*, 2005), and so can limit options for social-ecological system adaptation.

Climate models for the LMB predict continued warming across the region and a range of other threats, including decreased rainfall in the dry season, increased rainfall in the wet season, decreased soil water availability, and increased frequency and severity of extreme events (ICEM, 2014b). These changes will have far-reaching impacts on natural systems in the basin.

Projecting climate changes in the LMB for the year 2050 led to the identification of eight hotspot provinces and linked protected area clusters experiencing significant changes in temperature, rainfall and flooding. The provinces are: Chiang Rai and Sakon Nakhon in Thailand; Khammoune and Champassak in Laos; Mondulkiri and Kangpong Thom in Cambodia; and Gia Lai and Kien Giang in Vietnam. The five highest-ranked provinces and clusters are: Chiang Rai, Mondulkiri, Khammouane, Gia Lai and Kien Giang (Figure 10.2).

Projected changes in climatic conditions by 2050 vary across the LMB (ICEM, 2014g). In summary, changes in seven key indicators are expected (Figure 10.2):

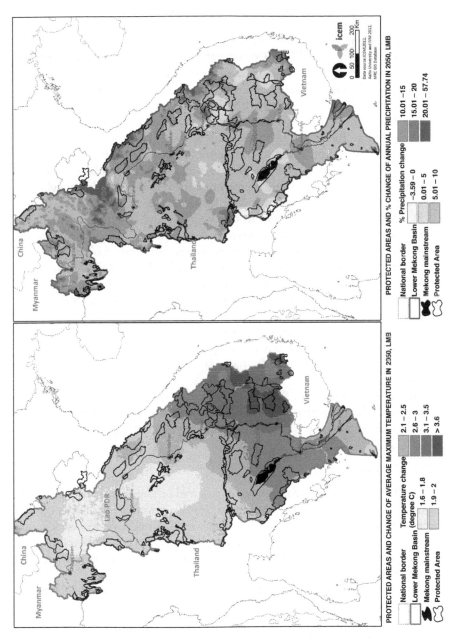

PROTECTED AREAS AND CHANGE OF AVERAGE MAXIMUM TEMPERATURE IN 2050, LMB

National border
Lower Mekong Basin
Mekong mainstream
Protected Area

Temperature change
(degree C)
1.6 – 1.8
1.9 – 2
2.1 – 2.5
2.6 – 3
3.1 – 3.5
> 3.6

PROTECTED AREAS AND % CHANGE OF ANNUAL PRECIPITATION IN 2050, LMB

National border
Lower Mekong Basin
Mekong mainstream
Protected Area

% Precipitation change
-3.59 – 0
0.01 – 5
5.01 – 10
10.01 –15
15.01 – 20
20.01 – 57.74

0 50 100   200
              Km

Data source: ICEM 2012,
Aalto University and IVM 2012,
MRC GIS Database

China
Myanmar
Lao PDR
Thailand
Vietnam

China
Myanmar
Thailand
Vietnam

Figure 10.2 Climate change in the Lower Mekong Basin (circles show hotspot provinces with their PA cluster)

- Temperature projection. In general, temperature in the LMB is expected to rise significantly over the next four decades. However, the level of warming will vary between PAs from 1 to 5°C. In many cases, temperature projections show that exposure to extreme temperatures will increase in duration and PAs will be exposed to conditions never before encountered.
- Rainfall projections. Both the quantity and seasonality of rainfall are projected to change. There is expected to be a reduction in rainfall in the dry season and, in many areas, an increase in the wet season.
- Water availability projection. Across most of the PAs, overall water availability is predicted to decrease with climate change. This modeled output is mostly a factor of higher temperatures and subsequent higher rates of evapotranspiration. Reduction in water availability is likely to be most severe in areas that are also projected to experience less rain during the dry season.
- Drought projection. Higher temperatures and decreasing relative humidity in the dry season are a common projection for PAs. As a result, droughts are projected to occur more often and for longer durations.
- Flooding projection. Increases in the intensity and volume of rainfall and runoff in the wet season are projected to lead to an increase in the frequency of flooding events. For PAs in more mountainous terrain, flash flooding will become a more frequent and destructive force. For PAs in the Mekong Delta and coastal zones, large-scale flooding is predicted to be more severe in duration and depth due in part to the additional effects of sea level rise and storm surge.
- Sea level rise projection. Sea level is projected to rise by approximately 15cm by 2030 and by approximately 30cm by 2050. Sea level rise will likely cause changes to flooding patterns, drainage, salinity exposure, coastal erosion and storm impacts.
- Storm surge projections. Projections regarding storm severity are uncertain; however, the frequency of very strong storms is expected to increase and individual occurrences of very high rainfall events are expected to increase.

Protected areas

In general, degraded PAs under significant external pressures are more vulnerable to climate change impacts. PAs (and particular PA zones) that are heavily utilized and relied upon for ecosystem services may enter a negative feedback cycle, whereby ecosystem services impacted by climate change will become more valuable and experience higher incidences of external exploitation pressure and, hence, higher vulnerability to climate change.

In its assessment of PA and natural system vulnerability in the LMB, ICEM (2014b) selected areas based on the following criteria:

- PAs located within priority climate change "hotspot" provinces;
- highly ranked PAs in terms of percentage change in climate conditions to 2050;

- PAs within the priority provinces that were representative of one or more ecozones;
- PAs within a province that formed part of a PA cluster or shared a contiguous boundary; and
- PAs with a substantial provisioning and servicing function for local communities as part of their farming ecosystem.

PA clusters were chosen in preference to individual PAs because they represent large areas of each ecozone and allow better demonstration of a range biological adaptation responses and strategies. Also, PA clusters have the potential to offer more secure and stable conditions for biodiversity in the face of climate change and other threats, mainly because of their size and diversity in habitats and biogeography. All individual and PA cluster vulnerability assessments are available in ICEM's protected areas report (ICEM, 2014g). In recognition of the large array of assets within each PA and the variability across all PAs in the LMB, climate change impacts and vulnerabilities were also categorized using an ecosystem services approach.

Provisioning services

Climate change impacts have negative consequences for a number of key provisioning services and assets of PAs. Provisioning services are most vulnerable in buffer zones, in community use areas and in the peripheries of PAs where NTFPs, suitable agricultural land and water sources are most heavily used, and where exposure to climatic change would be most pronounced. Services were found to be impacted in a number of ways, including:

- Decline in plant and animal productivity. Drought in the dry season, increased flooding and soil saturation in the wet season will lead to changes in pollination and flowering, spread and incidence of disease.
- Decline and loss of NTFPs. Reducing habitats and increased reliance and pressure on NTFPs in areas where agricultural production is impaired by climate change.
- Decline in water quantity and quality. Clean water is a key provisional service of PAs. Increased drought, reduced land cover, and changes in species composition and soil structure may affect the quantity and quality of water entering streams, ponds and lakes. The *trapeangs* throughout Mondulkiri Province of Cambodia, for example, are shallow ponds and lakes which will be exposed to droughts of great duration and severity, with increased temperatures and reduced rainfall during the dry season. Changes in surface water temperature, sedimentation and flood regimes will impact on water quality and aquatic systems generally.

Regulating services

Climate change impacts were found to affect many PA regulatory services, such as water cleansing, waste breakdown and nutrient recycling, soil erosion, climate (microclimate) control, pest control and flooding control. Identified vulnerabilities of regulatory services in the PAs included:

- Decreased regulation of erosion and sedimentation. An increased occurrence and decreased interception of sediment in water sources due to more extreme dry and wet conditions exposing and then flushing sediment into waterways.
- Decreased regulation of flash flooding and landslides. Projected increased precipitation in the wet season and increased runoff could lead to degradation of habitats, riverbank erosion, bank collapse and localized landslides.
- Decreased pest control functions. Biological control services are expected to decrease in reliability and effectiveness when various ecological linkages are disturbed, such as insect co-dependence and pest–predator relationships between insects, birds, amphibians, plants and mammals. Dramatic changes are likely to favor invasive species, displacing native predators, for example, and allowing corresponding native or introduced pests to flourish. In many cases increased ponding and stagnant water are likely in the wet season, thereby increasing breeding habitat for mosquitoes and other disease vectors. Likewise, drought conditions favor pests and have already been the cause of increasing damage to crops during drought years.
- Decreased nutrient recycling functions. Drier surface litter due to higher temperatures will lead to slower decomposition processes and buildup of otherwise recycled products. Losses or decline in certain insect activities may also reduce waste recycling, decomposition processes and, hence, nutrient recycling services.

Habitat or supporting services

Climate changes were found to be a cause of ecosystem shifts within PAs, reducing the supports for some species to thrive and survive. Identified vulnerabilities of habitat or supporting services in PAs included:

- Shifting/changes in habitat. Ecosystems are expected to shift or alter under climate change. The "movement" of certain species to higher elevations or latitudes because of changed temperature and rainfall regimes, for example, could change conditions and symbiotic linkages for other species, as well as lead to the appearance of competing species. In summary, key climate change-induced ecosystem shifts in LMB protected areas include:
  - geographic shifts in species ranges due to shifts in regular climate;
  - substantial range losses for individual species;
  - seasonal shifts in life-cycle events, such as advances in flowering due to changes in temperature;

- changes in animal migration patterns;
- changes in fish migration due to changes in the onset of the flood season;
- body-size changes, such as decreased body size due to higher temperatures;
- community composition changes – for example, species adapted to higher temperatures will become more predominant; and
- genetic changes, such as tolerance shifts.

Ecological shifts will lead to fundamental changes in the makeup of LMB PAs.

- Loss of habitat. Human settlement and infrastructure will limit the movement of ecosystems and, in some circumstances, prevent ecosystem shifts altogether, causing permanent loss of habitats. Also, increased risk and incidence of fire is likely to reduce habitat and disrupt support services.
- Reduction/degradation in biodiversity. In some PAs, some species are expected to disappear under climate change; others that are better adapted to new rainfall and temperature regimes will replace them. This will change the makeup of ecosystems and create opportunities for more hardy and aggressive native or exotic species, such as bamboo and other grassland species in degraded forests, or exotics in wetlands. Some species may experience a reduction in reproductive cycles and an altered period between flowering and maturing of seeds and propagules.
- Reduction in species population size. Higher levels of stress on PAs from a combination of influences, including climate change, are expected to lead to an overall loss in diversity and simplification of plant and animal assemblages. The number of species and the populations of some remaining species are expected to decline. Migratory species would seek other areas more suitable for breeding and nesting. Reduction in top soil moisture would reduce microflora and -fauna, suppressing decomposition and nutrient recycling, which will affect regeneration and plant growth.

Cultural services

Changes to ecosystems are expected to affect tourism, recreation, mental and physical health, and spiritual experiences associated with PAs, as well as the inspiration for culture, art and design. Identified vulnerabilities of cultural services in the PAs include:

- Declines in tourism. Reduced habitat is expected to cause losses in flagship species such as elephant, tiger and other cats – and subsequent losses in tourism. Increased intensity and regularity of flooding and storms may destroy tourist facilities and reduce access to tourist sites.
- Damage to infrastructure. Flooding, storms and sea level rise will damage infrastructure and cultural assets in and around the protected areas, including roads, bridges, temples and tourism facilities.
- Reduced community well-being and health. The overall losses in the condition of some PAs will have an impact on human well-being, especially

traditional communities with strong ties to the affected areas. Diseases may spread more easily in hotter and wetter climates unmoderated by diverse natural systems.

Wetlands

A small percentage of wetlands in the region fall within PAs. For this reason and because of their distinctive ecosystem characteristics, wetland vulnerability was assessed separately as an important category of natural system and source of livelihoods (ICEM, 2012). The climate change threats that are most significant for wetlands include changes in the magnitude, onset and duration of hydro-meteorological conditions which will alter the functioning of wetlands and their seasonal shift between aquatic and terrestrial environments. Direct climate threats include changes in precipitation, temperature, hydrology and sea levels, which will affect the source, transport and fate of water in the wetland system:

- Both local and upstream catchment *precipitation* are important factors in determining the nature and extent of wetland habitat. Mekong Basin rainfall is expected to increase in the wet season throughout the basin, while dry season rainfall will increase in northern and eastern areas and decrease in some parts of the vast Mekong floodplain–delta and the majority of the Sesan, Srepok and Sekong ("3Ss") river basins. At the basin level, mean annual precipitation is predicted to increase by 100–300mm/year, depending on the GCM used (ICEM, 2012). The highest increases are predicted in the central and northern Annamites to the east of the basin. This will result in greater variability in the Mekong moisture budget, with the highest level of exposure correlated with increasing elevation.
- *Temperature* plays an important role in wetland species productivity as well as surface water availability. There is considerable variability in temperature change throughout the LMB. The average annual maximum temperature will increase by 2–3°C with greater increases in the southern and eastern regions of the basin, while the largest change in temperature will occur in the 3Ss catchments, including a small area of the Srepok catchment, with an increase of over 4°C (ICEM, 2012).
- *The hydrological regime* plays a major role in shaping and maintaining wetland systems through the seasonal input of water and by determining the morpho-logical features of the wetlands. Alterations in these regimes will influence biological, biogeochemical and hydrological functions of the wetlands. Projected climate changes pose four key hydrological threats to Mekong wetlands:
    - increase in flood magnitude and volume;
    - increase in flood duration;
    - shortening of transition seasons; and
    - increase in dry-season water levels.

Based on these threats, of all Mekong wetland types, flooded forests are most exposed to climate change – especially those at higher elevations (ICEM, 2012). Riverine, freshwater, mangrove and peat wetlands in the LMB are all moderately exposed to climate change, with precipitation threats dominant for riverine and coastal wetlands and temperature dominant for peatlands. Grasslands, scrub and lakes/ponds of the LMB are on average amongst the least exposed wetlands to changes in temperature and rainfall.

Coastlines and deltas are among the most dynamic areas of river basins, with complex processes balancing marine encroachment with landmass expansion. Sea level rise will enhance the effectiveness of marine processes, shifting the balance and resulting in permanent inundation, and increased erosion over a greater proportion of the deltaic environment and an inland migration of coastal wetland environments where unimpeded. If movement inland is constrained by dykes and other developments, as is the case in most coastal areas of the LMB, then mangroves will become highly vulnerable, with large areas of forest lost (ICEM, 2012).

Capture fisheries

Capture fisheries in the LMB are buffered against climate change to an extent by exceptionally large aquatic ecosystem biodiversity (ICEM, 2014e). Some species would benefit from changing conditions, possibly maintaining the overall fisheries productivity, while other species will not fare well through direct effects and through losses in habitat, probably leading to an overall loss in biodiversity.[6]

Changes to habitat temperature will influence metabolism, growth rate, production, reproduction (seasonality and efficacy), recruitment and susceptibility to toxins and diseases affecting the natural ranges of some species, resulting in changes in biodiversity abundance in some areas.

Changes in rainfall patterns are likely to affect fisheries in a number of ways. Erratic rainfall will affect the flood pulse cycle of the Mekong River, affecting the hydrology of the Tonle Sap Great Lake and fish migrations, reproductive success and fish production that result. Shifting rainfall patterns, including longer dry periods, could affect the survival of fish through the dry season, particularly in Upper Mekong floodplain areas, already under pressure from hydropower development, overfishing and agriculture intensification.

Increased erosion in catchments will affect river floodplain water quality, reducing fish reproductive success and productivity. Increased runoff from inland areas could also result in the flooding of coastal lowlands, altering salinities and increasing fluvial deposition. Reduced rainfall during the dry season will reduce the capacity of upland streams to hold water for maintaining upland fish stocks. Flash flooding through increased rainfall during the wet season may affect habitats and the reproductive success of some species. Only those species able to handle these new extremes will proliferate.

Rising sea levels are likely to affect coastal fisheries through the migration of coastal mangrove areas northwards. Storm intensity and frequency will result in saline inundation of freshwater areas farther inland, resulting in periodic fish kills.

Upland fish species look particularly vulnerable to a wide range of pressures, including climate change. Black fish may be less vulnerable to wetland and river-ine fragmentation due to their limited migratory habits. As a result, this group of fish species may be less affected by climate change. White fish species are more vulnerable to declines in water quality (in terms of DO and alkalinity) and higher temperatures. The effects of climate change on some of these species may be severe. Estuarine species will be vulnerable to certain aspects of climate change, such as temperature increases in shallow coastal areas, but less vulnerable to others, such as sea level rise. Exotic species may benefit from the projected climate changes, increasing pressure on indigenous fish species.

In summary (ICEM, 2014e):

- upland fish will be most vulnerable to climate change;
- migratory white fish will also be vulnerable to climate change;
- black fish will be more "climate-proof" than other fish species; and
- invasive species will become more prevalent through climate change.

NTFPs and CWRs

An analysis of the potential climate change vulnerabilities of and impacts upon NTFPs and CWRs highlights considerable variation between different plant species and locations within the LMB (ICEM, 2014f).

Of all the climate changes considered, the increase in temperature is the most important for NTFPs and CWRs, particularly when it occurs during flowering, fruiting and seed dispersal periods. For example, wild rice species are highly vulnerable to projected increases in temperature. This CWR is already under threat from genetic erosion – mixing of genes with cultivated rice – and habitat loss with rapidly increasing pressure of land development for agriculture, urbani-zation, plantations or more general forest clearance. In some areas (such as Mondulkiri and Ratanakiri), climate change will increase the risk of forest fire through more intensive drying and accumulation of litter. If NTFPs and CWRs are not given the space and time to recover between forest fires, they will die out.

The province of Mondulkiri in Cambodia stands out as the most extreme in terms of its climate change and species vulnerabilities. Chiang Rai in northern Thailand stands out as the province where the vulnerability of NTFPs and CWRs to climate change is lowest. Gia Lai and Khammouan occupy intermediate positions. In Kien Giang, whilst the threat of sea level rise and storms affects makes the mangroves vulnerable, the vulnerabilities of aquatic plants and wild rice species seem to be low. The integrity of forest and wetland ecosystems is of critical importance to the natural resilience and adaptive capacity of most NTFP and CWR species – their vulnerability would be much greater without the protection that these provide.

Farming ecosystems and climate change

Natural systems are the foundation of agriculture and livelihoods in the LMB through the goods and services they provide and by building resilience to climate shocks. They regulate pests and diseases,[7] maintain soil productivity and improve water quality. They also directly provide significant nutrition, medicinal requirements[8] and income (for example, through NTFPs, including capture fisheries), as well as the long-term adaptability and productivity of agricultural systems and fisheries (for example, through supporting CWRs and wild fish species).

NTFPs lie at the heart of traditional farming systems in the Lower Mekong Basin and even have strong links with more commercial and industrial farming systems. Both NTFPs and CWRs rely on the intact[9] ecosystems of the LMB – forests, the Mekong River and wetlands – in order to survive. Natural system health is a determinant of their diversity and adaptability.

Figure 10.3 illustrates the linkages between NTFPs, CWRs and wild fish (as part of LMB natural systems) and agriculture, livestock, aquaculture and livelihoods. It shows why viable and sustainable NTFP and CWR systems (i.e. the forests, wetlands and river systems) are so important in the face of climate change (ICEM, 2014f). The traits of CWRs which make them important as resources for communities and natural systems to adapt to climate change and other man-made stresses include: their capacity to adapt to high altitudes, cold

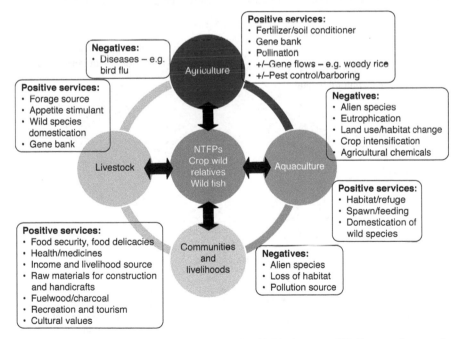

Figure 10.3 Links between NTFPs, CWRs and wild fish (as part of LMB natural systems) and agriculture, livestock, aquaculture and livelihoods

and xerophytic conditions; their resistance to fungal, bacterial and viral diseases, and to insect pests; their drought, flooding and salinity tolerance; and their potential for improved agronomic traits and grain quality.

Aquaculture farming is likely to be more vulnerable to climate change scenarios than capture fisheries (ICEM, 2014e). More intensive aquaculture systems will have much higher water quality and water quantity demands than more extensive systems in order to keep problems such as pollution or disease at bay (some of which may become more problematic with increased temperatures). This would be particularly damaging for the delta region, which has become economically dependent on aquaculture in recent years. Increased costs of farmed fish resulting from climate change adaptation costs could therefore have a serious effect on the quality of poorer people's diet. There are no other obvious animal protein alternatives. This is another example of alteration to a natural system leading to rising maintenance costs in order to maintain stability during climate change.

Farming ecosystems shifts

Farming ecosystems rely on climatic and ecological services. Climate and linked ecological shifts will lead to fundamental shifts in LMB farming ecosystems (Figure 10.4). Climate change shifts are spatial or temporal changes in regular or extreme climate. Changes in climate will lead to ecological shifts as species and habitats adapt to the new climate regime; an ecosystem shift occurs when the assemblage of species and habitats in a location changes to accommodate a new

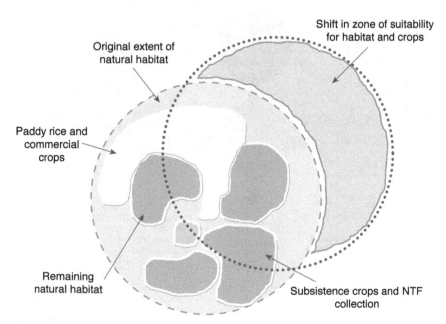

Figure 10.4 Effects of geographic shifts in climate

climate regime. The knock-on effects on LMB farming ecosystems will be profound (ICEM, 2014c).

Crops, NTFPs and CWRs that once flourished in an area may not be suited to new conditions. Changes in climate mean new crop species, cropping patterns, fishing activities and gathering and foraging habits become necessary, and a new balance in system components and inputs needs to be established in any one location. It can also mean that certain types of farming will need to shift to entirely new locations where conditions and natural system ingredients have changed to suit. This implies the need for flexibility and the ability of a system to adjust. Healthy, functioning and connected natural systems are much more flexible than degraded systems, and they have significant potential to assist farming systems to adjust and adapt.

Diminished or changing ecological provisioning may reduce availability and access to NTFPs and water. Weakening regulatory and habitat services may reduce pollination and pest control, as well as the soil's organic carbon content and micro-fauna and -flora. These ecosystem shifts may mean that in any location farming systems will require more intensive inputs and become more dependent on specialized, more resilient crops. Therefore, natural systems and ecological communities must be given the best possible opportunity to shift and adapt.

Integration of farming and natural systems for resilience

Farming systems have much to gain in terms of long-term productivity and resilience to climate change through greater harmonization and integration with natural systems. However, that shift will require a change from a focus on short-term productivity increases and profits to a broader, longer-term perspective on sustainability and multiple benefits.

During the past fifty years, the commercialization of farming systems brought about by technological advances and economic forces, including large agricultural subsidies, has increased food availability and decreased the real costs of agricultural commodities. Yet, the resulting agricultural practices have incurred costs related to losses in environmental quality and biodiversity, degraded or lost ecosystem services, emergence of pathogens, and the long-term instability of agricultural production (Tilman *et al.*, 2002). The shift from subsistence to commercial and industrial agriculture is gaining pace in the LMB and contributing to the degradation of natural systems. Figure 10.5 illustrates the continuum between subsistence farming and industrial farming and the changes in the level of inputs, productivity and ecosystem stability during climate change.

As farming systems move along the continuum they become less diverse, more intensive, and less resilient to climate change without substantial maintenance and inputs to keep them stable. Subsistence-based systems, while low in productivity, are inherently integrated with natural systems and benefit from their diversity and resilience to climate-related shocks (ICEM, 2014c).

Traditionally, this shift in farming ecosystems has taken place on a localized, gradual and incremental basis as individual farmers have sought more productive

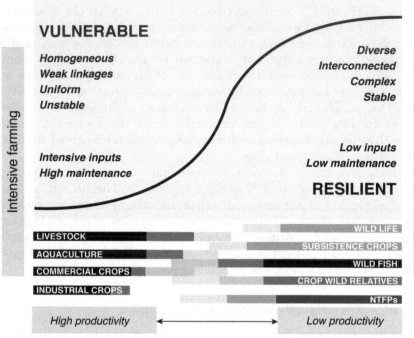

Figure 10.5 The agro-ecological systems and climate change continuum

and profitable crops and methods. Now, it is occurring rapidly and over large areas. Clear-felling of forests to make way for large industrial plantations and expansion of concession areas to cover large proportions of Cambodia and Laos are accelerating the shift along the continuum. These abrupt changes, or system leaps, greatly intensify the vulnerability of affected communities.

The transition from subsistence to more intensive forms of agriculture can bring significant benefits in terms of lifting people out of poverty, improving food security and relieving the pressure on natural areas by increasing incomes. Commercialization and industrialization of farming – that is, growing highly marketable and improved cultivars or varieties, together with an expansion of farm holdings, increased capital investment and additional hired labor – often increase farm incomes. However, in the process, large numbers of people may lose their land, livelihoods and food sources amidst increasing modification of natural systems.

The intensification of agricultural systems also faces sustainability issues, with erosion and loss of fertility due to the use of monocultures. A changing climate and higher climate variability will increase the vulnerability of this dynamic sector. Building more diverse and resilient cropping systems will be necessary for agricultural sector sustainability (ICEM, 2014d).

Local farming practices can sustain and enable poor families to adapt in a variable and challenging environment through manipulation of genetic diversity using indigenous knowledge and at low cost. Commercialization may exclude or reduce the role of local agro-diversity – for example, through increasing dependence on commercial seed supplies – which in turn may have longer-term implications for poverty and local resilience to climate change. This transition must therefore balance immediate economic efficiency and profits with local livelihoods, sustainability in agricultural systems and the conservation of natural systems.

Adaptation of natural systems in the LMB

The key strategy for coping with climate change in the Mekong region is to recognize that healthy natural systems are a foundation for the development and well-being of farming systems and for building resilience in communities, economic sectors and areas (ICEM, 2014c). Adaptation actions should contribute to ecological sustainability and social equity while also reducing climate change vulnerability. The corollary to that axiom is to ensure that adaptations do not contribute to environmental and biodiversity degradation.

Human actions reduce the resilience of natural systems where they degrade biodiversity, pollute ecosystems or disrupt natural disturbance regimes (Folke *et al.*, 2004). Healthy connected ecosystems are best placed to respond and adapt to a changing climate (Luck *et al.*, 2003). Given that natural systems in the LMB are subject to ongoing degradation and fragmentation, there is a need to address the existing management deficiencies when planning for climate adaptation. Determinants of the adaptive capacity of natural systems include (ICEM, 2014c):

- species diversity and integrity;
- species and habitat tolerance levels;
- availability of alternative habitat;
- ability to regenerate or spatially shift; and
- dispersal range and life strategy for individual species.

Protected areas

Protected areas are effective tools for optimizing the contribution of natural ecosystems in climate change response strategies (Dudley *et al.*, 2010). Protected areas enable adaptation through the maintenance of ecosystem integrity, buffering of local climate and the reduction of risks and impacts from extreme events. They provide the backbone for adaptation in LMB agriculture, fisheries, livestock and linked natural systems – and consequently for rural livelihoods. Yet, large and rapid transformations and simplification of biodiversity in the LMB is likely within and outside protected areas due to the impacts of climate change along with other threats. That transformation will tend to reduce overall productivity in linked farming ecosystems, with the poorest communities most affected.

Changes are inevitable. Some species and habitats will be lost from local areas, ecozones or from the whole basin. Others will shift from current locations to new areas. The structure and composition of ecosystems in some areas will change over time and require adaptation responses from protected area managers and farmers. Management should aim for the enhancement and maintenance of functions and values of the LMB protected area system as a whole. Basin-wide priorities for adaptation through protected areas and biodiversity conservation are set out below (ICEM, 2014c).

Expand and strengthen the protected area system

The protected area system should be expanded and strengthened to protect the full diversity of LMB habitats and increase opportunities for dispersal across the landscape. Protecting more habitats is one of the most effective ways to maintain viable populations of a wide range of species. Building a diverse protected area system of habitats is critical in the face of climate change. The size of an ecosystem is known to influence its resilience, with larger, less fragmented systems being more resilient (Zhou *et al.*, 2013). The role that forests play in supporting livelihoods and adaptation in agriculture (e.g. through NTFPs and CWRs) should be recognized and maintained wherever possible.

Strengthen the authority and capacity of protected area managers

The highest priority for PA adaptation in the LMB is the strengthening of capacities and processes for management planning, and for the effective implementation of a plan. At present, many PAs throughout the basin do not have management plans.

There are two basic principles which need to be embraced in PA legislation and strongly enforced at national and provincial levels: the principle of "*one area, one plan*" so that the PA management plan has authority over and is respected by all sectors and their individual development plans; and the principle of "*one area, one authority,*" which ensures that the PA managers are the recognized authority within their tenured territory and that they have a mandate that is respected across government. Realizing the second principle will require significant institutional reform in all Mekong countries to raise the status of PA managers within government. Standalone PA legislation and statutory authorities need to be considered.

Some adjustments, and new emphases and approaches, will be needed. These will include high-level political commitment and proactive interventions.

Management zones

Management zones are necessary to the effective management of activities within PAs of the LMB (ICEM, 2014g). Zoning defines what can and cannot occur in different areas of a PA in terms of human uses and benefits as well as PA development, maintenance and operations. While widely recognized and applied as a basic ingredient in PA management plans in the region, in practice they are not well defined on the ground or properly enforced.

Climate-induced shifts of ecosystems may require adjustments to PA zones to balance conservation and development around PAs. An important consequence of climate change is shifting habitat. That shift may force the migration of species to become more frequent and necessary for genetic exchange. Facilitating this migration requires long term planning of buffer zones and corridors, including extensive scientific research to shape management actions.

Key actions for zoning management include:

- *Establishment of a core zone of significant ecosystems or habitats.* This zone is the central area of the PA and the focus of strong protection measures. Defining the core zone's resources, as well as agreed control and safeguards therein, is crucial. In the context of climate change, this zone may need to be expanded or relocated to provide appropriate protection for significant habitat or species.
- *Establishment of a buffer zone.* This zone is an extension of a PA that surrounds the core zone. The intensity of human activities is greater than that allowed in the core zone. This zone creates a buffer wherein NTFP collection, tourism and other uses may occur. The existence of this area provides communities with an alternative to exploiting forest resources in the core zone and thereby protects the latter. Once again, it may be necessary to expand or relocate the buffer zone to increase the resilience of the core zone to climate change.
- *Establishment of a transition area.* This zone is similar to the buffer zone, except that even more intense development is permitted, such as agriculture.
- *Community participation and a flexible approach to management zones* are crucial to their success as a policy tool (ICEM, 2003). Climate change represents a major regime change and, therefore, must be integrated into ongoing monitoring and planning of management zones with PAs.

Building functional connectivity across the landscape

Climate adaptation capacity is generally increased through improving landscape-scale connections between core habitat patches. Maintaining connectivity involves establishing linkages between habitats to enable the movement of plants and animals, and to provide the supports that allow them to function. With climate changes, corridors of natural systems need to be available for organisms to move and relocate from one protected area to another. Creating corridors for adaptation is one of the most difficult but most important strategies facing the LMB countries in the decades to come.

Corridors to link PAs require extensive and long-term commitment to rehabilitation work. The local expressions of rehabilitation vary between ecosystems and habitats. Typically, rehabilitation brings back or creates benefits provided by PAs and increases their resistance to climate change impacts. Options for delivering corridors and rehabilitation are:

- Rehabilitation of degraded areas. Seeding and plantation activities on cleared land or critically degraded ecosystems may provide the basis of ecosystem rehabilitation. Ongoing monitoring of a range of biological processes is necessary for success.

- Enrichment planting. This is where valuable species are introduced or reintroduced to degraded forests or wetlands that complement existing species. In some cases, this may involve plantations of mixed native species for commercial purposes.
- Breeding programs. For some threatened or critical species breeding programs, including those in captivity may assist the reestablishment of breeding populations.

Build on and strengthen existing conservation management approaches

It is crucial to build on and strengthen existing conservation management approaches which are likely to continue to be important under climate change. Much of what is required is better and more proactive conservation management to meet adaptation requirements. After much support in the 1990s, protected areas and biodiversity conservation have been neglected by most donors who could not see substantial community development and poverty reduction benefits from their investments. Now there is a fresh and pressing need for increased and well-coordinated support for LMB protected areas as an essential adaptation and development strategy.

Integrated adaptation in protected area management planning

Climate change adaptation planning and strategies need to be a fundamental part of overall PA management planning. The priority for PA adaptation strategies is the reinforcement of conservation measures currently being planned and implemented in the LMB, and additional resources for and attention to new measures. A lack of management plans and inadequate implementation of existing plans are major constraints to climate change response. PA management planning requires support and strengthening with a precise and clear primary goal of biodiversity conservation. A lack of precision in the primary PA management goal and confusion between competing goals – for example, poverty reduction, community development and biodiversity conservation – has led to confusion and failure to sustain effective responses to any of them. Meeting that primary goal will enable protected areas to meet their subsidiary functions, including underpinning adaptation in farming systems.

Improve understanding of climate change impacts on biodiversity

Greater attention to and resources for improving scientific understanding of climate change impacts on species and ecosystems are required. Effective adaptation strategies need to be informed by scientific evidence. Research and monitoring are needed to understand which species and ecosystems are most sensitive to climate change and to distinguish climate change effects from those caused by other threats. Key areas of research and monitoring relate to:

- rates of ecological change, including early warnings of key changes in ecological processes;
- types of ecological change, such as in situ changes in species abundance, changes in interactions between species and distributional shifts;
- patterns of geographic range shifts over elevational and latitudinal gradients;
- changes in other threats to biodiversity and their interactions with climate change;
- detection of new invasive species at sites and changes in abundance and dynamics of species that may become problematic in the future; and
- the effectiveness of management actions (NSW Department of Environment, Climate Change and Water, 2010).

Building ecosystem resilience

There are clear links between ecosystem health and climate change. Higher levels of biodiversity provide more options for ecosystems to adapt, and therefore greater climate change resilience. Actions that maintain or expand biodiversity are conservation priorities in their own right but have become more important for their role in responding to climate change. These include three priorities (NSW Department of Environment, Climate Change and Water, 2010):

- *Understanding the maintenance requirements of key biological processes giving priority to "keystone" species* that play an important role in maintaining ecological processes, such as dispersal. Keystone species have a disproportionately large influence on community and ecosystem function relative to either their numbers or their biomass.
- *Identification, maintenance and proactive management of refugia and pockets of resilience.* Two examples of dry-season refugia that need to be maintained are: deep pools in the Mekong River and its tributaries which provide micro-habitats for many fish species; and deep valleys in mountainous terrain which are moister and cooler than the surrounding environment, with surface water that persists during severe droughts.
- *Mitigation of threats from invasive species and pests.* Monitoring of habitats supported by satellite imagery and ground surveys is needed to keep track of the spread of invasive species. This serious problem is gaining momentum as more areas are disturbed and the number of aggressive invasive species increases.

NTFPs and CWRs

In their natural environment, most species have clear seasonal patterns, especially in the climatic conditions of the Mekong, where there is a marked distinction between wet and dry seasons. They are well adapted to the extremes of the hot, dry season, when many species go into relative hibernation, shed their leaves, store food sources in tubers, or aestivate (earthworms) or migrate (honey bees).

They can often withstand the extremes of drought at the end of the dry season and decline in soil moisture availability. Some species' seeds can survive several years of dormancy and wait for the ideal conditions before they germinate. Fungi are excellent examples of this strategy, with some species able to survive periods of extreme climate and take advantage of climate variability to grow and reproduce when the conditions are less extreme. A number of species, especially grasses and herbs, climbers and aquatic plants, have vegetative reproduction and can grow from the rootstock and rhizomes as well as producing seeds. This means that they can also take advantage of different conditions to multiply and spread when seed production is limited.

Thus, there is a residual resilience in intact forest and wetland ecosystems to climate change. As conditions change there will be a gradual shift towards species or sub-species that can better tolerate the changes. The situation in stressed and modified natural environments will be different, and in these situations more rapid changes in the species and their ecology may be expected. Once a threshold is passed for the ecosystem as a whole there will be a more dramatic change, loss of key species, loss of forest cover and overall transformation of the ecosystem. Usually this threshold is not obvious until the loss has occurred.

Many of the NTFPs have several species or sub-species that are already distributed widely according to different existing conditions, and it is possible that these will "move" to take up new positions as the climate changes. Effective dispersal mechanisms are therefore an important aspect of species resilience.

Many NTFPs and CWRs are well adapted to climate extremes if undisturbed by other influences. Changes in rainfall patterns seem to have a less significant impact on NTFPs and CWRs. For most species in the hotspot provinces, the increases in annual rainfall are well within the normal range. Wetland plant species appear to be the least vulnerable of the NTFP plants assessed by ICEM (2014c). The increased temperature is moderated by the water habitat and may in fact induce growth in aquatic plants. They are well adapted to flooding, and most are also adapted to periods of drought, and will return when the floodwaters return.

The existing development threats to NTFP and CWR species are more immediate and significant than climate change. Habitat loss, changes in land use to agriculture, plantations, aquaculture, deforestation and over-harvesting are the most important stressors reducing the populations of many species. Climate change is an additional, but not necessarily the most important, stressor. Therefore, the most effective adaptation strategies for protecting certain species will be to reduce existing threats and at the same time enhance resilience to climate change. The options for enhancing NTFP and CWR resilience of species can be grouped under several broad adaptation strategies: habitat protection, rehabilitation and reforestation; management of sustainable NTFP harvesting, domestication and cultivation; and implementation of species conservation plans, including assisted movement of those at risk of habitat shift. So little is known about NTFPs – and even less about CWRs – that all adaption measures will benefit from more extensive research and monitoring.

Diversity for resilience in agricultural systems

Biodiversity is a product or outcome of co-evolving human and ecological systems over a very long time. Diversity in genetics, species and ecosystems is critical for the productivity, sustainability and adaptability of LMB farming ecosystems.

Genetic diversity

The loss of genetic diversity in farming ecosystems diminishes resilience and agricultural sustainability. To date, breeders of the major food staples (maize, rice and wheat) have been successful at improving resistance to pathogens and diseases in order to maintain yield stability despite low crop diversity in continuous cereal systems. However, any improvement in crop resistance to a pathogen is likely to be transitory because of their evolutionary interaction, requiring continuous intervention. For example, the useful lifetime of maize hybrids in the US – about four years – is now half what it was thirty years ago (Tilman *et al.*, 2002). It is uncertain if such conventional breeding approaches can work indefinitely. Similarly, agrochemicals, such as pesticides and antibiotics, are major selective agents – herbicide-resistant weeds were observed within about one or two decades from the introduction of each of the seven major herbicides (Palumbi, 2001).

Crop rotation and the use of spatial or temporal crop diversity can reduce the need to breed new disease-resistant crops and to develop new pesticides (Tilman *et al.*, 2002). This practice would also reduce costs. Both integrated pest management and biotechnology will be important in establishing durable resistance through multiple gene sources (Ortiz, 1998; DeVries and Toenniessen, 2001). Plant genetic diversity is the factor that enables adaptation (FAO, 2010, 2011). For example, the likelihood of crop failure due to drought decreases when traditional landraces with higher genetic diversity are adopted over modern crop varieties. One of the most effective ways of dealing with environmental changes is the ability to match crop genetics (i.e. variety) to a particular management system and region (Howden *et al.*, 2006). To retain this ability, local biodiversity must be protected. The coefficient of on-farm crop genetic diversity has a strong positive effect on crop yields and a negative effect on the variance of crop yields. This result is robust against different production function specifications, different types of crops, different scales of data (regional versus plot specific) and different measures of crop genetic diversity.[10]

In the Mekong Delta, there is a general trend toward the domination of cropping by a handful of high-yield rice varieties (Hashimoto, 2001). In the absence of ecological balance and natural controls, pest and disease outbreaks and fluctuations in environmental conditions will increasingly pose a threat to agricultural production (Hashimoto, 2001). Many traditional varieties of crop plants and livestock (as well as their ancestral or related species, which occur naturally within the Mekong Delta) have higher environmental tolerances than the more recently introduced varieties. Therefore, these could hold the key to the future development of more robust varieties for the local environment (Hirata, 2000). The absence of such genetic resources within the region will establish a feedback loop of perpetual reliance on costly inputs of agro-chemicals and new imported varieties, and continuing environmental and ecological degradation (Hashimoto, 2001).

Species diversity

A diversity of crop species can better adapt to environmental changes because the broader pool of different metabolic traits and metabolic pathways enables more effective resource use (such as water and soil nutrients) over a wide range of environmental conditions (Schläpfer *et al.*, 2002). Also, food insecurity is reduced when there is a diversity of plants, animals and microorganisms in and around the farm – an important principle for all farmers in the LMB but particularly those that are poorer and more isolated (Araya and Edwards, 2006; IFAD, 2011).

Ecosystem diversity

A significant opportunity to support and improve agricultural livelihoods while reducing environmental degradation and vulnerability to climate change is through the adoption of sustainable agricultural[11] practices. Rather than transition to large-scale, commercialized, high-input systems, the aggregate benefits in terms of productivity, provision of ecosystem services and resilience to climate change impacts may be much greater by improving existing management and productivity through sustainable agricultural practices. Expanding the area of land under agricultural production is not necessarily the solution, especially with changing climate. Often the short-term benefits of land conversion are outweighed by losses in ecosystem services and adaptive capabilities that intact natural systems provide. During the conversion period (say three–five years) sustainable agriculture can have lower productivity than high-input systems, but over the longer term it has comparable production levels plus positive effects on soil and water quality and reduced costs.

Integrating agricultural systems with adjacent natural, semi-natural or restored ecosystems through landscape-scale management holds significant potential for maintaining the benefits for individual farms of ecosystem services and reducing the negative effects of the farms on their natural system foundations (ICEM, 2014c). For example, surrounding cultivated fields with a diversity of trees and shrubs decreases soil erosion and can intercept nutrients that otherwise would enter surface or ground waters. Nutrient and silt loading from cultivated fields or pastures can be reduced with buffer zones along streams, rivers and lake shores. Crop pollination can be provided by insects and other animals living in nearby habitats or buffer strips, while other organisms from these habitats, such as parasitoids, can provide effective control of many agricultural pests (Tilman *et al.*, 2002).

Introducing biodiversity into farming systems improves soil fertility and, in the process, reduces net GHG emissions through carbon fixation. Agro-forestry through composting, for example, and organic agriculture significantly reduces greenhouse emissions compared with conventional practices (Ajayi *et al.*, 2011; Muller, 2009; Niggli *et al.*, 2009). Organic practices also contribute to reducing emissions through the use of organic fertilizers, crop rotations (including legumes, leys and cover crops) and through avoiding open biomass burning, synthetic fertilizers and the related production emissions from fossil fuels. In addition, systems that have and produce soil biodiversity foster carbon sequestration, good soil tilth and high fertility (Sperow *et al.*, 2003).

In general, the more a cultivation system mimics natural ecosystem functions, the fewer resource inputs are required and negative environmental effects felt (Folke and Kautsky, 1992). For example, the diversification of land use and the development of agro-forestry and silvofishery[12] systems in the Mekong Delta could contribute significantly to the creation of varied ecological habitats (Hashimoto, 2001). Such integrated systems that work with ecosystems without degrading the resource base on which they depend will be more sustainable and have positive contributions to the surrounding ecosystems and socio-economy (FitzGerald, 2002).

Conclusions

Despite the great diversity of livelihoods and ecosystems in the Lower Mekong Basin, the sectors of agriculture, livestock, fisheries, NTFPs and heath and rural infrastructure have one thing in common – their resilience to climate change is dependent on healthy, functioning natural ecosystems.

The foundation adaptation strategy for the Mekong region – that is, the set of measures that underlie all other responses to climate change – can be summarized as bringing diversity and complexity back into the agricultural landscape. Increased diversity in farming ecosystems means a broader range of species and a deeper genetic pool. Increased complexity means more mutually beneficial relationships and synergies between those components. The two foundational characteristics of a resilient farming ecosystem mean greater stability when confronted by climate change shocks. They also create more opportunities to recover by providing a broader range of adaptation options.

Farmers face many risks, such as climatic factors, pests and diseases, price uncertainties and changing government policies. Farm diversification is a response to avoid and minimize those risks. The main purpose of diversification in agriculture is to maintain an optimal level of overall production and return by selecting a mixture of activities which buffer the farmer against shocks affecting individual crops. Already smallholder farming in the LMB is extremely diverse and flexible. Yet, the trends are for consolidation of holdings and a shift to highly productive monocultures. Farm diversification is a key principle which needs to guide climate change adaptation in agriculture.

Another closely linked principle for adaptation is optimizing biodiversity in farming. That principle means more than increasing the range and number of crops grown on a farm (although that is critical for stability in output). It is about the overall enhancement and maintenance of ecosystem health on farms and their surrounding areas and catchments – that is, viewing farms as ecosystems which are integrated and managed with linked natural systems. The aim is steadily to increase the number of living components within the farm ecosystem in a way that is ecologically more diverse and stable and provides optimum growing conditions for a greater diversity of produce. Resilient farms include healthy soil micro-fauna and -flora, trees, shrubs, palms, bamboos and other woody perennials and well-managed and maintained waterways – all offering a range of produce and ecosystem services as an integral part of the crop and livestock system. A diverse and biodiversity-rich farming system builds on traditional practices with improvements from modern technologies and approaches.

Promoting diversity and complexity in farming ecosystems will require compromises on the nature, pace and scale of development across many sectors. It means taking a more cautious approach which avoids and compensates for degrading natural systems. Most important, it means all LMB governments giving priority to building natural capital as a way of ensuring long-term consistency in farm productivity and incomes in the face of climate change.

Notes

1   For example, in Lao PDR, 36.4 percent of farm holdings are medium sized (1.0–2.0ha), 27.4 percent are large (over 2ha), 23.8 percent are average (0.5–1.0ha) and only 12.5 percent are "small" (less than 0.5ha).
2   Climate change vulnerability refers to the degree to which an ecological system or species is likely to experience harm as a result of changes in climate. Vulnerability is a function of: *exposure* to climate change: the magnitude, intensity and duration of the climate changes experienced; the *sensitivity* of the species, community or asset to these changes; and the *adaptive capacity* of the system to adapt to these changes (ICEM, 2014a; IFAD, 2011).
3   Resilience is a stability property of ecosystems that reflects the capacity of a system to absorb shocks (Perrings, 1995) or the rate at which a system variable returns to the reference state after perturbation (Schläpfer *et al.*, 2002).
4   The genetic, species and habitat diversity of biota, including micro-organisms, plants and animals.
5   When humans modify an ecosystem to improve a service it provides, this normally results in changes to other ecosystem services. For example, actions to increase food production can lead to reduced water availability for other uses. As a result of such trade-offs, many services can be degraded, for instance fisheries, water supply and protection against natural hazards. In the long term, the value of services lost may greatly exceed the short-term economic benefits that are gained from transforming ecosystems (MEA, 2005).
6   Assessing the vulnerability to climate change "signals" in the Mekong fisheries is challenged by the "noise" from other factors (ICEM, 2014e). The largest single threat to the diversity and productivity of the Mekong fishery is the alteration of river morphology caused by hydropower projects. However, a wide range of other threats also exist. Physical barriers (even small-scale) which constrain the migration of fish species:

  • overfishing, resulting from increased numbers of fishermen and -women and increased size of gear;

- aggressive fishing methods, e.g. explosives;
- loss of productivity through habitat destruction/change;
- radical changes in land use patterns that change runoff patterns from upland areas; and
- establishment of exotic fish populations from aquaculture escapees.

7 There is a growing literature (e.g. Ash and Jenkins, 2007) documenting how biodiversity reduces the risk of exposure to several infectious diseases (such as malaria, Japanese encephalitis and rabies). "Biodiversity at the ecosystem level produces the appropriate balance between predators and prey, hosts, vectors and parasites which allows for appropriate controls and checks for both the spread of 'endemic' infectious disease as well as resistance towards invasive pathogens (from humans, animals or insects)" (Vira and Kontoleon, 2013: 69).

8 Biodiversity is an important source of traditional medicines (such as herbal medicines) for people in developing countries, especially where they have little (if any) access to formal healthcare (Vira and Kontoleon, 2013).

9 In intact ecosystems, "the majority of native species are still present in abundances at which they play the same functional roles as they did before extensive human settlement or use, where pollution has not affected nutrient flows to any great degree, and where human density is low" (Caro *et al.*, 2011: 1).

10 Just and Candler (1985); Smale *et al.* (1998, 2008); Widawsky and Rozelle (1998); Di Falco and Perrings (2005); Di Falco and Chavas (2008); and Heisey *et al.* (1997).

11 Agricultural sustainability means maintaining productivity while protecting the natural resource base. Possible actions include: improving low-impact practices, such as organic agriculture and providing incentives for the sustainable management of water, livestock, forests and fisheries. Science and technology should focus on ensuring that agriculture not only provides food but also fulfills environmental, social and economic functions, such as mitigating climate change and preserving biodiversity. Policy-makers could end subsidies that encourage unsustainable practices and provide incentives for sustainable natural resource management (IAASTD, 2008).

12 Silvofisheries are a form of low-input sustainable aquaculture where the cultivation of mangroves is integrated with brackish water aquaculture. They strive to mimic natural ecosystem functions in a culture system with reduced resource use; avoidance of chemicals and medicinal compounds; and the recycling of nutrients and materials to increase the efficiency of the system. This integrated approach to conservation and utilization of the mangrove resource allows for maintaining a relatively high level of integrity and biodiversity in the mangrove area while capitalizing on the economic benefits of brackish water aquaculture (FitzGerald, 2002).

References

Ajayi, O.C., F. Place, F.K. Akinnifesi, and G.W. Sileshi. 2011. Agricultural success from Africa: the case of fertilizer tree systems in southern Africa (Malawi, Tanzania, Mozambique, Zambia and Zimbabwe). *International Journal of Agricultural Sustainability* 9: 129–136.

Araya, H. and S. Edwards. 2006. *The Tigray Experience: A Success Story in Sustainable Agriculture.* Third World Network (TWN) Environment and Development Series 4. TWN, Penang.

Ash, N. and M. Jenkins. 2007. *Biodiversity and Poverty Reduction: The Importance of Biodiversity for Ecosystem Services.* UNEP–World Conservation Monitoring Center, Cambridge.

Caro, T., J. Darwin, T. Forrester, C. Ledoux-Bloom, and C. Wells. 2011. Conservation in the Anthropocene. *Conservation Biology* 26: 185–188.

Coates D., O. Poeu, U. Suntornratana, N. Thanh Tung, and S. Viravong. 2003. *Biodiversity and Fisheries in the Lower Mekong Basin*. Mekong Development Series No. 2. Mekong River Commission, Phnom Penh.

DeVries, J. and G. Toenniessen. 2001. *Securing the Harvest: Biotechnology, Breeding, and Seed Systems for African Crops*. CAB International, Wallingford.

Di Falco, S. and J.-P. Chavas. 2008. Rainfall shocks, resilience and the dynamic effects of crop biodiversity on the productivity of the agroecosystem. *Land Economics* 84: 83–96.

Di Falco, S. and C. Perrings. 2005. Crop biodiversity, risk management and the implications of agricultural assistance. *Ecological Economics* 55: 459–466.

Dudley, N., S. Stolton, A. Belokurov, L. Krueger, N. Lopoukhine, K. MacKinnon, T. Sandwith and N. Sekhran. 2010. *Natural Solutions: Protected Areas Helping People to Cope with Climate Change*. IUCN–WCPA, TNC, UNDP, WCS, World Bank and WWF, Gland.

European Union Delegation to Cambodia. 2012. *Country Environment Profile Royal Kingdom of Cambodia*. EU, Brussels.

Fischlin A., G.F. Midgley, J.T. Price, R. Leemans, B. Gopal, C. Turley, M.D.A. Rounsevell, O.P. Dube, J. Tarazona, and A.A. Velichko. 2007. Ecosystems, their properties, goods, and services. In M.L. Parry, O.F. Canziani, J.P. Palutikof, P.J. van der Linden, and C.E. Hanson (eds.), *Climate Change 2007: Impacts, Adaptation and Vulnerability. Contribution of Working Group II to the Fourth Assessment Report of the Intergovernmental Panel on Climate Change*. Cambridge University Press, Cambridge: 211–272.

FitzGerald, W.J. 2002. Silvofisheries: integrated mangrove forest aquaculture systems. In B. Costa-Pierce (ed.), *Ecological Aquaculture: The Evolution of the Blue Revolution*. Blackwell, Oxford.

Folke, C. and N. Kautsky. 1992. Aquaculture with its environment: prospects for sustainability. *Ocean and Coastal Management* 17: 5–24.

Folke, C., S. Carpenter, B. Walker, M. Scheffer, T. Elmqvist, L. Gunderson, and C.S. Holling. 2004. Regime shifts, resilience and biodiversity in ecosystem management. *Annual Review of Ecology Evolution and Systematics* 35: 557–581.

Food and Agriculture Organization (FAO). 2010. *Second Report of the State of the World's Plant Genetic Resources for Food and Agriculture*. FAO, Rome.

Food and Agriculture Organization (FAO). 2011. *Second Global Plan of Action for Plant Genetic Resources for Food and Agriculture*. FAO, Rome.

Gitay, H., S. Brown, W. Easterling, and B. Jallow. 2001. Ecosystems and their goods and services. In J.J. McCarthey, O.F. Canziani, N.A. Leary, D.J. Dokken, and K.S. White (eds.), *Climate Change 2001: Impacts, Adaptation, and Vulnerability. Contribution of Working Group II to the Third Assessment Report of the International Panel on Climate Change*. Cambridge University Press, Cambridge: 237–342.

Gopal, B. 2013. Future of wetlands in tropical and subtropical Asia, especially in the face of climate change. *Aquatic Sciences* 75: 39–61.

Hashimoto, T.R. 2001. *Environmental Issues and Recent Infrastructure Development in the Mekong Delta: Review, Analysis And Recommendations with Particular Reference to Large-Scale Water Control Projects and the Development of Coastal Areas*. Working Paper No. 4. Australian Mekong Resource Center, University of Sydney.

Heisey, P., M. Smale, D. Byerlee, and E. Souza. 1997. Wheat rusts and the costs of genetic diversity in the Punjab of Pakistan. *American Journal of Agricultural Economics* 79: 726–737.

Hirata, Y. 2000. Genetic resource diversity and hopeful future image in Mekong Delta especially focused on rice and soybean genetic resources. Paper presented at the First

Joint Environmental Conservation and Sustainable Agriculture, Cantho, Vietnam, 19–21 January.

Hooper, D.U., F.S. Chapin, J.J. Ewel, A. Hector, P. Inchausti, S. Lavorel, J.H. Lawton, D.M. Lodge, M. Loreau, S. Naeem, B. Schmid, H. Setala, A.J. Symstad, J. Vandermeer, and D.A. Wardle. 2005. Effects of biodiversity on ecosystem functioning: a consensus of current knowledge. *Ecological Monographs* 75: 3–35.

Howden, M., A. Ash, S. Barlow, T. Booth, S. Charles, B. Cechet, S. Crimp, R. Gifford, K. Hennessy, R. Jones, M. Kirschbaum, G. McKeon, H. Meinke, S. Park, B. Sutherst, L. Webb, and P. Whetton. 2006. *An Overview of the Adaptive Capacity of the Australian Agricultural Sector to Climate Change – Options, Costs and Benefits.* CSIRO, Canberra.

International Assessment of Agricultural Knowledge, Science and Technology for Development (IAASTD). 2008. *Agriculture at a Crossroad.* Island Press, Washington, D.C.

International Center for Environmental Management (ICEM). 2003. Regional report on protected areas and development. In *Review of Protected Areas and Development in the Lower Mekong River Region.* International Center for Environment Management, Hanoi.

ICEM. 2010. *Strategic Environmental Assessment (SEA) of Hydropower on the Mekong Mainstream: Summary of the Final Report.* Prepared for the Mekong River Commission by International Center for Environment Management, Hanoi.

ICEM. 2012. *Basin-Wide Climate Change Impact and Vulnerability Assessment for Wetlands of the Lower Mekong Basin for Adaptation Planning: Synthesis Paper on Adaptation of Mekong Wetlands to Climate Change.* Prepared for the Mekong River Commission by International Center for Environment Management, Hanoi.

ICEM. 2014a. *Climate Change Adaptation Methodology.* International Center for Environment Management, Hanoi. http://icem.com.au/portfolio-items/cam-infrastructure/ (accessed 12 March 2017).

ICEM. 2014b. *Mekong Adaptation and Resilience to Climate Change (Mekong ARCC): Main Report.* Prepared for the United States Agency for International Development (USAID) by International Center for Environmental Management, Hanoi.

ICEM. 2014c. *Mekong Adaptation and Resilience to Climate Change (Mekong ARCC): Synthesis Report.* Prepared for USAID by International Center for Environmental Management, Hanoi.

ICEM. 2014d. *Mekong Adaptation and Resilience to Climate Change (Mekong ARCC): Theme Report – Agriculture.* Prepared for USAID by International Center for Environmental Management, Hanoi.

ICEM. 2014e. *Mekong Adaptation and Resilience to Climate Change (Mekong ARCC): Theme Report – Fisheries.* Prepared for USAID by International Center for Environmental Management, Hanoi.

ICEM. 2014f. *Mekong Adaptation and Resilience to Climate Change (Mekong ARCC): Theme Report – NTFPs and CWRs.* Prepared for USAID by International Center for Environmental Management, Hanoi.

ICEM. 2014g. *USAID Mekong ARCC Climate Change Impact and Adaptation Study on Protected Areas.* Prepared for USAID by International Center for Environmental Management, Hanoi.

International Fund for Agricultural Development (IFAD). (2011) *Rural Poverty Report 2011.* IFAD, Rome.

Junk, W. J., P.B. Bayley, and R.E. Sparks. 1989. The flood pulse concept in river-floodplain systems. *Canadian Fisheries and Aquatic Sciences* 106: 110–127.

Just, R.E. and W. Candler. 1985. Production functions and rationality of mixed cropping. *European Review of Agricultural Economics* 12: 207–231.

Keesing, F., L.K. Belden, P. Daszak, A. Dobson, C.D. Harvell, R.D. Holt, P. Hudson, A. Jolles, K.E. Jones, C.E. Mitchell, S.S. Myers, T. Bogich, and R.S. Ostfeld. 2010. Impacts of biodiversity on the emergence and transmission of infectious diseases. *Nature* 468: 647–652.

Keskinen, M., S. Chinvanno, M. Kummu, P. Nuorteva, A. Snidvongs, O. Varis, and K. Vastila. 2010. Climate change and water resources in the Lower Mekong River Basin: putting adaptation into the context. *Journal of Water and Climate Change* 1(2): 103–117.

Koponen, J., D. Lamberts, J. Sarkkula, and A. Inkala. 2010. *Primary and Fish Production Report: DMS – Detailed Modelling Support Project.* Mekong River Commission and SYKE (Finnish Environment Institute) Consultancy Consortium, Vientiane.

Kottelat, M., I.G. Baird, S.O. Kullander, H.H. Ng, L.R. Parenti, W.J. Rainboth, and C. Vidthayanon. 2012. The status and distribution of freshwater fishes of Indo-Burma. In D.J. Allen, K.G. Smith, and W.R.T. Darwall (eds.), *The Status and Distribution of Freshwater Biodiversity in Indo-Burmai.* IUCN, Cambridge and Gland: 36–65.

Luck, G.W., G.C. Daily, and P.R. Ehrlich. 2003. Population diversity and ecosystem services. *Trends in Ecology and Evolution* 18: 331–336.

Lynam, A.J. 2010. Securing a future for wild Indochinese tigers: transforming tiger vacuums into tiger source sites. *Integrative Zoology* 5: 324–334.

MEA. 2005. *Ecosystems and Human Well-being: Synthesis.* Island Press, Washington D.C.

Mekong River Commission (MRC). 2009. *The Flow of the Mekong: Meeting the Needs, Keeping the Balance.* Management Information Booklet No. 2. Mekong River Commission for Sustainable Development, Vientiane.

MRC. 2010. *Basin Development Plan Programme, Phase 2.* Mekong River Commission for Sustainable Development, Vientiane.

Muller, A. 2009. *Benefits of Organic Agriculture as a Climate Change Adaptation and Mitigation Strategy in Developing Countries.* Discussion Paper Series. Environment for Development (EfD), Gothenburg.

National Agriculture and Forestry Research Institute (NAFRL), NUoL, SNV (2007) *Non-timber Forest Products in the Lao PDR: A Manual of 100 Commercial and Traditional Products.* National Agriculture and Forestry Research Institute, Vientiane.

Nekaris, K.A.I., G.V. Blackham, and V. Nijman. 2008. Conservation implications of low encounter rates of five nocturnal primate species (*Nycticebus spp.*) in Asia. *Biodiversity and Conservation* 17: 733–747.

Niggli, U., A. Fließbach, P. Hepperly, and N. Scialabba. 2009. *Low Greenhouse Gas Agriculture: Mitigation and Adaptation Potential of Sustainable Farming Systems.* Food and Agriculture Organization of the United Nations, Rome.

NSW Department of Environment, Climate Change and Water. 2010. *Priorities for Biodiversity Adaptation to Climate Change.* NSW Department of Environment, Climate Change and Water, Sydney. www.environment.nsw.gov.au/resources/biodiversity/1077 1prioritiesbioadaptcc.pdf (accessed 12 March 2017).

Ortiz, R. 1998. Critical role of plant biotechnology for the genetic improvement of food crops: perspectives for the next millennium. *Journal of Biotechnology* 1: 1–8.

Palumbi, S.R. 2001. Humans as the world's greatest evolutionary force. *Science* 293: 1786–1790.

Perrings, C. 1995. Biodiversity conservation as insurance. In T. Swanson (ed.), *The Economics and Ecology of Biodiversity Decline.* Cambridge University Press, Cambridge: 69–77.

Sarkkula, J., J. Koponen, H. Lauri, and M. Virtanen. 2010. *Origin, Fate and Impacts of Mekong Sediments: DMS – Detailed Modelling Support Project*. Mekong River Commission and SYKE (Finnish Environment Institute) Consultancy Consortium, Vientiane.

Schläpfer, F., M. Tucker, and I. Seidl. 2002. Returns from hay cultivation in fertilized low diversity and non-fertilized high diversity grassland. *Environmental and Resource Economics* 21: 89–100.

Smale, M. and A. Drucker. 2007. Agricultural development and the diversity of crop and livestock genetic resources: a review of the economics literature. In A. Kontoleon, U. Pascual, and T. Swanson (eds.), *Biodiversity Economics:Principles, Methods and Applications*. Cambridge University Press, Cambridge: 623–648.

Smale, M., J. Hartell, P.W. Heisey, and B. Senauer. 1998. The contribution of genetic resources and diversity to wheat production in the Punjab of Pakistan. *American Journal of Agricultural Economics* 80: 482–493.

Smale, M., J. Singh, S. Di Falco, and P. Zambrano. 2008. Wheat diversity and productivity in Indian Punjab after the Green Revolution. *Australian Journal of Agricultural and Resource Economics* 52: 419–432.

Sperow, M., M. Eve, and K. Paustian. 2003. Potential soil C sequestration on U.S. agricultural soils. *Climatic Change* 57: 319–339.

Stibig, H.-J., F. Stolle, R. Dennis, and C. Feldkötter. 2007. *Forest Cover Change in Southeast Asia: The Regional Pattern*. European Commission Joint Research Center, Brussels.

Sunderland, T.C.H., J. Sayer, and M. Ha Hoant. 2012. Introduction: evidence-based conservation from the Lower Mekong. In T.C.H. Sunderland, J. Sayer, and M. Ha Hoang (eds.), *Evidence-Based Conservation from the Lower Mekong*. Taylor and Francis, London.

Sunderlin, W.D. 2006. Poverty alleviation through community forestry in Cambodia, Laos, and Vietnam: an assessment of the potential. *Forest Policy and Economics* 8: 386–396.

Taylor, B.W., A.S. Flecker, and R.O. Hall Jr. 2006. Loss of a harvested fish species disrupts carbon flow in a diverse tropical river. *Science* 313: 833–836.

Tilman, D., K. Cassman, P. Matson, R. Naylor, and S. Polasky. 2002. Agricultural sustainability and intensive production practices *Nature* 418: 671–677.

Turner, W. R., K. Brandon, T. M. Brooks, C. Gascon, H.K. Gibbs, K. Lawrence, R.A. Mittermeier, and E.R. Selig. 2013. The potential, realised and essential ecosystem service benefits of biodiversity conservation. In R. Dilys, J. Elliott, C. Sandbrook, and M. Walpole (eds.), *Biodiversity Conservation and Poverty Alleviation: Exploring the Evidence for a Link*. John Wiley & Sons, London: 21–35.

United Nations Environment Program (UNEP) and Thailand Environment Institute (TEI). 2007. *Greater Mekong Environment Outlook 2007*. UNEP and TEI, Bangkok.

Vira, B. and A. Kontoleon. 2013. Dependence of the poor on biodiversity: which poor, what biodiversity? In R. Dilys, J. Elliott, C. Sandbrook, and M. Walpole (eds.), *Biodiversity Conservation and Poverty Alleviation: Exploring the Evidence for a Link*. John Wiley & Sons, London: 52–84.

Widawsky, D. and S. Rozelle. 1998. Varietal diversity and yield variability in Chinese rice production. In M. Smale (ed.), *Farmers, Gene Banks, and Crop Breeding*. Kluwer, Boston: 159–172.

WWF. 2013. *Ecosystems in the Greater Mekong: Past Trends, Current Status, Possible Futures*. WWF International, Gland.

WWF-Germany. 2011. *Rivers for Life: The Case for Conservation Priorities in the Face of Water Infrastructure Development*. WWF, Berlin.

Zhou, G.Y., C.H. Peng, Y.L. Li, S.Z. Liu, Q.M. Zhang, X.L. Tang, J.X. Liu, J.H. Yan, D.Q. Zhang, and G.W. Chu. 2013. A climate change-induced threat to the ecological resilience of a subtropical monsoon evergreen broad-leaved forest. *Global Change Biology* 19(4): 119–1210.

11 Participatory learning and management of biodiversity

The companion modeling approach

Pongchai Dumrongrojwatthana,
Chutapa Kunsook and Kobchai Worrapimphong

Introduction to the companion modeling approach

Participatory approaches have their origin in action research and have been used since the 1930s. These approaches emerged more strongly i response to top-down planning processes in rural development projects (Neef, 2005). Participatory approaches aim at involving diverse types of stakeholders (e.g. local resource users, NGOs, policy-makers, etc.) in all stages of research and/or development projects for shared learning and collective decision-making (Reed, 2008). In past decades, computer modeling and simulation have been integrated into participatory approaches because of their capacity for exploring scenarios that are difficult to test in reality. Moreover, they have been used to support collective learning and decision-making (Epstein, 2008).

In this chapter, a participatory modeling approach called "companion modeling" (ComMod) that is used for biodiversity in Thailand is presented. The approach was gradually built from a bundle of theories and ways to look at the collective management of renewable resources, including the science of complexity (Holling, 2001), resilience and adaptive management (Carpenter *et al.*, 2001; Walker *et al.*, 2002), constructivist epistemology (Kriz, 2010), post-normal science (Funtowicz and Ravetz, 1993), and patrimonial mediation (Ollagnon, 1991). The description of such key concepts linked to ComMod is presented in Trébuil (2008).

ComMod's main principles are to develop simulation models integrating various stakeholders' points of view and to facilitate dialogue and shared learning, and support negotiation and collective decision-making (Barreteau, 2003). Because most of the current problems in social-ecological systems are complex, rapidly evolving, and need to be addressed in more uncertain and less predictable environments, the main objectives of a ComMod process are:

- to better understand a complex agro-ecosystem through the collaborative construction and joint use of different types of simulation models integrating various stakeholders' points of view; and
- to use these models within communication platforms for collective learning to facilitate multiple stakeholders' coordination and negotiation processes, leading to the definition of agreed-upon collective action plans.

There are three iterative and evolving phases alternating between field and laboratory (Barnaud et al., 2006a) to be implemented with the stakeholders:

- initial diagnosis;
- sharing, adjustment and improvement of knowledge and perceptions on the problem through co-construction of a conceptual model; and
- collective discussions to generate acceptable scenarios to be tested in gaming and simulations, as well as collective assessment of these scenarios and decision-making on further action to be taken.

During the ComMod process, several complementary tools are implemented under the multi-agents system (MAS) concept, such as agent-based model (ABM), role-playing game (RPG) and geographic information system (GIS), to build a visual communication platform based on different kinds of knowledge (scientific and indigenous), and contributions from different (social and ecological) disciplines. The implementation of ComMod does not rely on a fixed tool; it may vary from case to case (Le Page et al., 2014).

Several studies have been conducted using the ComMod approach in Thailand (Barnaud et al., 2006b; Leteurtre, 2010; Vejpas et al., 2005). In this chapter, three case studies related to biodiversity management in Thailand are presented in order to demonstrate how ComMod processes are implemented to achieve the two main objectives: co-learning and setting up a biodiversity management action plan among stakeholders. We begin with a brief introduction of the study sites and biodiversity issues. Thereafter, the ComMod processes implemented at each site are described. The main findings are then presented, followed by a discussion of the limitations of the process and suggestions for how to improve the process in the future.

Case studies on biodiversity management: sites and contexts

The three cases deal with different levels of biodiversity management, including single species, multiple species, and ecosystem diversity. These cases are situated in three different regions of Thailand (Figure 11.1). Two cases are in coastal areas and the third is in a highland area. Summaries of the sites and contexts of the three cases are presented in Table 11.1.

This first case focuses on the collective management of single species, namely razor clam *Solen regularis*, in the Don Hoi Lord coastal area, Samut Songkhram Province, central Thailand. Don Hoi Lord is well known as the largest habitat for this clam in Thailand and was registered as the 1,099th Ramsar site in 2001. It has been an important site for razor clam fishery for local fishermen for more than eighty years (see harvesting details in Worrapimphong, 2010).

Don Hoi Lord (DHL) is nowadays a popular tourist destination because it is located not far from Bangkok. Numerous tourists visit the area, especially at weekends. They are not only sightseeing, but come to enjoy the seafood, particularly razor clam dishes. Scientific research into the razor clam population has shown that there were 65.51 clams/m^2 in 1989. By 2009, this had fallen to just

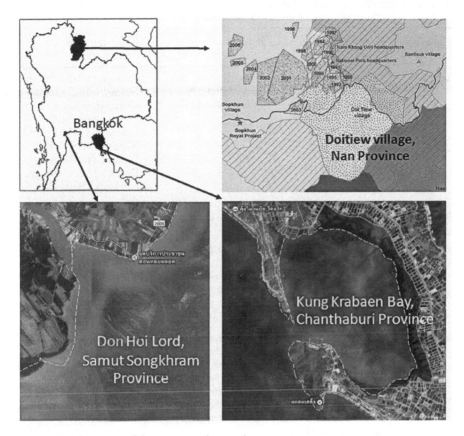

Figure 11.1 Location of three case studies in three provinces

Source: Modified from www.map.google.co.th.

0.51 clams/m² (Worrapimphong, 2010). Due to high demand from the market and heavier harvesting, sustainable management of this species is therefore urgently required.

Razor clam management at the Don Hoi Lord Ramsar site

Based on the literature, research into razor clams has been conducted since the. This research has focused on life history, environmental conditions, and the importance of razor clams for the local community. Although there have been many studies, razor clam management is still a topic of considerable debate. This is because all the studies were oriented to either a biological or a social perspective, with none of them offering integrative and participatory management approaches to improve management options. Hence the integrated study using the ComMod approach was carried out in this area. The goal of the study was to

Table 11.1 Summary of case studies using companion modeling approach for participatory management of biodiversity

Case study	Razor clam management	Bycatch management	Forest–farmland management
Location	Don Hoi Lord Ramsar site, Samut Songkhram Province, central Thailand	Kung Krabaen Bay, Chanthaburi Province, eastern Thailand	Ambiguous land among Doitiew village, Nam Khang Headwater Development and Conservation Unit, and Nanthaburi National Park, in Thawangpha District, Nan Province, northern Thailand
Biodiversity level of concern	Species diversity (single species)	Species diversity (multiple species)	Ecosystem diversity (farmland, grazing land, plantation, and forest)
Key issue	Decrease in razor clam population	Biodiversity due to lack of bycatch management from blue swimming crab fishery	Conflict on land use between Hmong farmers, especially herders, and two forest management agencies
Area	2,400 ha	640 ha	1,600 ha
Ecological constraints	Sand bar affected by many bio-physical factors Tidal effects on razor clam harvesting period More removal of razor clam from sand bar due to high market demand	More removal of bycatch due to the blue swimming crab fishery causing biodiversity degradation Seagrass bed and habitat degradation	Steep area Limited land for agriculture and cattle rearing Limited land for reforestation
Socio-economic constraints	More tourists visiting area More fishermen from outside area harvesting razor clams Lack of conservation supports from the local government	External force from international law to manage bycatch during blue swimming crab harvesting Local fishermen have limited information on this external force Local government did not communicate well with local people to help them adapt to the new regulation	High population growth More agricultural land required by local farmers Social tension between Hmong farmers and local government foresters, lack of trust, and limited dialogue due to different perceptions on the effects of cattle rearing and reforestation on forest regeneration
State of current research	Many scientific studies on razor clams and some socio-economic studies since the 1980s No integrated study	Many scientific studies on blue swimming crab Limited scientific and socio-economic studies on bycatch	No research carried out in this area

facilitate the development of suitable management plans for razor clam fishery (Worrapimphong, 2005; Worrapimphong et al., 2010).

Co-learning on bycatch species at Kung Krabaen Bay

The second case deals with co-learning on the diversity of bycatch species and their importance in the ecosystem in relation to the blue swimming crab *Portunus pelagicus* fishery in Kung Krabaen Bay (KKB), Chanthaburi Province, eastern Thailand. Before the start of the ComMod process on the bycatch issue, there was a ComMod process focusing on co-learning about the blue swimming crab's biology and sustainable management (Leteurtre, 2010). This initial ComMod process focused on this issue because the area faced a problem with respect to the reduction of size and yield of blue swimming crabs due to many factors, such as overharvesting of the ovigerous female crabs by local fishermen, habitat and nursing ground destruction, and inefficiency of crab management.

Numerous scientific studies on the blue swimming crab have been conducted by the Kung Krabaen Bay Royal Development Study Center: KKBRDSC (2014) and some independent researchers (Bhatrasataponkul et al., 2008; Kunsook et al., 2014). However, the crab situation in KKB is now in crisis, which has resulted in social tension, particularly between two groups of rival fishermen from two villages – Klongklud and Thaclang. Most of the fishermen from Thaclang harvest blue swimming crabs outside the bay, while the group from Klongklud harvests inside the bay. Each group blames the other for causing the decline in the crab population. This led to the implementation of the first ComMod process. An RPG called "IdeasFishery" was used as a tool. By placing locals from the two villages into collective learning and action through the simulation, the fishermen developed better self-organization ability. Some of them started taking about using crab banks at the household level to help conserve this crab species (Leteurtre, 2010). After the workshop, crab banks were established at both Klongklud and Thaclang, and Kunsook et al. are conducting ongoing scientific studies.

Currently, the area is facing a new biodiversity management issue. An international regulation on crab fishery means that the fishermen have had to adapt their fishing technique to conserve bycatch species. If fishermen cannot show that they are using good gear to reduce the number of harvested bycatch or have implemented strategies to conserve bycatch numbers, they might not be permitted to sell crabs to international markets. This led to the implementation of the second cycle in 2013, which focuses on the bycatch management issue.

Forest–farmland management in a Hmong highland village

The last case relates to conflict management between foresters and Hmong farmers on the use of forest–farmland in the highlands of Thawangpha District, Nan Province, northern Thailand (Dumrongrojwatthana, 2010). In general, deforestation and forest degradation are important problems in northern Thailand.

To solve such problems, top-down management practices have been implemented by the government, such as the establishment of forest and wildlife conservation areas. As a result, thousands of people, especially upland farmers, are occupying these areas illegally and condemned as forest encroachers (Roth, 2004). This created a conflict between government agencies and local communities due to their different perceptions and objectives in managing the land resources.

In this case, the conflict was heightened with the establishment of the Nam Khang Headwater Development and Conservation Unit (NKU) in 1990 and the Nanthaburi National Park (NNP) in 1996. These two government projects played important roles in the reforestation of degraded land and forest protection, respectively. On the other hand, since 1961, Hmong farmers have been practicing cropping and cattle-rearing (free-ranging) activities in the Doitiew (DT) area. In 2006, the park established its boundaries and started strict enforcement of the ban on farming activity and hunting within the park. Moreover, since then, the NKU has expanded its reforestation efforts into local farmers' fallows. Unsurprisingly, this resulted in social tension between farmers and foresters, particularly on land use for cattle grazing or forest plantation, due to contrasting perceptions. From the farmers' point of view, cattle rearing benefits biological conservation efforts as it prevents forest fires and accelerates forest regeneration. Meanwhile, the NKU and NNP argue that cattle rearing causes man-made forest fires and damages plant seedlings and saplings, especially in reforestation areas, and so degrades biodiversity.

A collaborative management process was eventually implemented to mitigate this land use conflict. The objectives of this participatory process were to facilitate dialogue between the farmers and the foresters, and to set up a collective landscape management plan.

ComMod process and participatory field workshop at each site

As is typical in ComMod processes, all three cases comprised iterative and evolving phases, including field data gathering, model implementation, and participatory simulation. At the beginning of the process, a scientific study was conducted at each site to improve understanding of researchers on the resource status or situation. In parallel, as ComMod takes into account a diversity of interests, interviewing of concerned stakeholders was carried out to understand their perceptions regarding the issue and their decision-making strategies in relation to the issue. This step is useful to identify key stakeholders that will be involved in the process (Barnaud et al., 2008). Thereafter, all scientific and socio-economic knowledge was used to create a conceptual model representing the system of study, which was then developed into a modeling tool based on the multi-agent systems concept, as mentioned earlier (Le Page et al., 2014). This tool was then co-improved with the active participation of stakeholders through gaming and simulation field workshops. Scenarios that were predefined by researchers or proposed by stakeholders were explored through the gaming sessions. Selected monitoring indicators were prepared for stakeholders to follow and discuss the results of different scenarios.

Table 11.2 summarizes the key characteristics of the ComMod process at each site. Figures 11.2 to 11.4 show the different steps of the ComMod process in the three cases, the gaming features, and their evolution by co-improvement with local stakeholders. Short descriptions of the various processes are presented below.

The ComMod process at Don Hoi Lord

The ComMod process was initiated at DHL in 2004. First field studies on razor clam population were conducted in 2004–2005. At the same time fishermen were interviewed to understand their decision-making process in relation to harvesting and selling razor clams. The data from field studies and interviews were used to build a first version of the ABM using the CORMAS (Common-Pool Resource and Multi-Agents System; see www.cormas.cirad.fr) simulation platform.

This ABM was used as game board and communication interface (Figure 11.2) with local stakeholders in the first field workshop in 2005. During that first workshop, four scenarios were proposed and explored with eleven fishermen from one village:

  i. baseline (freely harvesting among four zones, no regulation);
 ii. closing access to harvesting zone for three months per zone;
iii. permanent closure of one harvesting zone; and
 iv. closing access to harvesting zone for one year per zone.

The last three scenarios emerged from the discussion. All participants ultimately agreed with scenario ii because it provided good results in terms of razor clam population and seemed the most practical option.

However, some players and a government officer from the Tambon (District) Administrative Organization (TAO) who participated as an observer suggested conducting another workshop and inviting more players from another village, who were also harvesting razor clams in DHL. This led to the second field workshop, which was conducted with a total of ten fishermen from two villages and one razor clam trader. The activities and findings from the previous workshop were presented to the newcomers, followed by discussion on the additional scenarios to be tested. Finally, four scenarios were played in this workshop:

   v. baseline;
  vi. closing access to harvesting zone for three months per zone;
 vii. individual harvesting quotas (maximum of 3kg/harvester/day); and
viii. maximum harvesting effort.

At the end of the workshop, the fishermen agreed on scenario vii, but they requested a guarantee on razor clam price because the simulation suggested it was

Table 11.2 Characteristics of companion modeling processes at three sites

Case study	Razor clam management	Bycatch management	Cattle rearing and forest plantation management
Initial diagnostic	Razor clam population was studied using line transect technique Local fishermen, traders, and local government agencies were interviewed	Bycatch diversity data was collected from fishermen who use 2 fishing gear types – collapsible crab trap and crab gill net Local fishermen were interviewed	Effect of cattle grazing and reforestation on regeneration were investigated Hmong farmers, cattle traders, foresters, and livestock developers were interviewed
Conceptual modeling	A first version of the agent-based model was built and co-improved with local fishermen through small meetings at their houses	A first version of role-playing game (RPG) was produced by researchers and co-improved with fishermen during the gaming session	A first version of vegetation state transition diagram was created by researchers based on scientific evidence This was then co-improved with herders and foresters and used to build a first version of agent-base model
Participatory simulation through gaming and simulation in field workshops (WS) and main objective	Several informal workshops at fishermen's houses to co-improve and learn the ABM Two field workshops: 1st WS: To facilitate knowledge sharing and collective discussion on razor clam management 2nd WS: To present the results from the 1st WS to new stakeholders and facilitate collective discussion on razor clam management	One field workshop	Two informal sessions to validate vegetation state transition diagram and sensitize local stakeholders to the gaming and simulation tools Three field workshops: 1st WS: To stimulate communication, collective learning, and sharing of knowledge and perceptions between herders and foresters, and to improve trust among all stakeholders 2nd WS: To facilitate the co-management of the land and set up a joint action plan among stakeholders. 3rd WS: To facilitate learning about agricultural land and grazing land management among herders

Participants	1st WS: 11 fishermen (1 village) and 3 observers from Tambon (District) Administrative Organization (TAO) officer 2nd WS: 10 fishermen (2 villages), 1 trader, and 5 observers (4 TAO officers, and 1 fishery officer)	12 fishermen	1st WS: 14 herders and 3 Nam Khang Headwater Development and Conservation Unit (NKU) foresters 2nd WS: 5 herders, 2 NKU foresters, 3 National Park foresters, and 1 District Livestock Development officer 3rd WS: 14 herders (5 newcomers)
Facilitator(s) and assistants	1st WS: 2 researchers, 3 assistants 2nd WS: 1 researcher, 1 assistant	2 researchers and 8 assistants	1st WS: 3 researchers, 7 assistants 2nd WS: 1 researcher, 6 assistants 3rd WS: 1 researcher
Scenario(s) proposed in field workshop	1st WS: S-i: baseline (freely harvesting among 4 zones, no regulation) S-ii: closing access to harvesting zone 3 months per zone S-iii: permanent closure of 1 harvesting zone S-iv: closing access to harvesting zone 1 year per zone 2nd WS: S-v: baseline (freely harvesting among 4 zones, no regulation). S-vi: closing access to harvesting zone 3 months per zone S-vii: individual harvesting quotas (maximum of 3kg/harvester/day) S-viii: maximum harvesting effort	Baseline (free harvesting in different areas based on own fishing gear)	1st WS: S-i: 2 groups of herders manage cattle; no reforestation plots S-ii: 2 groups of herders manage cattle with reforestation plots of different ages established in the landscape by researchers S-iii (computer simulation): demonstration of vegetation dynamics with reforestation plots and without cattle in the landscape S-iv: herders and foresters manage a common landscape, negotiation is allowed, and different-age reforestation plots established in landscape by foresters. 2nd WS (on ruzi pasture technique): S-v/S-vii: herders manage cattle individually/collectively 3rd WS (requested by farmers): S-vii/S-viii: farmers manage cattle and upland rice land individually/collectively

(continued)

Table 11.2 Characteristics of companion modeling processes at three sites (*continued*)

Case study	Razor clam management	Bycatch management	Cattle rearing and forest plantation management
Debriefing/ analysis of the gaming results	Conduct at the end of each scenario by observing many indicators, such as numbers of razor clam and fishermen's income Plenary discussion at the end of WS to seek future management options	Compare results on numbers of bycatch and blue swimming crab, and harvesting trend Plenary discussion on the role of bycatch in ecosystem, future risks if fishermen do not manage bycatch, and identify best technique to manage bycatch	Conduct at the end of each scenario by observing many indicators, such as forest area, number of cattle, cattle status (fat, normal, thin), and herders' incomes Plenary discussion at the end of WS to seek future management options

likely to decrease. Later, all participants discussed the results of the workshop with the Provincial Governor and he agreed to support their decision.

A few months later, unfortunately, this process was terminated by the replacement of the Provincial Governor. However, the local fishermen are still working with researchers to monitor the razor clam population and co-improving the spatial representation of the razor clam harvesting areas in the ABM (Figure 11.2). In 2008–2009, razor clam data were collected again and used to improve the biological module in the model. Finally, a fully autonomous ABM representing fishermen's harvesting behavior was produced and used for effective scenario explorations in the laboratory and with local stakeholders (Worrapimphong, 2010). The key output of this DHL process has been the self-organization of local fishermen to manage the razor clam population in this area.

The ComMod process at Kung Krabaen Bay

The ComMod process on bycatch in KKB started relatively recently, and at the time of writing there had been only one field workshop, in March 2014. Therefore, this section will focus primarily on describing the RPG.

At the beginning of the process, data on bycatch diversity were collected from fishermen who use two fishing gear types – the collapsible crab trap and the crab gill net – in different zones both within and outside the bay (e.g. seagrass bed, mangrove, and open sea). Data were collected every month for one year. Species and number of individuals from each species were recorded. Relative abundance and species richness were then calculated. The results showed that different zones had different bycatch opportunities in terms of both species and numbers.

Thereafter, a set of game cards composed of bycatch pictures and names as well as blue swimming crab pictures was prepared (Figure 11.3). Twelve fishermen from Thaclang village attended the field workshop at the village meeting center. Participants were selected on the basis of the type of fishing gear they used – some used only trap or net, whereas others used both gears.

Before starting the workshop, participants were separated into six groups representing six households. Each group received fishing gear assigned by the researchers, as follows: 400 traps; 600 traps; three groups of eight nets; and eight nets plus 500 traps. Before starting the game, participants were asked to create the game board by drawing the KKB boundary with seagrass patches and mangrove area on an A0 sheet of paper (Figure 11.3). This helped the participants to relax and understand the gaming aspect of the workshop collectively. Once the game board had been completed, the gaming session began.

In one round, participants had to designate their harvesting locations and put the cards representing their fishing gear on the board. After all the players had finished, the next step was to draw chance cards of harvested blue swimming crabs and bycatches (species and number of individuals were predefined on the basis of field data related to fishing zones). Next, they were asked to separate the picked cards into blue swimming crab and bycatch cards. They then had to

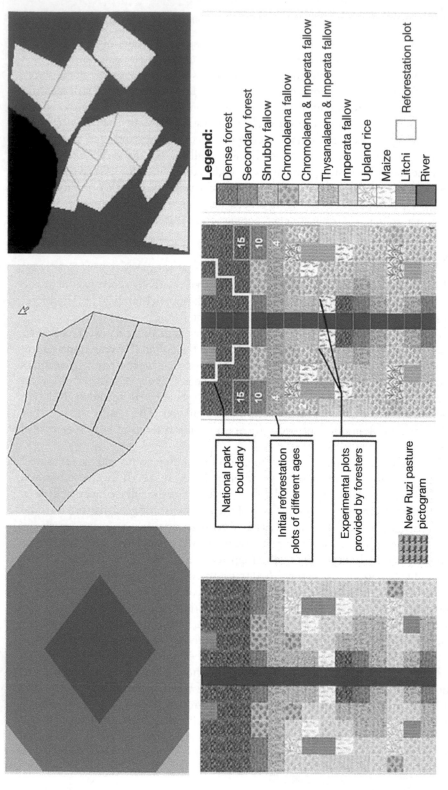

Legend:

- Dense forest
- Secondary forest
- Shrubby fallow
- Chromolaena fallow
- Chromolaena & Imperata fallow
- Thysanalaena & Imperata fallow
- Imperata fallow
- Upland rice
- Maize
- Litchi
- River
- Reforestation plot

National park boundary

Initial reforestation plots of different ages

Experimental plots provided by foresters

New Ruzi pasture pictogram

Figure 11.2 Examples of spatial representation of the system and its evolution after co-improvement with stakeholders. Top row: Don Hoi Lord case. Bottom row: Doitiew case

Figure 11.3 Examples of game cards (left), materials (middle), and spatial game board collectively drawn by fishermen with locations of fishing gear (right) for collective learning on bycatch management

continue to separate the bycatch cards into "for sale/consumption" and "throw away." These cards were posted separately on two feature boards. Finally, the participants "sold" their blue swimming crab cards to market traders and recorded the resulting income on record sheets. If any ovigerous females were harvested, these could be sold or released into the crab bank.

Four rounds were played with free choice of harvesting locations among players. At the end of the fourth round, a debriefing session was conducted with collective presentation of the bycatch boards. Diversity and number of individuals of the "for sale/consumption" and "throw away" bycatches were compared, as well as the number of blue swimming crabs. This allowed self-learning among the players. Then a group discussion touched on many topics, including the game's similarity to reality, the role and importance of bycatch in the ecosystem, future threats to the blue swimming crab fishery in the KKB area, and possible solutions for the bycatch issue. More participatory gaming and simulation activities will be conducted in this area.

The ComMod process in the Doitiew highlands

At the beginning of the ComMod process at DT in 2007, field studies on the effects of cattle raising and reforestation on forest regeneration were conducted. A total of thirty-two vegetation sampling plots were established across land use types and cattle-grazing intensities. Vegetation biomass in tree, sapling, seedling, grass, and herb layers were examined to represent forest succession (more succession equals more biomass, especially in sapling and tree layers). The results showed that cattle grazing in this area facilitated forest regeneration but at a slower rate than reforestation activity.

This conclusion was difficult to present to farmers and foresters because it proved that both groups' perceptions were correct. This led to the creation of a vegetation dynamic model integrated with reforestation and cattle-rearing activities. A vegetation state transition diagram was co-constructed with farmers and foresters, which in turn was used to build a computer-assisted RPG (cRPG). Prior to that, concerned stakeholders (i.e. farmers, NKU and NNP foresters, and

District Livestock Department (DLD) officers) were interviewed. Farmers were classified into four types: Type A have no cattle; Type B have two to fifteen heads of cattle and supplement their income from agriculture; Type C have similar herd sizes to Type B but supplement their income from small-scale trading in the village; and Type D have the largest herd sizes (up to forty heads) and generate most of their income from cattle. This socio-economic information was used to select participants to attend the field workshops.

Three gaming and simulation field workshops were conducted with diverse groups of stakeholders. The first was conducted over two days. Hmong farmers' understanding of the gaming features was tested on the first day, while the second day was devoted to discusssion of possible management plans among farmers and foresters from NKU. A new idea for landscape management emerged during the debriefing and plenary discussion sessions. This led to the modification of the cRPG to test the new management technique with more groups of stakeholders (NNP and DLD) in the second workshop. Once the second workshop had concluded, farmers requested access to this cRPG to learn about cattle and farmland management at the village level. Therefore, researchers modified the cRPG again. The role of computers was increased from the first version to the third version in the interests of saving time in future workshops. In the first workshop players drew their paddock boundaries on sheets of paper, while in the third workshop they played the game on computers (Figure 11.4). Moreover, the tool's purpose changed according to the requests of users (stakeholders); it was not dictated by the research team. A full description of the participatory process can be found in Dumrongrojwatthana and Trébuil (2011).

Lessons learned from the three cases

Gaming and simulation tools in the form of RPGs were used in all three cases as these are proven to be effective tools for learning (Crookall and Thorngate, 2009). In all three case studies, scientific knowledge and local knowledge on management were integrated through gaming and simulation tools within the iterative and evolving process. Participants in all cases used such tools for individual and co-learning, while new management action plans were discussed among diverse groups of stakeholders (Figure 11.5). The participants had no difficulty understanding or following the games, even when they had little formal

Figure 11.4 Evolution of computer-assisted tools used in the Doitiew case – from more human control (left) to more computer control (right)

Figure 11.5 Individual and collective learning through gaming and simulation sessions and discussion/negotiation among stakeholders

education, as in the case of the Hmong farmers. This was because all of the key elements and dynamic interactions were presented and linked to their daily lives.

The key lessons learned from the three cases are presented below.

Participants' individual and co-learning

In all cases, the ComMod activities stimulated the participants' awareness and adjusted their perceptions of the issue at stake. Such awareness is very important when persuading people to join a collective learning process (Röling and Wagemaker, 1998).

At the individual learning level, participants (players) learned by observing others' behavior during the gaming sessions and by observing the global trends of the system from the exploration of scenario(s). In the DT case, different types of farmers learned how to manage their herds by observing the decisions on the game board. They also discussed how to improve cattle breeding with each other. Moreover, they realized from the first field workshop that if they continued to practice their traditional cattle-rearing system while reforestation activities became increasingly constraining then the quality of their animals would decline in parallel with forest regeneration.

In the KKB case, participants admitted they had never realized that there were so many bycatch species, nor that they harvested three or four times as many bycatch species as blue swimming crabs. Another individual learning was linked to the game materials. By writing the name of each bycatch species on the game card, some participants acknowledged that they just knew their names. They also asked to keep the cards that were used in the game in order to educate their children at home.

In the DHL case, the ABM with modified spatial representation was used with small groups of fishermen in their own homes for individual learning on razor clam population dynamics. Moreover, each fisherman increased his awareness regarding the razor clam degradation issue. At the end of the process, this awareness was transformed into motivation and action.

Regarding co-learning, it was observed that each RPG played an important role as a learning and communication platform. By putting stakeholders together in different scenarios, they played the game with their own real experiences and observed the possible trends of the system collectively. They also understood that the results of the simulations were generated by their collective action. In all cases, participants said that they had never previously seen the system in its entirety, as they did in the game. This stimulated them to discuss the problem and possible solutions collectively.

In the DT case, during a gaming session in the first workshop, foresters were given thirty minutes to argue for the reforestation of a plot of land. At the end of this presentation, the farmers provided the foresters with a portion of land under the condition that they should be able to locate their cattle in this reforestation plot after three years of planting. This proved that the RPG helped both sides to learn from each other. At the end of the gaming sessions, all participants saw the future difficulty of a landscape covered by trees, and all agreed that the existing cattle-rearing system in the village had to be improved, so they started to think collectively on how this could be achieved. Another collective learning outcome in the DT case was a new way of managing rice growing in the village's territory. This was observed in the third workshop, when a group of players discussed and decided to pool their upland rice plots. They explained that this technique would help them to manage paddock and crop areas, as well as cattle rearing, more easily (Dumrongrojwatthana *et al.*, 2015).

In the KKB case, participants had a chance to draw a diagram representing a food web of the system. Through this, they collectively learned the important role of bycatch, especially as food sources for both blue swimming crabs and humans. Moreover, by sharing information on the impact of international fishery regulations with researchers, all of them agreed that they had to manage bycatch. They also wanted fishermen from nearby villages to learn about this bycatch issue.

Collective identification of scenarios to be explored and new management strategies

The ComMod processes in all three cases allowed participants to improve their understanding of the complexity of their respective systems and hear alternative viewpoints, but they also discussed and negotiated collectively to seek new management techniques. Scenario explorations and plenary discussions were important steps in this process. Baseline (current situation) and proposed scenarios from participants were explored in the field workshops. Sets of bio-physical and socio-economic indicators were prepared for different groups of stakeholders to observe system dynamics under changing conditions. At the end of each scenario, there was a short debriefing session when players discussed the results and exchanged opinions.

In the DHL case, there was collective identification in the first and second workshops, as described earlier. At the end of the second workshop, participants agreed upon a harvesting quota and guaranteed razor clam price.

In the DT case, new forest–farmland management techniques emerged from the discussion session in the first workshop. Participants came up with a new cattle management technique by agreeing to use "ruzi" (*Brachiaria ruziziensis*) pasture, with proposed land from the NKU. Similar to the DHL case, some participants suggested testing this new technique by inviting more stakeholders (i.e. National Park officers and livestock developers) to participate in the game before implementing it for real. In the second workshop, it was observed that the farmers were interested in establishing pasture in the virtual landscape. This led to the co-design of an action plan during the plenary discussion. Central to this plan, NKU agreed to provide a 10-hectare pilot plot of ruzi pasture, with DLD providing the ruzi seeds. NKU also agreed to provide funds for a fence and heavy machinery to prepare the land. The herders will provide the labor. Finally, cattle from two herders will be available for grazing this pasture when it is ready.

In the KKB case, although only a baseline scenario was played, the participants exchanged some ideas on bycatch management, such as spending more time during the blue swimming crab harvesting period removing bycatch from their fishing gear and releasing them back into the bay. However, some fishermen, especially gill net users, mentioned that murex sea snails (*Murex spp.*) are very difficult to remove from their nets. Indeed, the only practical way to do this is back at home, by which time most of the snails have died (and even then the nets usually have to be cut). The fishermen therefore asked the researchers to investigate potential practical uses for the snails' shells.

From gaming and simulation to implementation in reality

It is very important to turn a collective plan for natural resource management into concrete action, but this process is rarely easy. For instance, the new management strategies that emerged during the gaming sessions in the DHL and DT cases started to be implemented in real life. However, these implementations then stalled due to political factors at each site.

In the DHL case, a few months after the second workshop, participating stakeholders had a chance to present the collective agreement to the Provincial Governor. At that time, in 2006, he expressed an interest in implementing the suggestion to set up an individual harvesting quota for razor clams and to guarantee the buying price through the establishment of a provincial fund. However, a few months later, he was removed from office. As a result, implementation of the plan ceased because his successor was more interested in promoting DHL as a popular tourist destination, rather than conserving razor clams.

A similar situation developed in the DT case. The head of NKU who participated in the first two years of the process was moved to a new position, and his replacement was determined to stick rigidly to the government's line on forest management in headwater areas. As a result, no concrete action has been implemented.

There are no obvious solutions to these problems because they are both linked to personal perception and self-interest. However, one possible course of action is to integrate the new high-level decision-maker within the existing learning process. Another possibility is to use the ABM to create a new scenario that integrates the interests of the newcomers. However, such a solution must continue to recognize the original stakeholders' interests, too.

Self-organization on management and monitoring activity

Self-organization is very important for common-pool resource management, especially biodiversity (Ostrŏm, 2009). Among the three cases, the DHL case very successful in this respect. Its self-organization process took place in 2009 when there was a dramatic collapse in the razor clam population that had a significant detrimental effect on the local fishery system. Researchers were invited to a meeting with representatives of the Fishery Office of Samut Songkhram Province, the Department of Marine and Coastal Resource (DMCR), NGOs, and local fishermen to present the results of the ComMod process, which had been implemented since 2004. With financial support from the DMCR, a razor clam conservation group – composed of twenty-eight members – was instituted at DHL in June 2009. The local ComMod facilitator was selected to lead the group. After several informal meetings, the conservation group proposed a reserve area on a sand bar, in line with what they had learned and agreed during the scenario explorations. They asked researchers to provide technical support in the form of a GPS (Global Positioning System) survey and a map. Initially, a 3.5-hectare conservation area, co-created by the group, was established in September 2009. The razor clam population was then monitored by members of the group. Two years later, razor clam numbers were recovering, so the group agreed to expand the conservation area to 16.4 hectares and to continue to monitor the clam population.

Advantages of using ComMod in biodiversity management

Biodiversity can be considered as a common-pool resource issue dealing with multiple stakeholders (e.g. local, regional, and government agents) across spatial (e.g. small habitat/farm to landscape) levels and temporal (e.g. days to decades) scales. Therefore, biodiversity problems (e.g. biodiversity loss and degradation) are very complex and need a participatory and systemic approach to identify possible solutions. As presented in the three case studies, ComMod is suitable for use in biodiversity management research, especially in support of participatory learning on the importance of biodiversity in an ecosystem and collective decision-making among diverse types of stakeholders to conserve biodiversity collectively.

The tools used – RPG, ABM, or both – are flexible and easy for stakeholders to adjust, perceive, and understand because participants are engaged in the co-design and co-improvement of such tools throughout the process. Moreover, this increases the sense of ownership and the transparency of the model (Barreteau et al., 2001). Differences in the level of formal education were not serious obstacles to the use of these tools; indeed, the non-threatening gaming environment encouraged participants to express their opinions. The interactive and visual features of the simulation tools also minimized problems relating to participants who lacked the confidence to express their opinions in public and local language barriers.

The ComMod process supports knowledge integration from different sources. During the progressive development of the models, comments and requests from local stakeholders were taken into account and guided the modifications of the simulation tools in order to tailor them to the end users' changing needs. Along the way, this process allowed the integration into the models of "empirical/indigenous" knowledge (from fishermen and herders), "technical and institutional" knowledge (from NKU, NNP, DLD, DMCR, and TAO officials), and "scientific" knowledge (from the researchers).

Finally, the series of models developed during the process can be used as simulation tools to improve communication, envision the system, support co-learning on the interactions between stakeholders with respect to the apparent conflict, and resolve that conflict. Several biodiversity management scenarios proposed by stakeholders allow them to envision the future possibilities of their resources (Dionnet et al., 2008). The consequences (positive or negative) of their different decisions with respect to a common-pool landscape are represented on a simple but dynamic game board with a set of biological and socio-economic indicators. This visual representation is important for collective learning in a participatory modeling process (Horlitz, 2007). Stakeholders can observe and learn the effects of their decisions and they are stimulated to reflect on how to improve their management of the system.

Limitations of the process

Although all three cases enjoyed good outputs and outcomes, some limitations need to be pointed out.

Duration of the process

The ComMod approach is trans-disciplinary. It involves different kinds of expertise working together and takes into account the different points of view of local stakeholders to create a common representation of the system that is accepted by all of them. This process took almost two years in the DHL and DT cases before stakeholders could identify an acceptable solution. This limitation can be seen in many development projects because researchers have to build trust with all stakeholders before involving them in the collective process. The ComMod process could be accelerated through the use of existing field data.

Ability of participants to join the collective activities

In all three cases, working with different groups of stakeholders proved difficult. Farmers spend most of the day on the land, so they could attend collective activities only in the evening, whereas government staff preferred to attend workshops during office hours. Therefore, the process facilitator had to be flexible and open to modifying the schedule, including conducting the gaming and simulation sessions in the evening, if necessary.

Numbers of participants

The number of participants in each workshop was limited in order to maintain a high level of information exchange. In many ComMod cases, the number of participants in a participatory workshop is limited to ten–thirty people, not including the research team. Then, after the gaming session or at some later point in the process, new participants are often identified by local stakeholders. Therefore, process facilitators have to strike a balance between including the new stakeholders and continuing to accommodate the original participants. Regarding this limitation, out- and up-scaling are issues for all ComMod practitioners.

The need for more user-friendly tools

Local stakeholders in the KKB case found the simple game cards straightforward to use, while participants in the DHL and DT cases reported that the computer-assisted games were easy to understand. During the process, ABM simulators also worked with stakeholders to improve processing speeds in the gaming and simulation sessions. However, the models produced in the DHL and DT case studies were difficult for local users to handle by themselves, without assistance from researchers. Therefore, more user-friendly versions should be developed if they are to continue the sharing and joint learning process on their own.

Conclusion and ongoing research

The implementation of a ComMod process in three biodiversity management projects proved to be suitable and useful for achieving co-learning and problem mitigation in interactive fashion. Both the researchers and the local stakeholders benefited from the research through collaborative knowledge sharing during the iterative and progressively developing modeling process. Moreover, the ComMod process delivered a series of models that were co-developed and co-improved with stakeholders. These models were used to explore several scenarios with stake-holders and assess their decisions. They also played important roles as discussion platforms and for decision support on biodiversity management plans prior to implementation. Like several other participatory approaches, ComMod has both advantages and limitations that practitioners should take into account.

More studies relating to biodiversity are ongoing, such as adaptive management of the Melaleuca forest in southern Thailand, where diverse groups of local people are involved in the collection of timber and non-timber products, fishing and apiculture. Nowadays, the encroachment of oil palm plantations is having a negative impact on biodiversity in this forest. Moreover, the negative effects of climate change are already observable in the area, including more ground fires and thunderstorm events, both of which have damaged many trees. Therefore, ComMod research has been implemented to facilitate collective management among concerned stakeholders.

Two more ComMod processes are under way in northern Thailand: collaborative management between local farmers and a wildlife sanctuary on green peafowl (*Pavo muticus*), an endangered species in Payao Province; and participatory management of community forests in seven villages in Lainan Sub-district, Nan Province. Meanwhile, in east Kalimantan, Indonesia, there is now participatory management of the karst ecosystem, where many human activities – agricultural expansion (oil palm and para rubber plantations), legal and illegal logging, and hunting – have had a negative impact on fauna and flora.

Research is needed into many more aspects of biodiversity management, such biodiversity management in urban areas, climate change adaptation, food security, improvement of conservation and management networks, out- and up-scaling, etc. Indeed, this chapter may have provided insufficient information on the ComMod approach. Therefore, we point readers in the direction of two useful websites:

- Case studies relating to biodiversity conservation and management around the world can be found at: http://cormas.cirad.fr/ComMod/en/caseStudies/biodiv.htm.
- Further details of the ComMod approach are at: www.commod.org.

References

Barnaud, C., A.V. Paassen, G. Trébuil, and T. Promburom. 2006a. Power relations and participatory water management: lessons from a companion modelling experiment in northern Thailand. *Proceedings of 1st Challenge for International Forum on Water and Food (CPWF–CGIAT)*: 15.

Barnaud, C., P. Promburom, F. Bousquet, and G. Trébuil. 2006b. Companion modelling to facilitate collective land management by Akha villagers in upper northern Thailand. *Journal of World Association of Soil and Water Conservation* 1: 38–54.

Barnaud, C., G. Trébuil, P. Dumrongrojwatthana, and J. Marie. 2008. Area study prior to companion modelling to integrate multiple interests in upper watershed management of northern Thailand. *Southeast Asian Studies* 45: 559–585.

Barreteau, O. 2003. Our companion modelling approach. *Journal of Artificial Societies and Social Simulation* 6: 1.

Barreteau, O., F. Bousquet, and J.-M. Attonaty. 2001. Role-playing games for opening the black box of multi-agent systems: method and lessons of its application to Senegal River Valley irrigated systems. *Journal of Artificial Societies and Social Simulation* 4: 5.

Bhatrasataponkul, T., R. Yooman, J. Hachit, and K. Jiratchayut. 2008. Stock assessment of blue swimming crab and mud crab in seagrass habitat of Kung Krabaen Bay, Chanthaburi Province. Paper presented at Proceeding of Marine Science, Metropole Phuket Hotel, Thailand, 25–27 August.

Carpenter, S., B. Walker, J.M. Anderies, and N. Abel. 2001. From metaphor to measurement: resilience of what to what? *Ecosystems* 4: 765–781.

Crookall, D. and W. Thorngate. 2009. Acting, knowing, learning, simulating, gaming. *Simulation and Gaming* 40: 8–26.

Dionnet, M., M. Kuper, A. Hammani, and P. Garin. 2008. Combining role-playing games and policy simulation exercises: an experience with Moroccan smallholder farmers. *Simulation and Gaming* 39: 498–514.

Dumrongrojwatthana, P. 2010. Interactions between cattle raising and reforestation in the highland socio-ecosystem of Nan Province, northern Thailand: a companion modelling process to improve landscape management. Unpublsihed doctoral dissertation, Chulalongkorn University (Thailand) and Paris-Ouest Nanterre La Defense (France).

Dumrongrojwatthana, P., C. Le Page, and G. Trébuil. 2015. Designing a livestock rearing system with stakeholders in Thailand highlands: companion modelling for integrating knowledge and strengthening the adaptive capacity of herders and foresters. *Proceedings of the 5th International Symposium for Farming Systems Design, 7–10 September, Le Corum, Montpellier, France*: 305–306. http://fsd5.european-agronomy.org/documents/proceedings.pdf (accessed 14 March 2017).

Dumrongrojwatthana, P. and G. Trébuil. 2011. Northern Thailand case: gaming and simulation for co-learning and collective action; companion modelling for collaborative landscape management between herders and foresters. In A. Paassen, J. van den Berg, E. Steingrover, R. Werkman, and B. Pedroli (eds.), *Knowledge in Action: The Search for Collaborative Research for Sustainable Landscape Development*. Wageningen: Wageningen Academic Publishers: 191–219.

Epstein, J.M. 2008. Why model? *Journal of Artificial Societies and Social Simulation* 11: 12.

Funtowicz, S.O. and J.R. Ravetz. 1993. Science for the post-normal age. *Futures* 25: 735–755.

Holling, C.S. 2001. Understanding the complexity of economic, ecological and social systems. *Ecosystems* 4: 390–405.

Horlitz, T. 2007. The role of model interfaces for participation in water management. *Water Resources Management* 21: 1091–1102.

Kriz, W.C. 2010. A systemic-constructivist approach to the facilitation and debriefing of simulations and games. *Simulation and Gaming* 41: 663–680.

Kung Krabaen Bay Royal Development Study Center (KKBRDSC). 2014. *Researches under the Kung Krabaen Bay Royal Development Study Center*. www.fisheries.go.th/cf-kung_krabaen/abstract.htm (accessed 14 March 2014).

Kunsook, C., N. Gajaseni, and N. Paphavasit. 2014. Feeding ecology of blue swimming crab, *Portunus pelagicus* (Linnaeus, 1758) at Kung Krabaen Bay, Chanthaburi Province, Thailand. *Tropical Life Sciences Research* 25(1): 13–27.

Le Page, C., G. Abrami, O. Barreteau, N. Becu, P. Bommel, A. Botta, A. Dray, C. Monteil, and V. Souchère. 2014. Models for sharing representations. In M. Etienne (ed.), *Companion Modelling: A Participatory Approach to Support Sustainable Development*. Versailles: Quae and Springer: 69–101.

Leteurtre, E. 2010. Simulation & gaming to promote communication between researchers, managers and blue swimming crab fishery communities in Kung Krabaen Bay, Chanthaburi Province, Thailand: implementation of a participatory process using the companion modelling approach. Unpublished master's thesis, Université Paris Sud 11, Université Pierre & Marie Curie, AgroParisTech, ENS, and MNHN.

Neef, A. 2005. *Participatory Approaches for Sustainable Land Use in Southeast Asia*. Bangkok: White Lotus.

Ollagnon, H. 1991. Vers une gestion patrimoniale de la protection et de la qualité biologique des forêts. *Forest, Trees and People Newsletter* 3: 2–35.

Ostrӧm, E. 2009. A general framework for analyzing sustainability of social-ecological systems. *Science* 325: 419–422.

Reed, M.S. 2008. Stakeholder participation for environmental management: a literature review. *Biological Conservation* 141: 2417–2431.

Rӧling, N.G. and M.A.E. Wagemaker. 1998. *Facilitating Sustainable Agriculture: Participatory Learning and Adaptive Management in Times of Environmental Uncertainty.* Cambridge: Cambridge University Press.

Roth, R. 2004. On the colonial margins and in the global hotspot: park–people conflicts in highland Thailand. *Asia Pacific Viewpoint* 45: 13–32.

Trébuil, G. 2008. Companion modelling for resilient and adaptive social agro-ecological systems in Asia. Paper presented at the 4th National Agricultural Systems Conference "Agriculture for Community and Environment Ready to Handle Climate Change," Empress Hotel, Chiang Mai, Thailand, 27–28 May.

Vejpas, C., F. Bousquet, W. Naivinit, G. Trebuil, and N. Srisombat. 2005. Participatory modeling for managing rainfed lowland rice variety and seed systems in lower northeast Thailand: methodology and preliminary findings. In F. Bousquet, G. Trébuil, and H. Hardy (eds.), *Companion Modeling and Multi-Agent Systems for Integrated Natural Resource Management in Asia.* Los Baños, Philippines: International Rice Research Institute:141–163.

Walker, B., S. Carpenter, J. Anderies, N. Abel, G. Cumming, M. Janssen, L. Lebel, J. Norberg, and G.D. Peterson. 2002. Resilience management in social-ecological systems: a working hypothesis for a participatory approach. *Conservation Ecology* 6: 14.

Worrapimphong, K. 2005. Companion modelling for razor clam *Solen regularis* conservation at Don Hoi Lord, Samut Songkhram Province, Thailand. Unpublished master's thesis, Chulalongkorn University.

Worrapimphong, K. 2010. Integrated and collaborative ecological and socio-economic modelling for sustainable razor clam management at Don Hoi Lord Ramsar site, Thailand. Unpublished Ph.D. thesis, Chulalongkorn University.

Worrapimphong, K., N. Gajaseni, C. Le Page, and F. Bousquet. 2010. A companion modeling approach applied to fishery management. *Environmental Modelling and Software* 25: 1334–1344.

Part V

Policy, economics and governance of biodiversity

12 The shifting politics of biodiversity governance in Southeast Asia

A review and findings from Thailand

Witchuda Srang-iam

Introduction

The 1992 Convention on Biological Diversity (CBD) marked an important change in the global landscape of biodiversity politics and governance. The status of biological and genetic resources has changed from invaluable natural heritage of mankind to valuable resources that can be owned under national sovereignty or global intellectual property regimes. The new biodiversity regime thus involves maintaining a balance among three interrelated objectives: conservation, sustainable use and fair and equitable sharing of benefits derived from biological resources. These three objectives are mutually supportive in enhancing biodiversity. Conservation ensures that biodiversity is maintained, especially in natural settings. Sustainable use allows access to biodiversity for innovation. Benefit sharing is intended to reward the custodians of biodiversity, those who have enforced the conservation. These three dimensions comprise the CBD's objectives triangle, which can be seen as the resolution of conflicts between high-biotechnology developed nations and high-biodiversity developing nations (McAfee, 1999).

The conception of nature has evolved throughout the history of biodiversity politics. The concept of biodiversity has pointed to the variability among living organisms and its relations to the ecological complexity of nature. It also highlights the importance of cultural diversity that closely relates to biological diversity. The introduction of the ecosystem services concept by the Millennium Ecosystem Assessment (2005) has shifted attention further from biodiversity as facilitating natural functions to biodiversity as generating other services related to human well-being (Fisher et al., 2008). This conceptualization represents a human-centered orientation towards biodiversity as ecosystem services.

The new landscape of biodiversity politics has not only changed the conception of what biodiversity means for human society, but also how it should be handled. After the implementation of CBD, the world has witnessed a shift from the government of biodiversity commons to the governance of biodiversity resources. This shift reflects the retreat of states, traditionally the key actors, and the rise of communities and markets as non-state actors in managing biodiversity conservation and use. The incorporation of these new actors has led to the institutionalization of certain rights over biological and genetic resources, namely intellectual property and indigenous knowledge. For instance, the Nagoya Protocol to the CBD sets out clear obligations to ensure prior informed consent and fair and equitable benefit sharing of genetic resources, considering the fundamental role of customary laws on their use and exchange (Tobin, 2013). The adoption of the Strategic Plan for Biodiversity 2011–2020, including the Aichi Targets, has also mobilized "innovative financial mechanisms" for biodiversity conservation (Adenle et al., 2015), adding another layer of institutional complexity (Oberthür and Pożarowska, 2013).

The new governance of biodiversity and ecosystem services has allowed both state and non-state actors to exercise and contest their power in their own interests (Kull et al., 2015). The politics of biodiversity conservation and use has thus shifted from a single domain of governance towards "hybrid" governance platforms. Commonly known as community-based natural resource management, early hybrid forms of biodiversity governance devolve certain authority and rights to local and indigenous people as stewards of biodiversity. However, "devolved" natural resource management practices in developing countries reflect the continuation of central government control and the role of non-government organizations (NGOs), donors and external actors in mobilizing practices towards local interests (Shackleton et al., 2002).

Recent hybrid biodiversity governance practices represent a globally increasing trend in market-oriented governance, bridging between communities and markets (Büscher and Dressler, 2012) and between states and markets (Fletcher and Breitling, 2012; Martín-López et al., 2011). In Southeast Asia, the unfolding of market-oriented biodiversity governance initiatives varies depending on locally and contextually specific policies and practices (Roth and Dressler, 2012).

This chapter argues for a grounded understanding of how markets variously inform conservation and use of biodiversity. It draws on three case studies of market-oriented governance practices in Thailand, each differing with regard to key dominant actors: states, communities and markets. The within-case analyses look at how and why these actors argue for and engage in certain markets, as well as how they achieve their interests in biodiversity and ecosystem services, in particular. The cross-case analysis explores the emerging politics of biodiversity that determines biodiversity governance outcomes, in terms of access and benefit sharing.

The chapter first reviews three major paradigms in biodiversity governance as well as their unique assumptions about the role of states, communities and markets in managing biodiversity conservation and use. Then it describes case studies from Thailand and discusses how and why market-oriented governance practices become manifest across different settings. The chapter concludes by reflecting on the distinctive nature of biodiversity politics and its implications for the effectiveness and sustainability of market-oriented biodiversity governance in Thailand and other Southeast Asian countries.

Paradigms in biodiversity governance: the interplay among states, communities and markets

Throughout the history of global biodiversity governance, management policies and practices have followed three major paradigms: protected-area management; community-based resource management; and market-based conservation. These management paradigms have different assumptions regarding the role of states, communities and markets with their associated types of governance structure: command, cooperation and exchange. This chapter, however, does not take for granted the existing assumptions about the role of actors, since they could obscure the interplay of their power and knowledge in biodiversity governance. Rather, it considers how state, community and market actors realize and utilize these paradigms and associated claims instrumentally to pursue their own interests in biodiversity governance.

Protected area management emphasizes the key role of states in managing biodiversity under their jurisdiction. Before the 1980s, central governments were solely responsible for protecting biodiversity and the ecosystem services it provides. The command-and-control approach to protected area management was essential for maintaining biodiversity through conservation of in situ biodiversity. Protected areas were thus largely the domains of conservationists, ecologists and state officials. In the 1980s, the early expansion of protected areas, driven by growing concerns over the deforestation of rain forests, as well as increased funding and technical assistance programs from developed countries, led to significant expansion in parks and reserves in developing countries, including those of Southeast Asia.

Although the protected area approach is generally effective at maintaining forest conditions in the core areas, it fails to curb deforestation and land development in surrounding or buffer areas in developing countries (Naughton-Treves et al., 2005). Especially in the decline of public resources and support from international donors, governments in developing countries have been unable to enforce rules to prevent development activities that encroach on protected areas. Simultaneously, the systems of protected areas also contribute to the marginalization of local communities that live in poverty and are excluded from biodiversity as sources of their livelihoods (West et al., 2006).

The drawbacks of the state protecting biodiversity have led to an alternative approach, generally known as community-based natural resource management (CBNRM). This approach calls for a more active role for local communities in conservation efforts. It assumes that local people will do more to conserve biodiversity since they depend on it for their livelihoods. Since the 1980s, the scope of conservation programs has extended far beyond protected areas to include biodiversity on farms and in other landscapes. These programs recognize the value of local knowledge systems for both biodiversity conservation and sustainable use. In Southeast Asian countries, the role of NGOs in implementing these community-based conservation programs is notable, as they empower local communities to advance their claims over biodiversity resources (Bryant, 2002).

Along with the advocacy of the alternative, people-oriented approach, the systems of protected areas have reoriented towards people who were excluded from parks. During the 1980s and 1990s, the shifting development paradigm towards decentralization, participation and empowerment was translated into the need to incorporate local communities, as well as their concerns about liveli-hoods and other development-related goals, into conservation norms (Barrett *et al.*, 2001; Naughton-Treves *et al.*, 2005). The people-oriented protected areas have become institutionalized in the socio-economic agendas of international frameworks, including the World Parks Congress (1982) and the CBD (1992). The World Commission on Protected Areas has also developed new categories of protected areas that allow for sustainable use, in addition to those managed mainly for biodiversity conservation (IUCN, 1994). In developing countries, integrated conservation and development projects (ICDPs) have been imple-mented in many protected areas, where local communities are of primary concern (Brandon, 2000).

However, many of these community-based natural resource management pro-jects or people-oriented protected areas have failed to deliver win–win results on a wider scale. According to the protectionists, the people-oriented approach to conservation has focused too much on local development concerns and thus has largely failed to achieve the main goal of protecting biodiversity (Kramer *et al.*, 1997). Moreover, giving the community absolute authority to make conser-vation decisions seems questionable, particularly when such decisions go against development goals. Given continuing conservation–development conflict, there is ongoing debate about reinventing the protectionist paradigm to conserve biodiversity (Wilshusen *et al.*, 2002).

Since the 2000s, emerging biodiversity conservation policies and practices have turned towards the use of market-based instruments to resolve the conflicts between conservation and development, and to address the crises in the financing of biodiversity conservation (Milder *et al.*, 2010). The market-oriented approach emphasizes the role of the market in creating financial incentives to maintain biodiversity and its associated ecosystem services. Market-based interventions aim at fostering efficient management of biodiversity through creating a positive link between ecosystem services and rural development goals (Ingram *et al.*, 2012). In developing countries, many initiatives have emerged to remunerate the owners and managers of forests for their stewardship of biodiversity, including payments for biodiversity-conserving management or for allowing access to biodiversity resources and tradable rights (Jenkins *et al.*, 2004). In the Philippines, community-based initiatives have devised market-based mechanisms as new financial incentives for converting swidden agriculture into permanent forms of cultivation (Büscher and Dressler, 2012).

Market-oriented conservation seeks to create markets for biodiversity and its services in facilitating exchange between its providers and beneficiaries. To address failures in such markets, including negative externalities, public goods and ill-defined property rights, the calculated values of ecosystems provide incentives to conserve biodiversity, or compensations for damage, as well as disincentives to pollute or reduce ecosystem services (Van Hecken and Bastiaensen, 2010).

The main criticisms of market-oriented conservation include "neoliberalization of nature" literature on the retreat of the state and increased authority of the private sector over the management of biodiversity (Bakker, 2010; Castree, 2008). Central to the criticism is the commodification process through which nature is converted into standard units of exchange and assigned with values. The quantification and valuation of ecosystem services could lead to disappointing outcomes in the absence of appropriate direct pricing methods (Spangenberg and Settele, 2010). Economic measures of ecosystem services are often indirect and may not represent the true values of nature (Spash and Aslaksen, 2015), for example, by estimating the amount people are willing to pay for received benefits or the costs to society in terms of lost benefits (Pagiola *et al.*, 2004). The monetary valuation of ecosystem services has important implications for social and environmental justice, fostering unequal access to and distribution of benefits among users and providers (Bulte *et al.*, 2008; Matulis, 2014). Thus, the commodification could shift constructions of nature – from what nature means to local communities to what it should mean to them in commodity terms (Büscher and Dressler, 2012).

Nevertheless, the implementation of market-oriented instruments in Southeast Asia and other developing countries is associated with a lower degree of commodification (Hahn *et al.*, 2015). Many market-oriented governance practices in the developing world do not represent market deregulation; rather, they are planned by specific actors, such as governments, NGOs or private firms, with a view to achieving desirable biodiversity outcomes (Vatn, 2010). In countries where markets for biodiversity are unavailable or not widespread, governments, NGOs and corporations have been either core intermediaries or buyers of market-oriented conservation programs (Fauzi and Anna, 2013; Fletcher and Breitling, 2012; Pham *et al.*, 2010). Under such circumstances, these actors could play a key role in creating markets for biodiversity, and in so doing they could determine their rights and responsibilities as well as the distribution of costs and benefits from biodiversity. These "hybrid" market-oriented governance practices could thus result in divergence from market ideals – either efficient use or fair benefit sharing (Corbera *et al.*, 2007).

As case studies in Vietnam show, market-based conservation programs have neither enhanced ecosystems nor benefited local communities and the poor as providers of ecosystem services (McElwee, 2012; Suhardiman *et al.*, 2013). Evidence from Thailand and the Philippines suggests an instrumental use of markets to rearticulate hierarchies of state control through community-based conservation (Dressler and Roth, 2011). The payments for ecosystem services (PES) cases from Cambodia comfirm that empowering local institutions and reinforcing intrinsic motivations are necessary to make these programs more sustainable (Clements *et al.*, 2010).

Hybrid market-oriented biodiversity governance: theory and practice in markets for biodiversity in Thailand

This chapter uses the environmental governance framework (Lemos and Agrawal, 2006) to examine "hybrid" forms of market-oriented biodiversity governance practice through which the interplay of state, market and community actors determines governance outcomes. It makes a conceptual distinction between the hybrid model of market-oriented governance, which is the focus of the chapter, and the purely market-based model that is found in the neoliberal literature. The hybrid model differs in that some intermediaries outside the market are involved in the creation of biodiversity and other environmental markets. Additionally, in the absence of a strong environmental market economy, hybrid biodiversity governance schemes could become a new arena for the state–community–market contestation shaped by the protectionist, community and neoliberal paradigms. This chapter seeks to understand the new politics of biodiversity by examining hybrid practices at the local level, where the powers of states, communities and markets interact.

The payments for ecosystem services (PES) concept represents a typical market-oriented biodiversity governance approach in both developed and developing countries. The original idea of PES, which draws on the Coase theorem (Coase, 1960), defines the concept as "voluntary exchange of well-defined ecosystem services between a minimum of one buyer and a minimum of one provider who must secure the service provision" (Wunder, 2005: 5). Theoretically, payments should be determined voluntarily in the market through demand from ecosystem service buyers and supply by ecosystem service providers. In practice, the applications of the PES concept vary significantly along the voluntary–regulatory continuum (Sattler and Matzdorf, 2013). In the developing world, most PES practices require government either to pay itself or to make others pay for ecosystem services, thus following the Pigouvian theorem (Pigou, 1932). The broad definition of PES (e.g. Muradian et al., 2010) serves to incorporate non-market forces in determining payments or regulating the flow of environmental goods and services. The payments can be predetermined to cover the opportunity costs of service provisions, or to distribute the benefits fairly. In practice, the PES approach shows a combination of the cost-allocation, polluter-pays principle and the benefit-sharing, beneficiary-pays principle, rather than a shift from the former to the latter, as asserted (e.g. Mauerhofer et al., 2013).

This chapter extends existing definitions of PES to include a broader range of practices in the creation of markets or quasi-markets for non-market ecosystem services. The following sections examine three cases of emerging markets for biodiversity in Thailand, representing differing PES approaches across scales and domains for governance – states, communities and markets, respectively. Table 12.1 presents a comparison of the cases. The first case is a state scheme that seeks payments for forests and their services to finance reforestation programs. The second is a nationwide, community-based program that pays farmers for planting trees. The third case concerns the implementation of small pilot PES projects at the local level under corporate social responsibility (CSR) programs.

Table 12.1 Three emerging markets for biodiversity in Thailand

Markets	Intermediaries	Service providers	Environmental goods/services	Payers/buyers
Markets for reforestation	Department of National Parks, Wildlife and Plant Conservation	Department of National Parks, Wildlife and Plant Conservation	Timber and forest products Soil fertility Increased rainfall Cooling weather	Farmers and developers who destroy forests in protected areas
Markets for tree conservation	Network of Civil Organizations for Tree Bank	Small-scale farmers nationwide	Timber and forest services Agro-biodiversity Carbon sequestration	Governments and corporations
Markets for biodiversity-based economy	Biodiversity-based Economy Development Office	Mangrove Forest Conservation Club	Eco-tourism resources Aesthetics	Tourism-related business operators
		Indigenous upland community	Water availability	Provincial Waterworks Authority

Markets for reforestation

The 1992 Enhancement and Conservation of National Environmental Quality Act (B.E. 2535) imposed civil liability for "destruction, loss or damage to natural resources owned by the State or belonging to the public domain." Section 97 concerns the liability to pay "compensation to the State representing the total value of natural resources." Under the law, the design and implementation of civil liability are the responsibility of state authorities with respect to types of natural resources under their jurisdiction. The Department of Natural Parks, Wildlife and Plant Conservation (DNP), under the Ministry of Natural Resources and Environment, is responsible for the conservation of Thailand's biodiversity, in particular the management of protected areas.

In its enforcement of civil liability, the DNP adopted two compensation rates based on the costs of deforestation: 150,000 baht per rai (US\$1 = c. 30 baht; 1 hectare = 6.25 rai) for deforestation in headwater areas (designated as Watershed Classes 1 and 2); and 68,244.22 baht per rai outside headwater areas. The calculated costs of deforestation comprise the direct loss of timber and forest products, the approximate loss of soil nutrition, and the costs of reforestation, using market prices in 1993 and 1996. Adoption of these two compensation rates was highly criticized, since the rates did not consider contextual factors that could affect the values. In addition, the value of deforestation should vary according to forest conditions.

In response to this criticism, DNP experts developed a general computing model and procedure for evaluating and calculating values of environmental loss from forest destruction. In calculating the value of ecosystem structure and

Table 12.2 Losses that make up the total value of deforestation in a protected area

Direct loss of timber and forest products	40,825.10 baht/rai/year
Loss of evergreen forest	61,263.36
Loss of mixed deciduous forest	42,577.75
Loss of deciduous dipterocarp forest	18,634.19
Direct loss from forest ecosystem structure	59,264.15 baht/rai/year
Loss in soil fertility	4,064.15
Loss in soil infiltration	600
Loss of soil	1,800
Loss of water from solar evaporation	52,800
Indirect loss of local climate regulation	50,853.45
Loss from decreased rainfall	5,400
Loss from increased temperature	45,453.45
Total loss	150,942.70 baht/rai/year

services, the model applies the direct replacement cost approach using available market price. For instance, the calculated value of water/soil loss includes the cost of buying and transporting the amount of soil/fertilizer/water for replacement. Similarly, the calculation of loss in terms of warmer weather is based on the cost of electricity used in cooling the air. The model was applicable only to deciduous dipterocarp forest, mixed deciduous forest and evergreen forest in 2004, and to any type of forest from 2010 onwards.

The DNP model expands the scope of the calculated impacts that encompass three components of ecosystems: structure, function and service. For each deforestation site, the model requires data inputs, such as species richness, slope and soil texture, collected in both deforested and nearby forest areas, as well as basic data, such as annual rainfall and geographical profiles. The calculated amount of compensation serves only as a guideline for the DNP; the amount of "reasonable" compensation remains open for determination by the courts. For instance, all losses that make up the total value of deforestation in a protected area in southern Thailand are listed in Table 12.2.

By adopting the model, the DNP has created a market for reforestation. For each particular site of deforestation, farmers pay the total value of forests and their services, including timber and forest products, forest ecosystem structure and its regulating services. As a result, they pay not only for the damage they have done but also for the value for which beneficiaries should pay. These payments are much higher than the cost of reforestation. Since its adoption of the model, the DNP has charged developers and farmers who cut down the forests equally.

However, some NGOs, including Thai Land Reform Network, have accused the DNP of imposing excessive liability costs on farmers who usually have no ability to pay. They argue that the demanded payments amount to "global warming fines" and that these are unfair tariffs on farmers who, as a group, have contributed far less to global climate change than the energy and transportation sectors. A case finally came to the Administrative Court, but it was dismissed because the model itself is not an administrative order, only a guideline. DNP experts argued that the model considers the forest's regulating service only at the local climate scale. Additionally, the same model is used for developers and farmers who encroach on protected areas. However, according to DNP officials, farmers are often disadvantaged because they lack the knowledge to dispute assessed liabilities through the civil court system.

Markets for tree conservation

In 2005, the head of the Pa Toh Watershed Management Unit, under the DNP in Ranong, initiated an idea of creating markets for tree conservation. These markets can provide incentives in the form of payments to farmers who plant or conserve trees, in addition to the long-term benefits of tree planting, such as supplies of timber. The payments for tree conservation seek to address the ineffectiveness of state reforestation programs, particularly in terms of survival rate, and the limited success of local conservation activities that often compromise local livelihoods. The idea and its implementing organization were later established as the Tree Bank and the Network of Civil Society Organizations for Tree Bank, respectively. The Tree Bank argues that the value of living trees should be recognized, just as the value of dead timber is recognized. Throughout its development, the Tree Bank scheme has continued to debate the rights and scope of environmental service provisions as well as the sources and valuation of payments. As of 2014, the scheme was still searching for buyers for tree services.

Initially, the Tree Bank proposed that the government should redirect part of its reforestation budget to pay farmers for planting specific tree species on their land. The eligible trees, according to Tree Bank rules, included all timber types with the exception of eucalyptus, para-rubber, fruit trees, agarwood and other business-oriented or monoculture plantations. Payments for tree conservation would compensate farmers for the opportunity costs of planting and conserving trees as well as reward tree environmental services, such as agricultural biodiversity. The payments were in form of interest based on the value of the trees the farmers grew. The calculation of tree values would cover the annual costs of investment for the first ten years, and thereafter would be equal to the price of timber on the market. The scheme would pay each farmer for a maximum of 1,000 trees in order to prevent exploitation by large landholders.

Although the early Tree Bank scheme gained support among more than a thousand communities nationwide who wished to act as tree service providers, it could not persuade the government to become a buyer. Indeed, the government dismissed the Tree Bank proposal, saying that it was incompatible with the interests of the state in many respects. First, the type and scale of the claimed benefits from the Tree Bank did not correspond to the state policy – that is, small individual agro-forestry versus large patches of tropical forest. The definition of tree rights for payments also seriously conflicted with state property regimes. Specifically, the Tree Bank recognized farmers' rights over the trees they grew, even if the trees were planted on state property. Furthermore, a government feasibility study found that payments for biodiversity were unnecessary, given that the market for specific types of timber provided sufficient incentive for planting more trees. Following the study, the Bank of Agriculture and Agricultural Cooperatives (BAAC), a state enterprise, incorporated the Tree Bank idea into its CSR program in 2012. This new Tree Bank serves as a special loans program through which farmers can secure loans based on the timber values of trees on mortgaged land. Meanwhile, not a single community registered with the original Tree Bank received any payments.

After the scheme failed to obtain government support, the Network of Civil Society Organizations for Tree Bank reorganized the Tree Bank in a bid to attract corporations as buyers. In anticipation of interest from both private and corporate buyers, pilot projects were implemented to produce concrete, measurable outcomes, such as a certain number of trees. The network was able to match a few corporations with communities in relation to their environmental damage. For example, an electricity generating company paid communities affected by its power plant operations. Similarly, a coffee production firm provided support to communities where farmers maintained the coffee agro-forestry system. The support was in the form of a lump-sum payment from the firm's annual CSR budget, paid to a number of farmers in the pilot communities. However, it ceased when the coffee company learned that a majority of these farmers were no longer growing coffee.

Recently, carbon sequestration has become a new claim of tree environmental services, to address some drawbacks of existing service provision. For example, the services from agro-biodiversity can be doubtful and they are hardly measurable. The appeal of carbon sequestration is not only due to its policy salience, but also because the markets for carbon are available internationally. The availability of these carbon markets facilitates connections between community carbon service provisions and potential buyers. The new targeted buyers for the Tree Bank thus include corporations and individuals who are interested in offsetting carbon through tree conservation. However, this entails huge transaction costs for the network associated with credible monitoring, reporting and verifying procedures.

Markets for biodiversity-based economy

The Biodiversity-based Economy Development Office (BEDO) is another public organization that is interested in using the PES instrument to convert biodiversity services into main or supplementary income sources for local communities. The main approach is to acknowledge that biodiversity produces significant monetary value in terms of ecosystem services. In 2010, the BEDO, in cooperation with the DNP, launched a pilot PES project at the Khao Ang Ruenai Wildlife Sanctuary. However, this project failed to identify beneficiaries who were willing to pay for ecosystem services, even though the realizable value of the benefits was significant. Hence, the BEDO has pursued a new strategy by harnessing typical CSR activities into the PES channel. Using this strategy, it has concluded two PES deals, one in Krabi and the other in Chiang Mai.

In Krabi, the BEDO has introduced the PES principle into eco-tourism on a small island called Koh Klang. In 2012, local communities, tourism-related business operators and the sub-district administrative organization reached an agreement to launch the first ever PES project in Thailand. As part of the agreement, groups of tourism-related business operators, including the Passenger Boat and Motorcycle Taxi Club, the Islanda Eco Village Resort and local restaurants, pay a monthly fee of 500 baht for the environmental services they have enjoyed. These fees go to the established PES fund, managed by the Mangrove Forest Conservation Club, for conservation activities, including mangrove rehabilitation and garbage collection. As of April 2014, the PES fund had accumulated a surplus of 25,000 baht which was yet to be utilized. The payments received are negligible in comparison to the assessed value of Koh Klang's ecosystem services – 196,029 baht per rai per year. The calculated ecosystem services include fishery products, coastal erosion, firewood, tourism resources and carbon sequestration.

The other PES deal concerns the forested watershed ecosystem of Mae Lao–Mae Sae Wildlife Sanctuary, Chiang Mai. According to the DNP model, the value of this area's forest ecosystem services is 171,585 baht per rai. The Region 9 office of the Provincial Waterworks Authority (PWA) agreed to allocate 2013's organizational social responsibility budget (170,000 baht) to compensate for upstream activities that provide water downstream. In this regard, the BEDO identified a Karen community as service providers. As part of the twelve-month agreement, the villagers agreed to contribute their labor in building six check dams for water retention in return for daily wages. The total payment for their labor was 10,425 baht, distributed among seventy-five villagers who participated in the construction, with a majority of the budget used to purchase building materials. The remainder was used to pay the villagers for monitoring and preventing forest fires through the use of fire barriers. It is uncertain whether the PWA will continue to support the village or extend the scheme to other villages in the watershed area, so the sustainability of the project remains questionable.

In both of these PES projects, the nature of the CSR activities does not relate directly to the ecosystem services as claimed. For example, the PWA paid the villagers to build check dams for themselves, which has little connection to the water PWA is exploiting. The buyers determined the terms and values of their services and payments, whereas the providers had no bargaining power. Meanwhile, in Koh Klang, conservation of the mangrove forests continued regardless of the payments. The cost of provision is usually low, consisting of small expenses and often voluntary labor from the villagers. Moreover, there were high transaction costs associated with reimbursement of payments made from the CSR budget. As a result, payments for environmental services are relatively low compared to the latter's assessed values.

In addition to these two pilot projects, the BEDO has established a network of large business operators that are committed to converting their CSR into PES programs. The new plan for PES includes, for instance, conversion of an upland maize-farming area into more sustainable agricultural practice in Ban Mae Sanan, Nan Province. This land conversion plan will be sponsored by Charoen Pokphand Produce, a leading maize-seed production company and a subsidiary of the Charoen Pokphand (CP) Group, the largest agro-industrial and food conglomerate in Thailand.

Understanding the politics of hybrid market-oriented biodiversity governance: a theoretical framework

The creation of markets for biodiversity has generated new opportunities for broader negotiations among state, community and market actors, as well as broader governance of related ecosystem services. These actors have been active in political processes through which the emerging markets for biodiversity have developed. The creation of biodiversity markets has thus deviated from ideal market practices in certain respects. These deviations display some similarities across case studies. In most cases, buyers and providers generally have unequal power, with the providers often having less bargaining power. State, community and market actors exercise and compete for power in defining the terms and rights of provision as well as setting the price. In the absence of markets for biodiversity, intermediaries often play significant roles in determining the outcomes of these power competitions. Moreover, they themselves exercise direct regulatory or moral power to promote market transactions, as in the cases of the DNP and the Tree Bank, respectively. They also indirectly arrange the markets (e.g. designate the buyers and sellers, and oversee the negotiation process), as in the case of the BEDO. The interplay among these political actors shapes governance outcomes that may not be effective and may exclude the marginalized and the poor from biodiversity benefits.

Based on a cross-case comparison, three modes of politics can be identified in relation to the nature of ecosystem services and governance. Each case represents a particular setting that determines the potential of markets for biodiversity, power relationship and practices of buyers and providers. Three modes of politics elaborate a coherent theoretical framework that reveals the distinctive nature of market-oriented biodiversity politics, as well as its implications for the effectiveness and equity of biodiversity governance in Thailand and other Southeast Asian countries.

In the case of degraded ecosystems where there is little or no assigned value to ecosystem services, the creation of markets for biodiversity is possible only under hierarchical governance. States or governments usually perform the key role in assigning and enforcing the values that are necessary for reinvestment in biodiversity rehabilitation over the long term. The roles of community and private actors are often limited to negotiating the markets. Under these circumstances, the costs of ecosystem restoration are much higher than the benefits derived from biodiversity and alternative developments. As in the DNP case, the state forest agency has thus created markets for deforestation that account for lost benefits, calculating the direct replacement costs on the basis of market values. This calculation results in excessive compensation that seems unfair to poor farmers vis-à-vis land investors and developers who engage in deforestation activities, especially in light of the benefits the latter derive from land development.

For biodiversity that falls under community regimes, the recognized values of biodiversity conservation are limited and therefore usually fail to attract potential buyers. The focus of market creation thus extends beyond biodiversity-centered conservation to include human well-being and other ecosystem services. The Tree Bank scheme has sought to create incentives for farmers to plant or maintain trees, which traditionally have had market value only as timber. But the values of timber and other direct uses over the long term are usually lower than shorter-term benefits derived from alternative use of the land – in other words, the opportunity costs of tree planting and conservation. The Tree Bank has thus created markets for agro-biodiversity, and recently shifted towards boarder ecosystem services as carbon sequestration. However, the establishment of markets for non-marketed ecosystem services incurs high transaction costs associated with, for example, clarifying resource claims and environmental service provision rights, estimating values of services, and matching buyers with sellers. In this case, securing potential buyers poses a significant challenge for intermediaries, as the markets are generally unsustainable without continued support.

The ideal case for creating markets for ecosystem services is when the benefits derived from biodiversity are much higher than the total provision costs. This case includes, for example, eco-tourism or where a tourism-based economy relies heavily on ecosystem services. In such circumstances, provision of biodiversity may be managed efficiently under private regimes. Corporations may be persuaded to fund existing or new community conservation programs because, while they obtain significant benefits from high biodiversity, they need to pay only small amounts to service providers for their conservation activities. However, these payments may be so low that they provide insufficient incentive for biodiversity conservation, and certainly fail to alleviate poverty at the local level. Corporations generally have bargaining power which enables them to decide on the nature and value of activities for compensation or service provision. Paradoxically, the existence of community-based conservation often reduces the power of communities when negotiating contracts. As we have seen, ecosystem service stewardship has tended to continue regardless of payment. Markets for ecosystem services become instrumental in determining CSR activities. After all, rich ecosystem services do not necessarily translate into rich livelihoods for local communities.

These case studies from Thailand demonstrate some trade-off relationships between the sustainability of markets and the effectiveness of biodiversity and ecosystem service provisions. Where markets for biodiversity are possible under private governance, the negotiations in the markets may not lead to effective management of biodiversity and may also exclude potential beneficiaries and providers. The nature of conservation activities is geared towards the interests of corporations, and often relates to their costs on biodiversity and livelihoods, with a narrower scope and a lower value than their benefits from biodiversity. Meanwhile, the sustainability of biodiversity markets with desirable terms and conditions for effectiveness is uncertain, given excessive transaction costs. Changing biodiversity markets through the introduction of a new regulatory scheme or the enforcement of existing rules may cause legitimacy problems, and even lead to unfairness in the implementation.

Conclusion

This chapter has reflected on emerging market-oriented biodiversity policies and practices as hybrid forms of biodiversity governance in Thailand. The creation of markets for biodiversity and more broadly ecosystem services has brought up new constituencies and expanded new possibilities for biodiversity governance outside biodiversity-rich protected natural areas and communities. Through assigning utilitarian or economic values to biodiversity, the new domain for hybrid governance has opened up not only new opportunities but also challenges relating to the ineffectiveness and inequality of biodiversity markets.

Based on comparative case studies, the chapter has discussed three modes of political practice in the creation of biodiversity markets that occur in different settings. The findings have led to some suggestions regarding the design of policies and practices to achieve effectiveness and equity in biodiversity governance. For degraded ecosystems, the scope and valuation of payments could be based on the costs of reforestation rather than the benefits. The application of the polluter-pays principle in market creation should apply to the costs of cleaning up, which will prevent government rent seeking. Inclusion of non-marketed ecosystem services when determining the payments is unfair to alleged polluters who are not voluntary "buyers." Unlike the principle of beneficiaries pay, a selection of buyers could voluntarily and collectively pay for ecosystem benefits.

The case of community-based biodiversity governance poses a real challenge for the use of a market-based approach. Successful market creation relies upon clear definition of terms and payments for service provision that can draw the attention of potential buyers. In the absence of buyers who will directly benefit from such services, states or governments can step in. If that is the case, transparency and credibility of ecosystem service provision claims are crucial to the legitimacy of community market-oriented regimes.

Where a market for biodiversity is possible, intermediaries tend to match low-cost providers with high-benefit buyers in order to establish and sustain operation of the market. While this acknowledges the importance of market efficiency, the intermediaries ignore power relations underlying the development of markets for biodiversity and ecosystem services. Under such circumstances, they could create unequal bargaining power in favor of potential buyers. Consequently, the providers, with very little power to negotiate, may not reach the deals that are essential to improve their livelihoods and reduce poverty. Instead, intermediaries should empower the providers so they have equal standing when negotiating the creation of markets for biodiversity and ecosystem services. For example, they could exert influence in determining payments and sharing of benefits that at least compensate the opportunity costs of the providers, or improve the livelihoods of local communities as biodiversity stewards. Alternatively, they could strategically select provisions over which providers have some control for enabling or facilitating user benefits.

References

Adenle, A.A., C. Stevens, and P. Bridgewater. 2015. Global conservation and management of biodiversity in developing countries: an opportunity for a new approach. *Environmental Science and Policy* 45: 104–108.

Bakker, K. 2010. The limits of "neoliberal natures": debating green neoliberalism. *Progress in Human Geography* 34: 715–735.

Barrett, C.B., K. Brandon, C. Gibson, and H. Gjertsen. 2001. Conserving tropical biodiversity amid weak institutions. *BioScience* 51: 497–502.

Brandon, K. 2000. Moving beyond integrated conservation and development projects (ICDPs) to achieve biodiversity conservation. In D.R. Lee and C. Barrett (eds.), *Tradeoffs or Synergies? Agricultural Intensification, Economic Development and the Environment*. Wallingford: CAB International: 417–432.

Bryant, R.L. 2002. Non-governmental organizations and governmentality: "consuming" biodiversity and indigenous people in the Philippines. *Political Studies* 50: 268–292.

Bulte, E.H., L. Lipper, R. Stringer, and D. Zilberman. 2008. Payments for ecosystem services and poverty reduction: concepts, issues, and empirical perspectives. *Environment and Development Economics* 13: 245–254.

Büscher, B. and W. Dressler. 2012. Commodity conservation: the restructuring of community conservation in South Africa and the Philippines. *Geoforum* 43: 367–376.

Castree, N. 2008. Neoliberalising nature: the logics of deregulation and reregulation. *Environment and Planning* 40: 131–152.

Clements, T., A. John, K. Nielsen, D. An, S. Tan, and E.J. Milner-Gulland. 2010. Payments for biodiversity conservation in the context of weak institutions: comparison of three programs from Cambodia. *Ecological Economics* 69: 1283–1291.

Coase, R. 1960. The problem of social cost. *Journal of Law and Economics* 3: 1–44.

Corbera, E., K. Brown, and W.N. Adger. 2007. The Equity and legitimacy of markets for ecosystem services. *Development and Change* 38: 587–613.

Dressler, W. and R. Roth. 2011. The good, the bad, and the contradictory: neoliberal conservation governance in rural Southeast Asia. *World Development* 39: 851–862.

Fauzi, A. and Z. Anna. 2013. The complexity of the institution of payment for environmental services: a case study of two Indonesian PES Schemes. *Ecosystem Services* 6: 54–63.

Fisher, B., K. Turner, M. Zylstra, R. Brouwer, R.D. Groot, S. Farber, P. Ferraro, R. Green, D. Hadley, J. Harlow, P. Jefferiss, C. Kirkby, P. Morling, S. Mowatt, R. Naidoo, J. Paavola, B. Strassburg, D. Yu, and A. Balmford. 2008. Ecosystem Services and economic theory: integration for policy-relevant research. *Ecological Applications* 18: 2050–2067.

Fletcher, R. and J. Breitling. 2012. Market mechanism or subsidy in disguise? Governing payment for environmental services in Costa Rica. *Geoforum* 43: 402–411.

Hahn, T., C. McDermott, C. Ituarte-Lima, M. Schultz, T. Green, and M. Tuvendal. 2015. Purposes and degrees of commodification: economic instruments for biodiversity and ecosystem services need not rely on markets or monetary valuation. *Ecosystem Services* 16: 74–82.

Ingram, J.C., K.H. Redford, and J.E.M Watson. 2012. Applying ecosystem services approaches for biodiversity conservation: benefits and challenges. *S.A.P.I.EN.S* 5(1). https://sapiens.revues.org/1459 (accessed 15 March 2017).

International Union for Conservation (IUCN). 1994. *Guidelines for Protected Area Management Categories*. Cambridge and Gland: International Union for Conservation.

Jenkins, M., S.J. Scherr, and M. Inbar. 2004. Markets for biodiversity services: potential roles and challenges. *Environment: Science and Policy for Sustainable Development* 46: 32–42.

Kramer, R., C.V. Schaik, and J. Johnson. 1997. *Last Stand: Protected Areas and the Defense of Tropical Biodiversity*. Oxford: Oxford University Press.

Kull, C.A., X. Arnauld de Sartre, and M. Castro-Larrañaga. 2015. The political ecology of ecosystem services. *Geoforum* 61: 122–134.

Lemos, M.C. and A. Agrawal. 2006. Environmental governance. *Annual Review of Environment and Resources* 31: 297–325.

Martín-López, B., M. García-Llorente, I. Palomo, and C. Montes. 2011. The conservation against development paradigm in protected areas: valuation of ecosystem services in the Doñana social–ecological system (south western Spain). *Ecological Economics* 70: 1481–1491.

Matulis, B.S. 2014. The economic valuation of nature: a question of justice? *Ecological Economics* 104: 155–157.

Mauerhofer, V., K. Hubacek, and A. Coleby. 2013. From polluter pays to provider gets: distribution of rights and costs under payments for ecosystem services. *Ecology and Society* 18: 41.

McAfee, K. 1999. Selling nature to save it? Biodiversity and green developmentalism. *Environment and Planning: Society and Space* 17: 133–154.

McElwee, P.D. 2012. Payments for environmental services as neoliberal market-based forest conservation in Vietnam: panacea or problem? *Geoforum* 43: 412–426.

Milder, J.C., S.J. Scherr, and C. Bracer. 2010. Trends and future potential of payment for ecosystem services to alleviate rural poverty in developing countries. *Ecology and Society* 15: 4.

Millennium Ecosystem Assessment. 2005. *Ecosystems and Human Well-being Synthesis*. Washington, D.C.: Island Press.

Muradian, R., E. Corbera, U. Pascual, N. Kosoy, and P.H. May. 2010. Reconciling theory and practice: an alternative conceptual framework for understanding payments for environmental services. *Ecological Economics* 69: 1202–1208.

Naughton-Treves, L., M.B. Holland, and K. Brandon. 2005. The role of protected areas in conserving biodiversity and sustaining local livelihoods. *Annual Review of Environment and Resources* 30: 219–252.

Oberthür, S. and J. Pożarowska. 2013. Managing institutional complexity and fragmentation: the Nagoya Protocol and the global governance of genetic resources. *Global Environmental Politics* 13: 100–118.

Pagiola, S., K. von Ritter, and J. Bishop. 2004. *Assessing the Economic Value of Ecosystem Conservation*. Environment Department Paper No. 101. Washington, D.C.: World Bank and IUCN.

Pham, T.T., B.M. Campbell, S. Garnett, H. Aslin, and H.M. Ha. 2010. Importance and impacts of intermediary boundary organizations in facilitating payment for environmental services in Vietnam. *Environmental Conservation* 37: 64–72.

Pigou, A.C. 1932. *The Economics of Welfare*. London: Macmillan.

Roth, R.J. and W. Dressler. 2012. Market-oriented conservation governance: the particularities of place. *Geoforum* 43: 363–366.

Sattler, C. and B. Matzdorf. 2013. PES in a nutshell: from definitions and origins to PES in practice: approaches, design process and innovative aspects. *Ecosystem Services* 6: 2–11.

Shackleton, S., B. Campbell, E. Wollenberg, and D. Edmunds. 2002. Devolution and community-based natural resource management: creating space for local people to participate and benefit? *Natural Resource Perspective* 76. www.odi.org/sites/odi.org.uk/files/odi-assets/publications-opinion-files/2811.pdf (accessed 15 March 2017).

Spangenberg, J.H. and J. Settele. 2010. Precisely incorrect? Monetising the value of ecosystem services. *Ecological Complexity* 7: 327–337.

Spash, C.L. and I. Aslaksen. 2015. Re-establishing an ecological discourse in the policy debate over how to value ecosystems and biodiversity. *Journal of Environmental Management* 159: 245–253.

Suhardiman, D., D. Wichelns, G. Lestrelin, and C.T. Hoanh. 2013. Payments for ecosystem services in Vietnam: market-based incentives or state control of resources? *Ecosystem Services* 6: 64–71.

Tobin, B.M. 2013. Bridging the Nagoya compliance gap: the fundamental role of customary law in protection of indigenous peoples' resource and knowledge rights. *Law, Environment and Development Journal* 9: 142–162.

Van Hecken, G. and J. Bastiaensen. 2010. Payments for ecosystem services: justified or not? A political view. *Environmental Science and Policy* 13: 785–792.

Vatn, A. 2010. An Institutional analysis of payments for environmental services. *Ecological Economics* 69: 1245–1252.

West, P., J. Igoe, and D. Brockington. 2006. Parks and peoples: the social impact of protected areas. *Annual Review of Anthropology* 35: 251–277.

Wilshusen, P.R., S.R. Brechin, C.L. Fortwangler, and P.C. West. 2002. Reinventing a square wheel: critique of a resurgent "protection paradigm" in international biodiversity conservation. *Society and Natural Resources* 15: 17–40.

Wunder, S. 2005. *Payments for Environmental Services: Some Nuts and Bolts*. CIFOR Occasional Paper No. 42. Jakarta: Center for International Forestry Research.

13 Biodiversity ecosystems and human health

A complicated but important science policy challenge

*Hans Keune, Suneetha M. Subramanian,
Unnikrishnan Payyappallimana,
Sate Ahmad, Mohammed Mofizur Rahman,
Mohammed Nurul Azam, Zita Sebesvari,
Joachim H. Spangenberg, Serge Morand,
Conor Kretsch, Pim Martens and
Heidi Wittmer*

Introduction

In this chapter we will mainly focus on biodiversity challenges in relation to human health. We will do this from a science–society interface perspective. What are the important challenges regarding this topic when aiming for policy and society practice relevant research and action? First, we will briefly introduce the main biodiversity–human health linkages, in terms of both health benefits and risks. Second, we will address the main international challenges regarding these linkages for science and society. We will specifically focus on the need for integration and collaboration and the need for dealing with complexity in a societally relevant manner. Third, we will introduce several case studies that will illustrate the above within practice and within the local Southeast Asian context. Finally, we will draw some conclusions from the case studies and briefly reflect on a way forward for these important challenges for society, both in Asia and around the world.

Biodiversity and human health

According to McMichael (2009), human population health should be the central criterion, and is the best long-term indicator, of how we are managing the natural environment. Environmental philosophers and others may question the valuation of human life over all other life in this assertion, as well as the implicit assumption that "we" can truly manage the natural environment. Human population health nonetheless is an important criterion for efforts by institutional and other actors to guide human action in the natural environment. More than ten years ago, the Millennium Ecosystem Assessment synthesis report for the health sector

(WHO, 2005) outlined several linkages between ecosystems and health. A recent state-of-knowledge review under the auspices of the Convention on Biological Diversity (CBD) and the World Health Organization (WHO) provides a wide-ranging overview of well-established insights (WHO and CBD Secretariat, 2015).

Health benefit examples, or ecosystem services (ES) to human health, are the importance of biodiversity to traditional and modern medicinal practice, and the utility of various species for medical research. Genetic and species diversity is functional to food production, and can play an important role in addressing issues of nutrition security, including certain disease risks (e.g. obesity, diabetes) through dietary improvements. Biodiversity also plays a role in safeguarding air quality and access to fresh water, disaster risk reduction, and supporting emergency responses and climate change adaptation. Furthermore, diverse natural environments may enhance experiences that reduce stress, support the development of cognitive resources, stimulate social contacts, attract people for physical activity, and support personal development throughout an individual's lifespan. Moreover, recent studies show that declining contact with some forms of (microbiotic) life may contribute to the rapidly increasing prevalence of allergies and other chronic inflammatory diseases among urban populations worldwide. Biodiversity thus can make an important contribution to public health-related ecosystem services and the reduction of health risks.

However, recent discussions also recognize that biodiversity can produce potential nuisances to our health, such as adverse impacts on infectious diseases, allergies or pests. For example, the relationship between biodiversity and infectious diseases is not straightforward: the disease dynamics are complex and system-dependent. Biodiversity may reduce the risk of infectious disease emergence or spread (dilution effect), while its loss or unsustainable exploitation can increase disease transmission. Land use change and ecosystem disruption are widely recognized as drivers of infectious disease emergence. In contrast, however, areas of naturally high biodiversity may serve as source pools for emerging pathogens. The emergence and spread of certain pathogens from wildlife to livestock and/or humans, and related social and economic costs, have been well documented. These diseases include HIV, hantavirus, avian influenza, Lyme disease, malaria, dengue fever, leishmaniasis, Nipah virus and Ebola. Hence, high biodiversity does not necessarily reduce disease risk in all situations.

The need for integration

The major challenges facing humanity in the twenty-first century are closely interlinked and increasingly globalized. At local, regional and global scales the crises in health, food and nutrition security, water, energy, climate change, biodiversity loss and poverty frequently overlap in both origins and proximate causes, in their various impacts, and in terms of the policy and practical approaches needed to address them and to effect sustainable long-term solutions. International initiatives such as the Sustainable Development Goals of the United Nations (UN) and the Aichi Targets of the CBD seek to overcome these, and require more collaboration and cross-linking between actors in civil society, business and governance. Integration is a fundamental challenge for sustainable development, demanding improved dialogue between sectors and the integration of perspectives, policies and strategies between social, environmental, economic and cultural arenas, as well as the creation of new institutional frameworks.

Against this background the linkages between the biosphere and human health and well-being have become of increasing importance in international science and policy in the past two decades, with new interdisciplinary and "transdisciplinary" fields emerging to address the gaps in knowledge and action based on ecosystem approaches to health (or ecohealth). This includes the concepts of "One World, One Health" and conservation medicine, and initiatives such as the EcoHealth Alliance and Cooperation on Health and Biodiversity. The WHO–CBD state-of-knowledge review (WHO and CBD Secretariat 2015) advocates more integration, recognizing the diverse integrating concepts and communities and propagating the One Health concept as one overarching ideal. Earlier, the WHO, the World Organization for Animal Health (OIE) and the Food and Agriculture Organization (FAO) issued a similar call for integration in a tripartite concept note entitled *Sharing Responsibilities and Coordinating Global Activities to Address Health Risks at the Animal–Human–Ecosystems Interfaces* (FAO, OIE and WHO, 2010).

These approaches build on the concept of the ecosystem approach to biodiversity conservation promoted by the UN's Convention on Biological Diversity (CBD), which aims to account for the interactions between various levels of biological complexity and recognizes that "humans, with their cultural diversity, are an integral component of ecosystems" (CBD, n.d.: COP Decision V/6). As such, ecosystem approaches to health are systemic approaches to population health that recognize intimate links between the health of the biosphere and the health of human communities, and frequently incorporate perspectives of ecology, human and veterinary medicine, agriculture, economics, sociology, and aspects of risk assessment, engineering and conflict resolution. To date, much of this work has, for the most part, been led from within the environment and conservation biology disciplines, fronted by environmental research institutes and NGOs, alongside intergovernmental environmental agencies. However, as the science on these issues has progressed, so too has the understanding within the medical science and healthcare communities of how major public health issues and emerging health threats may be associated with global environmental change, and how interactions with ecological systems affect disease risks, health outcomes and the efficacy of public health management strategies (WHO and CBD Secretariat, 2015).

There are numerous constraints to systemic and cross-sector approaches. These include cultural, resource and political barriers, knowledge gaps, and differing temporal scales of operation. Campbell-Lendrum (2005) identified three core difficulties that hinder such engagement by the health sector on environmental issues: a lack of awareness of the relevance to health; a methodological approach focused on discrete cause–effect relationships rather than systemic issues; and little input to processes addressing the environmental root causes of health problems. Therefore, realization of these approaches requires development of a strong evidence base, a mutual understanding of perspectives between sectors, common frameworks for assessment, and practical collaborative strategies based on an appreciation of shared risks and opportunities. The concept of ecosystem goods and services (referring to the benefits which ecosystems provide to society, and often considered together simply as "ecosystem services") has been important in helping to bridge these gaps, and serves as a framework on which to build ecosystem approaches to health and well-being. Whilst the precise relationship between biodiversity and the delivery of ecosystem services is not always clear and is often contentious, not least in terms of the precise connections between biodiversity and services that are relevant to health, in a general sense it is widely accepted that biodiversity is important for the key traits of resistance and resilience in ecological systems and as such underpins ecosystem services. Although the attribute of "diversity" is not necessarily essential to the delivery of certain ecosystem services in every scenario – for example, a monoculture forest plantation might supply as much timber as a native mixed woodland of similar size – diversity helps to secure the sustainability and flow of multiple ecosystem services, and supports adaptation to environmental change. (That same monoculture might be more susceptible to disease or drought, and provide less in terms of other services, such as pollination and food resources; see, for example, Gamfeldt *et al.*, 2013.)

Whilst still an emerging and evolving field, the study of ecosystem services is progressing rapidly, with particular attention paid in recent years to the economic aspects of ecosystems, and the economic policy implications of ecosystem change and biodiversity loss. Despite the growing awareness of linkages between biodiversity, ecosystem services and public health, the health dimension remains comparatively undervalued in policy and practical contexts, even within recent economic studies.

However, in recent years there have been some significant developments in international policy as well as in local, regional and global responses towards the integration of biodiversity and human health, of which the WHO–CBD collaboration is one of the most promising examples (WHO and CBD Secretariat, 2015). At the 2015 meeting of the CBD's Subsidiary Body on Scientific, Technical and Technological Advice (SBSTTA) the WHO–CBD recommendations were supported and CBD parties were requested to report on progress on those recommendations in their national reports and national biodiversity strategies and action plans. Of specific relevance are Aichi Target 14 ("By 2020, ecosystems that provide essential services, including services related to water, and contribute to health, livelihoods and well-being, are restored and safeguarded, taking into account the needs of women, indigenous and local communities, and the poor and vulnerable") and Sustainable Development Goals 2 (food security/nutrition), 3 (ensure healthy lives) and 12 (sustainable consumption).

Scientific challenges

Whilst the need for integrating biodiversity and ecosystem approaches into the health sector is evident, it is equally important that the biodiversity sector for its part recognizes the potential impacts – positive and negative – of conservation policies and activities on human well-being. This is important, for example, when in urban planning incorporation of nature is considered for improving citizen health (Lõhmus and Balbus, 2015; Keune *et al.*, 2013). Interdisciplinary science can provide options for such management, and society at large can then make informed decisions when choosing among those options. Ernstson (2013) states that these decisions must also account for issues of social justice, such as the need to ensure equity in decision-making, accounting for varying costs and benefits of resource management that may be experienced by different communities, and has highlighted issues of conflicting societal choices in terms of how urban ecosystems and associated benefits may be valued. Jax *et al.* (2013) have also highlighted the issue of equity and justice in terms of access to ecosystem services. Considering how specific patterns of design can restrict access to ecosystems that may support health, or make their availability more important, this is surely another important dimension to be addressed. Additional scientific research will inevitably be needed to better understand the interplay between biodiversity and public health. Hartig *et al.* (2014) in their review discuss the growing amount of research on health benefits from nature, yet also highlight the considerable scientific challenges that remain. Keesing and Ostfeld (2015) draw a similar conclusion regarding the complex relationship between biodiversity and infectious diseases. Addressing these challenges will require further research, greater use of conceptual models that straddle disciplinary agendas, and greater involvement of practitioners in urban planning and associated assessments.

New institutional science–society arrangements: the need for integrated transdisciplinary approaches

Due to their complexity, both biodiversity and public health science have to cope with limited understanding. The relationship between biodiversity and public health is even more complicated and of great societal concern: public health is high on the agenda and its relation with biodiversity has enormous potential. The combination of societal importance and a vast decline in biodiversity worldwide also challenges the evidence-based culture of policy-making: by the time science is able to prove the importance of biodiversity for public health, important biodiversity will probably be lost. Thus scientific practice is also challenged: do the traditional methods of scientific proof suffice here? If we take environmental health as an example of combining public health with environmental issues, over the last few years there has been considerable debate about this problem of limited knowledge and great social importance. Traditional scientific routines searching for complete knowledge and strict statistical proof have been criticized and precautionary approaches are advocated (Grandjean, 2008). There is increased demand for more participatory approaches to knowledge assessment and policy interpretation in which experts, policy-makers and stakeholders collaborate. Public health can bring different actors together as it is an important concern in many policy domains. As such, it can be powerful as an integrator of a diverse range of policy objectives.

Whilst specific barriers hinder the natural flow of ideas between various scientific disciplines concerned with health–biodiversity linkages, there are also particular issues at the interface of science and public policy. How should scientific assessments be translated into effective public policy? How should societal choices account for issues of scientific complexity and uncertainty? How can valuation methodologies (e.g. of the importance of biodiversity to health and well-being) account for differing cultural perspectives, differing policy goals, and a variety of potentially conflicting community needs? The challenge at the science–society interface consists in developing adequate interfaces, but also in dealing with its intrinsic complexity as a social interface. Long (2001: 243) defines a social interface as "a critical point of intersection between different life worlds, social fields or levels of social organization, where social discontinuities based upon discrepancies in values, interests, knowledge and power, are most likely to be located." Interfaces lead to realities that have to be recognized as complex by practitioners, scientists, policy-makers and funding bodies.

An important aspect of ecosystem governance and a strong rationale for the further development of science–policy interfaces is the need for policy to incorporate different types of knowledge in decision-making processes. A "science–policy interface" (SPI) can be defined as: "relations between scientists and other actors in the policy process which allow for exchanges, co-evolution, and joint construction of knowledge with the aim of enriching decision-making" (Van den Hove, 2007: 807). The most prominent example in the field of biodiversity is the Intergovernmental Science-Policy Platform on Biodiversity and Ecosystem Services (IPBES). The plea for including a broader range of knowledge and stakeholders beyond "'elite actors', from natural scientists to national governments," has been voiced repeatedly, recently by Turnhout *et al.* (2012: 454).

Study cases

Biodiversity, community health and traditional knowledge (Suneetha M. Subramanian and Unnikrishnan Payyappallimana)

Biodiversity and related traditional knowledge have been actively used in natural product drug discovery (Newman and Cragg, 2007). Beyond mainstream healthcare systems, they form the basis for health and nutritional security to people in insufficiently connected and marginalized regions of the world. These regions are marked by richness in medicinal resources, traditional medical practitioners with knowledge of their use, and an inadequacy of public healthcare infrastructure and personnel.

To address improvement in healthcare access for people, there is a renewed interest in strengthening the potential of traditional medical knowledge and health practitioners to fulfill this role. At the moment the focus of multilateral organizations such as the World Health Organization (WHO) is predominantly on ensuring quality, safety, efficacy, rational use and national regulations related to traditional medicine within a market-oriented perspective, with limited attention towards existing knowledge and practices of local communities and their revitalization for achieving health goals at the community scale.

Below, we highlight some important areas that require focus in this context, and list some related inspiring case studies from South and Southeast Asia, where attempts have been sustained at integrating traditional medicine, resources, practices and practitioners.

Traditional and local knowledge

Identifying local health priorities and supplementing them with ecosystem and community-specific traditional medical knowledge and resources through primary health programs is critical (see Box 13.1).

ABHAIBHUBEJHR HOSPITAL FOUNDATION (CAF), THAILAND

The Chao Phya Abhaibhubejhr Hospital Foundation (CAF) in Prachin Buri, Bangkok, was founded in 1983, focusing on conservation and the revitalization of traditional local practices of food, healthy lifestyles, and cultural traditions in addition to clinical care. A self-reliant community healthcare model has been established through the use of: a database of documented traditional knowledge,

---

Box 13.1 Integration and mainstreaming of traditional medicine

In India, as part of the UNDP–GEF–CCF program, a major conservation program was initiated with the State Forest Departments. A key pilot component in this project was to link biodiversity conservation with the official healthcare delivery systems. This involves a unique model addressing related issues to integrate local medicinal plants, related traditional knowledge and practitioners into the primary healthcare system in twenty-four primary healthcare centers (PHCs), including twenty allopathic centers and four Ayurveda dispensaries. The work forms part of the efforts of a voluntary organization called Karuna Trust and is implemented through twenty-five health volunteers (*arogyamitras*) from selected PHCs. Priority health problems, health practices and local medicinal plants for these conditions are identified, and their safety and efficacy are assessed through a participatory approach, complemented by a traditional medicine manufacturing unit and home and community herbal gardens

(Payyappallimana and Subramanian, 2012)

networking of health practitioners and community knowledge holders on the Thai–Myanmar border, northeastern and southern Thailand; scientific validation of community-based assessment; promoting conservation as well as organic cultivation and semi-processing of medicinal plants; product development (such as herbal medicines, food supplements) based on local knowledge with strict quality control as per good manufacturing practices and ethical commerce. For example, Thao Wan Prieng, a capsule used for reducing muscle pain and stiffness, is widely used by the communities. To foster awareness among younger generations, the foundation organizes youth camps in forests on local health traditions and biodiversity (interview with Dr. Supaporn Pitiporn, Abhaibhubejhr Hospital Foundation, Bangkok, 26 November 2011).

TRADITIONAL MEDICINES IN PRIMARY HEALTHCARE, JAPAN

A popular and successful model of including traditional medicines in primary healthcare services is haichi (*okigusuri*) – a traditional public healthcare model based on the concept of "self-help medication" that has been practiced since the seventeenth century. This model originated in Toyama, the third-largest prefecture in terms of pharmaceutical production in Japan, one third of which specializes in haichi medicines. Haichi is based on a "use first, pay later" principle. The distributor visits each household at set times, providing a medicine box with ten to twenty essential drugs. During subsequent visits, the distributor collects his or her fee for the medicines used. Remarkably, all such medicines are regulated for household use by the Ministry of Health, Labor and Welfare and thus cannot contain medicine that is usually prescribed by a doctor. While the medicines are distributed, they are classified according to efficacy and risk. The contra-indications or special precautions are explained to the households, whose information is documented in an electronic database.

LESSONS LEARNED

The Indian case shows that a participatory and community-based program can lead to increased confidence among community members in their practice. It also helped to integrate them into conventional primary healthcare. The Thai case demonstrates an approach that comprises strengthening various links in a value chain of traditional medicine preparation to reinforce and revitalize such traditions. The Japanese case is used as a best-practice model in some East Asian countries due to its success.

CAPACITY-BUILDING

In order to build on the natural resources, namely ecosystems and biodiversity, as well as the knowledge, skills and capabilities of the populations who live in close proximity to biological resources, it is also important to focus on appropriate capacity development and educational activities (see Box 13.2).

Box 13.2 Community knowledge and capacity development for education

CAPTURED is a unique collaborative programme of universities in Ghana and Bolivia as well as NGOs such as FRLHT and ETC/COMPAS. In total, it brings together sixteen universities from four continents. It was developed as a response to a situation where education systems in Africa, Asia and Latin America could not articulate their relevance to their own societies.
CAPTURED's work includes:

- Developing innovative forms of higher education and research that contribute to revitalize traditional knowledge based on a dialogue between academic and endogenous scientific communities.
- Building theories on endogenous development and strengthening the institutional capacities of the related universities in endogenous research and training.

LESSONS LEARNED

CAPTURED's approach has proven to be a successful model to promote education models that integrate experience-based subjective knowledge and reinforce intergenerational transfer of knowledge and skills in communities through appropriate pedagogical systems that are sensitive to traditional approaches. Its work highlights the need to foster receptivity among traditional healers and integrate their knowledge systems into mainstream education systems.

Public–private partnership

As biological resources, such as medicinal plants, and other resources form important aspects of the health, nutrition–biodiversity rubric, it is critical to ensure their survival through effective conservation and sustainable use (Hamilton, 2004). One in five of the world's plant species is estimated to be threatened with extinction in the wild, with unsustainable harvesting being a major driver. Where they exist, efforts to promote sustainable management, transparency and increased benefit sharing have to contend with widely distributed harvest communities and highly complex trade chains. This calls for coordinated action among different actors, including the private sector, to address the issue of sustainable use of medicinal plant resources (see Box 13.3). The situation is even more complex with regard to faunal resources that are used for medicinal purposes.

Enterprise development using biological resources such as medicines, intermediate products for commercial utilization and dietary interventions fostering enhanced livelihood security can facilitate contextually relevant conservation and access to medicinal resources (see Box 13.3).

Box 13.3 Public–private partnership for medicinal plant conservation

Studies suggest that 81 percent of 960 medicinal plant species are in active trade and entirely or largely sourced from the wild in India (Ved and Goraya, 2008). Alarmingly, more than 70 percent of these are collected through destructive practices as stem, wood, bark, roots or even whole plants are used (FRLHT, 1999, 2009). In addition, the demand for plant resources from modern industries, including pharmaceuticals, botanicals and cosmetics, is on the rise (Laird and Wynberg, 2008). This, along with various other factors, such as the loss of forests, encroachment and conversion, and unsustainable practices, has led to the destruction of genetic diversity and habitats of several valuable natural resources.

To address these pressing issues, the Foundation for Revitalization of Local Health Traditions (FRLHT) in India initiated the establishment of Medicinal Plant Conservation Areas (MPCAs) as an integrated approach of in situ and ex situ conservation programs in India. The aim is to conserve and study medicinal plants through a participatory approach in their natural habitats, preserve their gene pools and develop strategies for the management of rare, endangered and vulnerable species. Between 1993 and 2012, FRLHT jointly with the State Forest Departments established 112 MPCAs across 13 Indian states in a globally unique model of public–private partnership (PPP). In addition, there has been a recent move to recognize these locations as biodiversity heritage sites. A related initiative is the establishment of Medicinal Plant Conservation Parks (MPCPs). This is a community-based ex situ conservation initiative aimed at the sustainable use of medicinal plant resources and preserving knowledge associated with their use. Within each geographical region, communities have been mobilized to create ethno-medicinal forests and resource centers housing herbaria and drug collections as well as local pharmacopoeia databases based on community knowledge.

LESSONS LEARNED

The initiative affirmed the role of cultivation as a complementary approach to conservation, emphasizing the need to consider local biological and cultural diversity as well as local development priorities.

Conservation and health, local knowledge

Implicit in such decentralized conservation measures is the need to strengthen local innovation through livelihood programs and local enterprises (see Box 13.4). A further requirement is to develop the capacities of traditional health practitioners

---

Box 13.4 Community livelihoods: linking conservation with community health

An example of good practice for linking conservation with development objectives is the program to enhance local livelihoods among highland communities developed by the International Center for Integrated Mountain Development (ICIMOD). Its overall goal is to promote the cultivation, value addition and marketing of specific medicinal plants of high value found in countries such as Nepal, Bhutan and Bangladesh.

(see Box 13.5). Mechanisms for protection of such traditional knowledge resources (see Box 13.6), prevention of their erosion and establishing links with scientific research are related areas that also demand further attention.

LESSONS LEARNED

The ICIMOD program highlights that promoting community-based enterprises through traditional medicinal resources and products is a successful model for both improving livelihoods and fostering sustainable natural resource management. The case further highlights a best-practice model to streamline policies related to access to resources and equitable sharing of benefits (ABS) while at the same time including value-addition activities at the local level.

Another good example of an NGO facilitating capacity development for traditional healers is Friends of Lanka, based in Sri Lanka. It has promoted the documentation of practices, research and networking of traditional health practitioners. Around seventy-five healers have been identified from a total population of 8,000, treating various conditions such as snake and insect bites and certain food or natural poisons which are considered as leading causes of

---

Box 13.5 Capacity development for traditional health practitioners

Since 1993, FRLHT has been promoting local healer activities in various states of India. As such, it established the Medicinal Plant Conservation Network (MPCN) with a number of NGOs that work with different rural communities. As part of this effort, traditional healers' associations have been formed at different administrative levels. These associations hold regular meetings that facilitate exchanges among the healers and act on the basis of self-regulatory guidelines. Along with NGOs and state forest departments, they have been actively engaged in supporting medicinal plant conservation programs in various states (FRLHT, 1999).

---

morbidity and mortality in rural areas of developing countries (Payyappallimana and Subramanian, 2012).

LESSONS LEARNED

Networks such as the MPCN and the associations of healers established within its framework are a successful approach to facilitate knowledge and experience exchange among traditional healers nationally, regionally and internationally.

More recently, communities have been articulating their rights over their knowledge and resources by developing their own bio-cultural community protocols (Bavikatte et al., 2010). Defined by communities, these highlight the legal rights that are vested in communities by virtue of international and national laws and provide self-descriptions of each community's profile, resources, rights and responsibilities. They also outline the terms of engagement with external agents.

Box 13.6 Sustainable protection of traditional knowledge

Health-related knowledge has been more commonly accessed for developing new medicines and continues to be profiled as the primary sector where misappropriation of knowledge, practices and resources occur. Hence, searchable databases such as the Traditional Knowledge Digital Library (TKDL) in India, which are pertinent to health-related traditional knowledge (TK) and ensure the protection of related resources and knowledge, are being developed. By providing information on TK in languages and formats that are comprehensible to patent examiners at International Patent Offices (IPOs), the database makes an immense contribution to preventing the granting of wrong patents. In parallel, various organizations are undertaking similar exercises to document oral knowledge or knowledge in informal domains through Community Knowledge Registers (CKRs). Chiefly led by NGOs, these registers encourage community members to discuss and document their knowledge and practices in different categories of resource use, eventually reinforcing such uses as strong social traditions.

LESSONS LEARNED

The positive outcomes and impact of the TKDL case and the Community improvement of CKRs and bio-cultural protocols and link them with national databases for protection. In addition, they show that it is necessary to build on and upscale good practices of ethical and equitable agreements with international organizations and industries related to the use of TK and natural resources for research or commercial purposes.

Box 13.7　Using traditional medicinal practices for preventive and curative health interventions

The Research Initiative on Traditional Antimalarial Methods (RITAM) was established in 2001. Initiated by a group of international researchers to explore ways to increase the relevance of including traditional medicine in the repertoire of choices available for prevention and cure of malaria, RITAM is working on traditional antimalarials with more than two hundred members in over thirty countries. Data from the field in India has shown statistically significant positive outcomes in malaria prevention.

(Payyappallimana and Nagendrappa, 2007)

Community health programs

Addressing public health through effective traditional medical knowledge and locally available biological resources is important. There have been a number of successful programs addressing community health and sustainable natural resource management. A couple of examples are discussed below.

HOME HERBAL GARDENS AS A SELF-RELIANT COMMUNITY HEALTH PROGRAM

Home herbal gardens, as conceived by FRLHT, include fifteen to twenty prioritized medicinal and nutritional plants that are designed to provide self-reliant primary healthcare in a community. In addition to aiding the conservation of medicinal plants, these gardens also help to address nutritional challenges. Today, approximately 200,000 home gardens across ten states in India are meeting the primary healthcare needs of some of the poorest households, while reducing their healthcare expenditure. A majority of participants are now contributing fully to meet the costs of raising their own medicinal plants. Studies show that there are substantial healthcare cost savings when home remedies are used (Hariramamurthi *et al.*, 2006).

THAI–BURMA BORDER REFUGEES – TRAINING BACKPACK HEALTH WORKERS

Healthcare access for populations in special circumstances, such as refugees, is a subject of keen interest to policy-makers. While a basic human right, access to healthcare is often stymied by political factors, a lack of personnel who are qualified in modern medicine, and poor infrastructure. Burmese refugees on the Thai–Burma border faced all of these problems. Efforts to establish a clinic (the Mae Sot Project) under the strong leadership of one of the refugees (Dr. Cynthia Maung), which aimed to offer both in-patient and out-patient care, were challenged by a lack of personnel and medical resources. Working with the Global Initiative for Traditional Systems of Health (GIFTS of Health), a program of health intervention including traditional healers – who were among the refugees and connected to each other through an informal network – was designed. It was found that the refugees preferred to consult with traditional healers for their primary healthcare needs, but preferred Western medicine for acute conditions. It was also observed that cooperation between Western clinical services and traditional health practitioners was closely linked to their well-being, including cultural continuity and refugee identity. The program was funded via a combination of grants from local NGOs and patient contributions (Bodeker and Neumann, 2012).

LESSONS LEARNED

The first case indicates that utilizing traditional medical knowledge through community-based participatory approaches is both feasible and urgently needed to find solutions to the continuing high incidence of preventive and curative diseases, such as malaria, in regions where they are endemic. This also requires consideration of ethical factors, such as free, prior and informed consent (Payyappallimana and Nagendrappa, 2007). FRLHT's home herbal gardens scheme is a successful model to promote access to healthcare through sustainable natural resource management of medicinal plants. It has not only reduced poverty in rural areas but also revived local knowledge of medicinal plants and traditional health practices.

Validation of community knowledge and practices

There is an urgent need to develop tools related to the validation of community knowledge and practices relevant to biodiversity conservation, health and nutrition security (see Box 13.8).

Box 13.8 Documentation and assessment of local health traditions

To assess and promote best practices of local health traditions while also protecting the intellectual property rights of community knowledge, FRLHT developed the Documentation and Participatory Rapid Assessment of Local Health Traditions (DALHT). This approach brings together local traditional health practitioners, allopathic and Ayurvedic physicians, botanists and field workers. It is used by local NGOs and community-based organizations to:

1. Identify and document repeatedly used remedies from local knowledge holders.
2. Prioritize health conditions.
3. Identify and document literature related to symptoms and remedies for specific conditions from codified systems as well as modern research studies.

Apart from assessing the efficacy of practices, it is helpful to differentiate and identify sound traditional health practices, taking into consideration the world-views and epistemologies of folk practitioners and those who possess house-hold knowledge. Based on consensus, these key stakeholders further identify and promote best practices within the community. Additionally, incomplete practices are supplemented with knowledge from other sources, while ineffective practices are discarded.

Nearly 70 percent of Indian practices have supporting evidence from Ayurveda and modern pharmacology on their prescribed use (Santhanakrishnan *et al.*, 2008). Based on the local pharmacopoeia, simple users' manuals are prepared in the vernacular with details relating to their use.

LESSONS LEARNED

Proving the efficacy of every community-based practice through various stages of clinical studies would demand considerable amounts of time and money. This is often impractical for a rapid community health intervention. In this context it is important to employ locally relevant but still scientifically rigorous assessment methods through participatory approaches. Methods such as DALHT are both cost effective and practical.

Several community-based approaches to validation have been introduced, with participatory clinical studies used alongside as the likes of DALHT. One problem with these semi-formal methods is a lack of capacity among the stakeholders who conduct the studies to share their findings with the scientific and policy-making communities in an acceptable form. Addressing this would facilitate knowledge sharing with other communities and adoption of similar efforts elsewhere, influence national and international policy-making, and foster rigorous research among research institutes and scientists.

The state of biodiversity and health linkages in Bangladesh (Sate Ahmad, Mohammed Mofizur Rahman, and Mohammed Nurul Azam)

The populous deltaic country of Bangladesh, which lies in the Ganges–Brahmaputra–Meghna river basin in South Asia, is rich in biodiversity. This is due to its geographical location in the subtropical belt at the confluence of two major biotic sub-regions of the Oriental Region: the Indo-Himalayas; and Indo-China. It is home to a total of 121 species of mammals (including the famous royal Bengal tiger and various cetaceans), 690 birds, 158 reptiles and 53 amphibians (Khan *et al.*, 2008) and 3,611 floral species (Ahmed *et al.*, 2008). The richness of species in Bangladesh is well recognized when compared with other areas in the region. The diversity of birds is a great example: the total number of bird species in Bangladesh is equivalent to that of Europe. However, biodiversity in Bangladesh is declining at a rapid pace even though a large proportion of its citizens depend on biodiversity and related ecosystem services for their lives and livelihoods.

Owing to its low per capita income and poor health outcomes, Bangladesh has concentrated most of its policy efforts in the areas of economic development, poverty alleviation, education and public health. Since the 1980s, coastal populations in Bangladesh have made some progress in all of these areas, but often at the cost of environmental degradation. Biodiversity conservation has been largely neglected and has featured only as an afterthought in most policy discussions, even though a large percentage of the economy depends on diverse natural resources and flows of ecosystem services. Recently the government has paid more attention to biodiversity, although its policies in this respect are rarely based on sound scientific research and they usually lack proper implementation.

The country has made significant progress in the arena of public health, with a steep decline in rates of fertility and mortality in both urban and rural areas (NIPORT *et al.*, 2011). Health science research has been gaining pace in its own way and has received international recognition. Research in the field of biodiversity has also progressed but with little or no explicit focus on the link between biodiversity and health. Nevertheless, such linkages can be drawn from various studies, such as those involving indigenous medicinal plants, exotic species, water quality, mangroves as a habitat for cholera vectors and as means of protection against natural hazards such as cyclone and storm surge, genetic diversity in agriculture, and so on. Only recently have some research institutes started looking explicitly at the relationship between ecosystem services and human health. Only a handful of these studies have been undertaken thus far, such as the project on "Assessing Health, Livelihoods, Ecosystem Services and Poverty Alleviation in Populous Deltas" under the ESPA program (ESPA Delta, 2012). Biodiversity, once perceived as a luxury of the few, is now an articulated demand of the common citizen as this has linkages with income generation, poverty reduction and improving human well-being. Similar multidisciplinary research should focus on the relationship between biodiversity and human health, and pave the way for developing countries like Bangladesh to prioritize biodiversity conservation and place greater emphasis on this neglected area.

Bangladesh has practiced both in-situ and ex-situ conservation to maintain the country's biodiversity. The declaration of protected areas, ecologically critical areas (ECAs), World Heritage sites and Ramsar sites are a few examples of in-situ conservation while botanical gardens, preservation plots, gene banks and arboretums are examples of ex-situ conservation (Mukul, 2007). In the international arena of biodiversity conservation, Bangladesh has signed and ratified most of the treaties and conventions (Mukul, 2007).

Significant progress has been made in Bangladesh's legislative regime in terms of biodiversity conservation. The flagship convention of the UN on biodiversity – the Convention on Biological Diversity (CBD) – was signed by Bangladesh in 1992 and ratified in 1994. The government also ratified the Convention on the International Trade of Endangered Species of Wild Fauna and Flora (CITES) and the Ramsar Convention on Wetlands in 1982 and 1992, respectively (Ministry of Environment and Forest, n.d.). The National Biodiversity Strategy and Action Plan (NBSAP) of Bangladesh was prepared in 2005 to fulfill the country's commitment to the CBD (Government of Bangladesh, 2005). It outlines several pieces of environmental legislation that are important for the protection and conservation of biodiversity. These include the Environment Conservation Act 1995; Environment Conservation Rules 1997; the Forest Act 1927 (amended in 1990 and 2000); the Protection and Conservation of Fish Act 1950; the Marine Fisheries Ordinance 1983; the Wildlife (Preservation) Order 1973; the Wildlife Preservation (Amendment) Act 1973; and the Bangladesh National Conservation Strategy (NCS).

Notwithstanding these policies and strategies, the government and research organizations have fallen short of addressing biodiversity in a comprehensive and integrated manner. Thus the association between biodiversity and human health is absent in most cases. However, in 2005, for the first time, the NBSAP explicitly acknowledged the importance of maintaining biodiversity for human health and stated that one of its objectives was to "ensure that long-term food, water, health and nutritional securities of the people are met through conservation of biological diversity" (Government of Bangladesh, 2005: v). The role of science in policy-making is a necessary but not sufficient condition. In Bangladesh, significant gaps exist not only between science and the formulation of policies, but also between policy formulation and implementation. There has been a significant lack of integration in the wide range of legislation mentioned above, and coordination among relevant government institutions and ministries has been absent for the most part. Thus implementation of these policies and strategies has been erratic, at best, and largely ineffective.

In addition to the science–policy gap, there is disciplinary fragmentation and a compartmentalized institutional attitude. Essential data on biodiversity and health are usually not shared by institutions or made publicly available, sometimes leading to the duplication of effort. Research prioritization in Bangladesh is primarily donor driven, because that is where the money is, and extremely volatile, with constantly shifting focus as the donors' priorities change. Thus, research is never truly demand driven. Moreover, funding often tends to be sector-specific, which leaves less money for sound and integrated research.

Pesticide use, biodiversity loss and farmers' health risks in Vietnam (Joachim H. Spangenberg and Zita Sebesvari)

In rural settings with a strong focus on agricultural production, such as the Mekong and Red River Delta in Vietnam, farmers face both direct and indirect exposure to pesticides: direct during application and contact or inhalation while handling them; and indirect when consuming polluted food items and water. Large-scale pesticide use puts ecosystem services at risk (Chagnon *et al.*, 2015; Settele *et al.*, 2015) and chronic exposure to a mixture of pollutants can lead to a large variety of health impacts (Prüss-Üstün and Corvalán, 2006). This is an issue all over rural Southeast Asia, where governance is weak or absent and pesticide use, driven by market promotion, is way above technically necessary levels (Heong *et al.*, 2015). Furthermore, pesticides are sold without comprehensive application information, promoted through misguiding advertising from the chemical companies, available in almost every corner shop, stored in unsafe places, such as kitchens, and often applied to crops without protective gear. Even pesticide dealers complain about the lack of regulation and control. Low awareness of pesticide risks for both human health and the environment often coincides with both unquestioned habitual routines of spraying, perpetuated by a lack of biological or ecological knowledge (Spangenberg et al., 2015), and high reliance on natural resources such as clean water. Natural clean water is indispensable for irrigation, fishing and aquaculture, and for domestic use (drinking) from surface water bodies in the dry season, yet there is only basic treatment which does not remove the large majority of pesticide pollutants (Toan et al., 2013; Chau et al., 2015).

Linkages between biodiversity and agrichemical use

Pesticide contamination can cause significant losses in both species and family richness of stream invertebrates, even at concentrations that are below the level of concern in European regulations (Beketov et al., 2013). Some of the affected invertebrates, such as dragonflies, contribute to biological pest control. Reducing them therefore likely leads to higher pest infestation rates and – if prevailing plant protection mechanisms do not change – ultimately to increased use of pesticides. This vicious cycle was shown to prevail throughout Southeast Asia, where every year significant losses of rice harvest occur due to infestations by plant-hoppers, in particular the brown plant-hopper (*Nilaparvata lugens*) and the white-backed plant-hopper (*Sogatella furcifera*). Most plant-hoppers, and all those which are categorized as pests, are rapidly reproducing organisms, adapted to thrive in environments undergoing perturbations (Heong, 2009). The usual reaction to hopper infestation – especially in intensive wet rice agricultural areas such as central Thailand, Vietnam and parts of China – is to intensify insecticide spraying to combat the insects. This is based on a mental model which associates every infestation with significant harvest losses and conceives spraying pesticide as the natural, best-practice solution (Spangenberg et al., 2015). However, spraying often increases the rice crop's vulnerability to pests in the event of subsequent infestations, as it indiscriminately destroys natural hopper enemies and the ecosystem services they provide (Yang et al., 2014). While the method of choice seems to be of limited effectiveness, so far intensification of spraying rather than trialing alternative means of reducing hopper-induced losses has been observed, even though the principles of sustainable rice management are well established (Savary et al., 2012). Obviously a problem with the feedback mechanism is preventing effective learning.

The long-term change to integrated pest management (IPM) practices remains weak despite several training campaigns. The reasons for this are manifold but certainly linked to the lack of learning, persistent trust in pesticide application, and weak understanding of the relationship between pests and natural enemies. Farmers themselves have mentioned concerns about employing IPM in their fields when it is not applied in surrounding fields, pressure from pesticide retailers, and low net profits when employing IPM (Toan *et al.*, 2013). Effectively, the landscape managers (farmers, planners, etc.) have maximized one ecosystem service (yield) at the expense of many others (biocontrol, water purification, etc.), triggering a feedback loop that undermines their intended maximization.

Linkages between agrichemical use and health in farming households

Farming family members and hired farm workers are exposed to pesticides during application (mainly inhalation) and handling (mainly skin contact) and through ingestion of contaminated food or water. The exposure in the course of handling and application would be easily reduced if protective measures were applied. Th farmers' behavior is influenced not only by limited knowledge but also by their social situation and their basic world-view. When interviewed regarding the health impacts they felt after spraying insecticides, medium-sized farmers in the Philippines pointed out that they do not do the spraying themselves but hire relatively well-paid workers for the purpose, and that the latter can be easily replaced in the event of pesticide-induced health problems, which the farmers attributed to a lack of care. Small-scale farmers in Vietnam described symptoms such as headaches, stomach problems, breathing problems and sometimes even coordination problems after spraying, but they neither changed their modus operandi nor complained to the manufacturers or authorities. Instead, they simply accepted these issues as normal professional risks, normal aspects of farmers' lives and working routines. In their view, the pesticides' effectiveness and low price almost always outweighed the risks for human or environmental health.

The consumption of polluted water is likely to have more serious health consequences for rural households than for urban dwellers due to farmers' use of pesticide-polluted surface water for drinking and other domestic purposes (Toan *et al.*, 2013; Chau *et al.*, 2015). Furthermore, the consumption of contaminated fish and vegetables poses a significant risk for subsistence farmers as their diets include a high proportion of these products (Hoai *et al.*, 2011).

The linkages between water quality and health in the Mekong Delta, Vietnam, are not comprehensively monitored or assessed; water quality in smaller rivers and canals is largely unknown. Moreover, the authorities have failed to monitor a number of agrichemicals that are now in widespread use in the region. In the course of the WISDOM project, fifteen of these recently introduced pesticides were monitored, most of them categorized as moderately or slightly hazardous; their concentrations in surface water and household drinking water occasionally exceeded health guidelines. However, Toan *et al.* (2013) argued that the year-round co-occurrence of a wide range of pesticides in the region's water supply likely causes chronic exposure among aquatic organisms as well as the rural population. They therefore called for more effective treatment methods, monitoring of a wider range of pesticides, and the development of health guidelines that address mixture toxicity.

For more sustainable and strategic solutions beyond application rates and waste management at the household level (Berg and Tam, 2012; Sebesvari *et al.*, 2011, 2012), measures should attempt to tackle all of the important aspects of pesticide risk to human health. One approach could be ecological engineering – arguably the least known of the methods that enable us to cross the biodiversity–human health bridge.

Ecological engineering

An alternative to pesticide spraying is offered by ecological engineering (Gurr *et al.*, 2012), which emphasizes not the suppression but the deliberate exploitation of existing biological structures and mechanisms, such as food webs, to provide effective biocontrol and pollination services (Gagic *et al.*, 2012). Ecological engineering includes reduced and delayed spraying to avoid disturbing the available biocontrol potential, and actively supporting it by planting suitable, nectar-rich plants on the paddy dykes, which provide shelter and food for biocontrol agents.

When trialed in China, Thailand and Vietnam, the ecological engineering approach reduced both harvest losses and input costs (seed, insecticides and fertilizer) and saved farmers' working time (Gurr *et al.*, 2016). It also reduces health risks for both producers and consumers. Nevertheless, as yet, there has been little enthusiasm among Southeast Asia's farmers to adopt the approach.

Challenges

Given the complexity of cause–effect mechanisms, it is often difficult to prove that a given environmental concentration of pesticides is having a direct impact on the health status of individuals. Precaution supported by epidemiology is a possible solution, but this requires reliable and accessible public health data, which is lacking in many Southeast Asian countries.

One of the major reasons for concern is that we know relatively little about the additive effects of chemicals – their mixture toxicity – in an environment and human health context. While current pesticide regulations consider lethal doses of single pesticides, evidence is growing that low-level exposure to a combination of pesticides can be more harmful than high exposure to individual pesticides. However, it is often difficult for farmers (including well-informed European farmers) to resist the temptation to mix several pesticides, as doing so significantly reduces labor costs and the equipment needed for spraying (Dänhardt and Spangenberg, in press).

Individual goodwill or marginal management changes will not alter this situation. A transformation of agricultural practice must encompass all elements of land management, and must be supported by adjusted framework conditions. Positive health impacts can be an important argument for such a change.

BiodivHealthSEA *(Serge Morand)*

The BiodivHealthSEA (Local Impacts and Perception of Global Changes: Biodiversity and Health in Southeast Asia) project aims to investigate the local impacts of global changes on zoonotic diseases (60 percent of emerging diseases), with a focus on rodent-borne diseases, in relation to biodiversity changes.

Southeast Asia is a unique location in which to investigate the perception and effects of global changes, including global governance, on the interaction between biodiversity and health. It is a hotspot of infectious emerging diseases that have the potential to become global pandemics (Coker *et al.*, 2011) as well as a hotspot of biodiversity that is at threat due to land use and climate changes (Sodhi *et al.*, 2014; Wilcove *et al.*, 2013). It therefore attracts the attention of international organizations, developmental agencies and non-governmental conservationist organizations due to its global significance in terms of biodiversity and health.

Rodents are used as models for investigation of biodiversity changes and for their important roles as reservoirs of zoonotic diseases (leptospirosis, scrub typhus, plague, viral hemorrhagic fever). Southeast Asia is a diversification center for murid rodents, which appear to be good indicators of habitat changes, carriers and reservoirs of numerous diseases, and agricultural pests. These animals are traditionally hunted, and some of them are transnationally traded. They are particularly suitable topics for researchers who are studying the relationship between humans and the environment, and those who are assessing health due to diseases transmitted by rodents.

The objectives of BiodivHealthSEA are to analyze: the local impacts of global changes on zoonotic diseases in relation to biodiversity modification and environmental change; and the local perceptions of global environmental and biodiversity changes in connection with the "global governance architecture" and national public policies and actions of NGOs in the sectoral domains of health, conservation biology and land planning. Cross-correlate analyses were performed using various data at the regional and national levels to determine the socio-environmental determinants (climate change, biodiversity loss, change in forest cover) of endemics and epidemic infectious diseases. It has been shown that, within Asia-Pacific countries, the overall richness of infectious diseases is positively correlated with the richness of birds and mammals (Morand *et al.*, 2014). However, the number of zoonotic disease outbreaks is positively correlated with the number of threatened mammal and bird species, and the number of vector-borne disease outbreaks is negatively correlated with forest cover. It therefore appears that biodiversity is a source of pathogens, but also that the loss of biodiversity, as measured by forest cover or threatened species, is associated with an increase in outbreaks of zoonotic and vector-borne diseases (Morand *et al.*, 2014).

The project includes partners from France (CNRS, CIRAD, IRD, Montpellier University), Thailand (Kasetsart University, Mahidol University, Chulalongkorn University, Thammasat University, Maha Sarakham University, Udon Thani

University), Cambodia (LR Mérieux at the University of Health Sciences), Lao PDR (National Institute of Health, NAFRI, CC Mérieux Vientiane), Singapore (National University) as well as collaborators in Vietnam, the Philippines and Malaysia.

At a local level, the effects of rapid deforestation on host–parasite interactions in Southeast Asia were investigated using an extensive data set on rodents, their pathogens and their parasites, acquired from seven localities in the region for which land-cover changes were developed (Dupuy *et al.*, 2012). Unpublished analyses suggest that rapid fragmentation does not affect parasite species richness per se but does strongly impact host–parasite interactions, as investigated using network analysis. Local parasite and pathogen diversities seem to be affected by habitat fragmentation, suggesting a similar trend to the one observed at the regional level (i.e. a decrease of zoonotic diseases is linked with biodiversity loss). However, synanthropic rodents are favored in human-dominated habitats (and disturbed habitats), which may favor the spread of the pathogens they carry to humans – a development that some researchers have already observed (Blasdell *et al.*, 2012; Herbreteau *et al.*, 2012).

The investigation of international initiatives and national policies in the domains of health, climate regulation and biodiversity conservation will assess the roles of institutional actors and the coordination (or lack thereof) between sectors (e.g. health and conservation sectors are establishing a partnership through the One Health initiative). (This is investigated at the regional and national levels and is presented in Chapter 14 of this volume.)

Ongoing analyses of several local situations by a multi-disciplinary team (social scientists, environmental scientists, ecologists and epidemiologists) will provide data on biodiversity, rodent-borne diseases, land uses, local practices and health surveillance as well as information on perceptions of biodiversity/health/environment interdependence, global changes and their impacts among various stakeholders, from international decision-makers to natural resource users and scientists. Using participatory approaches as well as questionnaires and in-depth interviews, the project will attempt to answer a number of important questions (Binot and Morand, 2014):

- How do the various actors understand the patterns of zoonotic disease emergence and the interactions between environment and health?
- How does this understanding interfere with cultural background, scientific disciplines and/or traditional knowledge?
- How do different social groups perceive the notion of "zoonotic disease," merging animal and human health?

The answers to these questions could provide a useful basis for implementing ecosystem services for disease regulation and monitoring in specific cultural, political, ecological and socio-cognitive contexts.

The complicated case of malaria (Hans Keune)

Even when scientific findings seem relatively straightforward, for example in the combined promotion of biodiversity and public health, societal responses may be complicated and even painful. The use of insecticides such as DDT in malaria control is evidence of this (Parmesan *et al.*, 2009). DDT is known for its effectiveness in controlling the spread of malaria. Unfortunately, it is also known for its non-target effects. There are good reasons why DDT has been forbidden in Western countries for many years: it affects the food chain and remains in the environment for many years, as was extensively described in 1962 in Rachel Carson's famous book *Silent Spring* (Carson, 1962). In addition to its severe effects on animals, DDT is believed to have severe human health impacts: it is a probable human carcinogen; it damages the liver; it temporarily damages the nervous system; and it reduces reproductive success (ATSDR, 2002). Nevertheless, relatively recently (2006), the World Health Organization actively supported the short-term use of DDT in tropical regions, notwithstanding its longer-term public health and ecological risks (WHO, 2011). Still, the WHO clearly struggles with the pros and cons of this chemical. On the one hand, it acknowledges the importance of the Stockholm Convention on Persistent Organic Pollutants (http://chm.pops.int/default.aspx). On the other, it acknowledges that not all communities have the financial resources to utilize safer alternatives, which makes the use of DDT a poverty issue (WHO, 2011). It is worth investigating whether substituting another chemical for DDT, or employing alternatives such as traditional or modern ecological approaches, would be the most sustainable solution.

Conclusions

In this chapter we focused on challenges of biodiversity conservation in relation to human health policy, and on research and action in South and Southeast Asian countries. These cases confirm the challenges identified in this topic area and at the same time illustrate some entry points for options on how to address them.

The complex relationship between biodiversity and health: a difficult starting point for policy

The case studies confirm the complex relationship between biodiversity and public health with regard to how ecosystem change and biodiversity loss may affect outbreaks of infectious – especially zoonotic – disease; how pesticide use can damage both biodiversity and public health; and the distinct logics of traditional biodiversity-based medicinal care and modern health sciences. The BiodivHealthSEA project shows that, also among countries of the Asia-Pacific region, it appears that biodiversity is a source of pathogens, but that the loss of biodiversity, as measured by forest cover and/or threatened species, seems to be associated with an increase in outbreaks of zoonotic and vector-borne diseases. The entry point chosen for improving biodiversity conservation for public health (i.e. improving the "ecosystem service" for disease regulation) consists of detailed analysis of how the different parties involved understand the interactions between environment and health – that is, how the different scientific disciplines but also the different social groups involved perceive the notion of "zoonotic disease."

The pesticide use case also illustrates the challenge of complex cause–effect mechanisms: it is often hard to prove the causality of a given environmental concentration of pesticides for the health status of individual human beings. One of the main reasons for concern is a lack of research into the additive effects of chemicals both for the environment and for human health, their mixture toxicity. Nonetheless, results show that pesticide use could be lowered considerably without affecting agricultural yield, while at the same time improving natural pest control and reducing human health impacts – a clear win–win. The main factor impeding such a change in agricultural practice lies in the beliefs of farmers, who have been taught over decades that pesticides are the only reliable means to address pest control and have come to accept symptoms of intoxication as "normal aspects" of farm life. Studies on the cumulative effects of pesticide use and demonstrations of alternative management practices are needed if these beliefs are to be changed. Epidemiology might be a solution, but this requires reliable and accessible public health data, public information and improved governance. Similarly, farmers need to be shown the effectiveness of alternative pest management strategies, and they will need more knowledge if they are to apply these more complicated strategies effectively.

Another example of the complexity of the biodiversity and health nexus is the example of malaria. It shows that, even when scientific findings seem relatively straightforward, complicated policy trade-offs may emerge. In this case, the trade-off is between short-term vector control and the long-term human health and environmental impacts of DDT. Part of the complexity here also has to do with poverty issues (poor populations seem to have no alternative to DDT), which (from a WHO perspective) result in the acceptance of DDT as a means of malaria control. This approach confines the framework of policy action to the use of pesticides, excluding alternatives such as traditional antimalarial approaches or ecological engineering, which may offer more sustainable solutions to some health-related problems, as was shown in the cases of traditional knowledge and pesticide use.

The traditional knowledge case introduces yet another socio-economic driver of unsustainability which complicates sustainability efforts – projects that promote effective conservation and the sustainable use of biological resources, such as medicinal plants, have to compete with complex trade chains. One in five of the world's plant species is estimated to be threatened with extinction in the wild, with unsustainable harvesting being a major driver. Conservation practices need to take account of this and involve healers and emerging businesses.

The need for integration

The linkages between the biosphere and human health have gained prominence in international science and policy over the past two decades, with new interdisciplinary and transdisciplinary fields emerging to address the gaps in knowledge and action in ecosystem approaches to health. Still, a number of bridge-building challenges remain in this field. Most of the cases in this chapter address these challenges and present relevant practical examples.

Several cases exemplify the need for interdisciplinary approaches. For example, in the BiodivHealthSEA project, social scientists, environmental scientists, ecologists and epidemiologists study local impacts and the perception of global changes related to biodiversity and health in Southeast Asia. Meanwhile, in Bangladesh, disciplinary fragmentation and a compartmentalized institutional attitude have hampered research on biodiversity and health. Essential data for research into biodiversity and health is rarely shared by institutions and not publicly available.

Several of the projects discussed in this chapter try to build bridges to the understanding of local actors regarding biodiversity–health linkages. For example, the BiodivHealthSEA project uses participatory approaches, questionnaires and in-depth interviews in the hope of obtaining a useful basis for disease regulation and monitoring adapted to the specific cultural, political, ecological and socio-cognitive contexts.

The traditional knowledge case showcases several successful attempts to integrate traditional medicine, resources, practices and practitioners. The Indian case shows that the participatory and community-based program led to increased practice confidence among community members and the integration of traditional practices into conventional primary healthcare. It was also shown that utilizing traditional medical knowledge through community-based participatory approaches requires consideration of ethics, including voluntary, prior and informed consent. Integration with business by promoting community-based enterprises through traditional medicinal resources and products has been another successful model, addressing both improving community livelihoods and sustainable natural resource management. Networks such as the MPCN and associations of healers established within its framework have facilitated knowledge and experience exchange among traditional healers nationally, regionally and internationally.

At the same time, this case underlines the importance of developing mechanisms for the protection of traditional knowledge resources, prevention of their erosion and the promotion of linkages with scientific research: there is a need to develop tools to validate community knowledge and practices relevant to bio-diversity conservation, health and nutrition security. Here the development and improvement of community knowledge registers and bio-cultural protocols, as well as linking them with national databases for protection, have achieved positive results and merit further encouragement.

Proving the efficacy of every community-based practice through various stages of clinical study requires a considerable amount of financial as well as trained human resources and time. This is often impractical for a rapid community health intervention. Hence, in this context, it is important to have locally relevant and at the same time scientifically rigorous assessment methods through participatory approaches. Apart from assessing the efficacy of practices, it helps to identify sound traditional health practices, taking into consideration the world-views and epistemologies of folk practitioners and those who possess household knowledge. Based on consensus, these key stakeholders are able to identify and promote best practices within the community. Where such practices are found to be incomplete, they may be supplemented with knowledge from other sources, while ineffective practices are discarded.

Several community-based validation approaches have already been introduced. The main issue with these semi-formal methods is a lack of capacity among the stakeholders who conduct the studies to share their experiences the scientific and policy-making communities in an acceptable form.

The need for more research into the relationship between biodiversity and human health

Interest is rising in the relationship between biodiversity and human health among policy-makers and scientists in the region. Several of the cases in this chapter describe research projects with a clear focus on biodiversity and health. Moreover, these projects try to connect to and thereby share and strengthen local, traditional knowledge and practices that are at risk of disappearing.

Nonetheless, it is often difficult for this topic to find a place on the political agenda, alongside other pressing issues relating to poverty and public health, as is clearly the case in Bangladesh. Even so, research in the field of biodiversity has progressed in that country, mainly with little or no explicit focus on the link between biodiversity and health; only recently have some research institutes started looking into this relationship explicitly. While significant progress has been made in Bangladesh's legislative regime in terms of biodiversity conservation, and the National Biodiversity Strategy now acknowledges some of the links between biodiversity and health, significant gaps remain, not only between science and policy formulation, but between policy formulation and implementation.

Overall, it is clear that there is huge potential to improve human health via better understanding and integration of biodiversity conservation and health knowledge and practices in Southeast Asia. Nonetheless, one significant challenge remains: despite the growing awareness of the links between biodiversity, ecosystem services and public health, the health dimension remains comparatively undervalued in both policy and practical contexts. We hope that this chapter contributes to addressing this challenge by sharing our experiences from the literature and case studies.

References

Agency for Toxic Substances and Disease Registry (ATSDR). 2002. *Public Health Statement for DDT, DDE, and DDD*. www.atsdr.cdc.gov/ToxProfiles/tp35-c1-b.pdf (accessed 16 March 2017).

Ahmed, Z.U., Z.N.T. Begum, M.A. Hassan, M. Khondker, S.M.H. Kabir, M. Ahmad, A.T.A. Ahmed, A.K.A. Rahman, and E.U. Haque. 2008. *Encyclopedia of Flora and Fauna of Bangladesh*, Vol. 1: *Bangladesh Profile*. Dhaka: Asiatic Society of Bangladesh.

Allen, H.K., J. Donato, H. Huimi Wang, K.A. Cloud-Hansen, J. Davies, and J. Handelsman. 2010. Call of the wild: antibiotic resistance genes in natural environments. *Nature Reviews Microbiology* 8: 251–259.

Bavikatte, K., H. Jonas, and J. von Braun. 2010. Traditional knowledge and economic development: the biocultural dimension. In S.M. Subramanian and B. Pisupati (eds.), *Traditional Knowledge in Policy and Practice: Approaches to Development and Human Well-being*. Tokyo: UNU Press: 294–326.

Beketov, M.A., B.J. Kefford, R.B. Schäfer, and M. Liess 2013. Pesticides reduce regional biodiversity of stream invertebrates. *Proceedings of the National Academy of Sciences USA* 110: 11039–11043.

Berg, H. and N.T. Tam. 2012. Use of pesticides and attitude to pest management strategies among rice and rice-fish farmers in the Mekong Delta, Vietnam. *International Journal of Pest Management* 58: 153–164.

Binot, A. and Morand, S. (2014) Implementation of the "One Health" strategy: lessons learnt from Community-Based Natural Resource Programs (CBNRP) for communities' empowerment and equity within an "Ecohealth" approach. In S. Morand, J.-P. Dujardin, R. Lefait-Rollin, and C. Apiwathnasorn (eds.), *Socio-Ecological Dimensions of Infectious Diseases*. Singapore: Springer: 325–335.

Blasdell, K., J.F. Cosson, Y. Chaval, V. Herbreteau, B. Douangboupha, S. Jittapalapong, A. Lundqvist, J.P. Hugot, S. Morand, and P. Buchy. 2012. Rodent-borne hantaviruses in Cambodia, Laos PDR and Thailand. *EcoHealth* 8: 432–443.

Bodeker, G. and C. Neumann. 2012. Revitalization and development of Karen traditional medicine for sustainable refugee health services at the Thai–Burma border. *Journal of Immigrant and Refugee Studies* 10: 6–30.

Campbell-Lendrum, D. 2005. How much does the health community care about global environmental change? *Global Environmental Change* 15: 296–298

Carson, R. 1962. *Silent Spring*. Boston: Houghton Mifflin Harcourt.

CBD. n.d. *Convention on Biological Diversity*. www.cbd.int/decisions/ (accessed 16 March 2017).

Chagnon, M., D. Kreutzweiser, E.D. Mitchell, C. Morrissey, D. Noome, and J. van der Sluijs. 2015. Risks of large-scale use of systemic insecticides to ecosystem functioning and services. *Environmental Science and Pollution Research* 22(1): 119–134.

Chau, N.D.G., Z. Sebesvari, W. Amelung, and F.G. Renaud. 2015. Pesticide pollution of multiple drinking water sources in the Mekong Delta, Vietnam: evidence from two provinces. *Environmental Science and Pollution Research* 22(12): 9042–9058.

Chow, J. 2015. Spatially explicit evaluation of local extractive benefits from mangrove plantations in Bangladesh. *Journal of Sustainable Forestry* 34: 651–681.

COHAB. n.d. *Co-operation on Health and Biodiversity*. www.cohabnet.org/ (accessed 16 March 2017).

Coker, R., R. Rushton, S. Mounier-Jack, R. Karimuribo, P. Lutumba, D. Kambarage, D. Pfeiffer, K. Stärk, and M. Rweyemamu. 2011. Towards a conceptual framework to support one-health research for policy on emerging zoonoses. *Lancet Infectious Diseases* 11: 326–331.

Cutts, B.B., K.J. Darby, C.G. Boone, and A. Brewis. 2009. City structure, obesity, and environmental justice: an integrated analysis of physical and social barriers to walkable streets and park access. *Social Science and Medicine* 69: 1314–1322.

Dänhardt, J. and J.H. Spangenberg. In press. Valuation of pest suppression potential through biological control: insights to farmers perceptions supports the necessity of non-monetary approaches. *International Journal Biodiversity Science, Ecosystem Services and Management*.

Dearing, J. A., R. Wang, K. Zhang, J. Dyke, H. Haberl, S. Hossain, P. Langdon, T. Lenton, K. Raworth, S. Brown, J. Carstensen, M. Cole, S. Cornell, T. Dawson, C. Doncaster, F. Eigenbrod, M. Flörke, E. Jeffers, A.W. Mackay, B. Nykvist, and G. Poppy. 2014. Safe and just operating spaces for regional social-ecological systems. *Global Environmental Change* 28: 227–238.

Dupuy, S., V. Herbreteau, F. Feyfant, S. Morand, and A. Tran. 2012. Land cover dynamics in Southeast Asia: contribution of object-oriented techniques for change detection. *Proceedings of the 4th Geobia, Rio de Janeiro, Brazil:* 217–222.

El Arifeen, S., K. Hill, K. Z. Ahsan, K. Jamil, Q. Nahar, and P.K. Streatfield. 2014. Maternal mortality in Bangladesh: a countdown to 2015 country case study. *Lancet* 384: 1366–1374.

Ernstson, H. 2013. The social production of ecosystem services: a framework for studying environmental justice and ecological complexity in urbanized landscapes. *Landscape and Urban Planning* 109: 7–17.

ESPA Delta. 2012. *Assessing Health, Livelihoods, Ecosystem Services and Poverty Alleviation in Populous Deltas.* www.espadelta.net (accessed 28 February 2014).

FAO, OIE and WHO. 2010. *Sharing Responsibilities and Coordinating Global Activities to Address Health Risks at the Animal–Human–Ecosystems Interfaces.* www.who.int/influenza/resources/documents/tripartite_concept_note_hanoi_042011_en.pdf (accessed 16 March 2017).

Foundation for Revitalisation of Local Health Traditions (FRLHT). 1999. *The Key Role of Forestry Sector in Conserving India's Medicinal Plants: Conceptual and Operational Features.* Bangalore: FRLHT.

FRLHT, 2009. *Overview.* http://envis.frlht.org/overview.htm (accessed 4 June 2012).

Gagic, V., S. Hänke, C. Thies, C. Scherber, Ž. Tomanović, and T. Tscharntke. 2012. Agricultural intensification and cereal aphid–parasitoid–hyperparasitoid food webs: network complexity, temporal variability and parasitism rates. *Oecologia* 170: 1099–1109.

Gamfeldt, L., T. Snäll, R. Bagchi, M. Jonsson, L. Gustafsson, P. Kjellander, M. Ruiz-Jaen, M. Fröberg, J. Stendahl, C. Philipson, G. Mikusiński, E. Andersson, B. Westerlund, H. André, F. Moberg, J. Moen, and J. Bengtsson 2013. Higher levels of multiple ecosystem services are found in forests with more tree species. *Nature Communications* 4:1340.

Government of Bangladesh 2005. *National Biodiversity Strategy and Action Plan for Bangladesh 2005.* Dhaka: Ministry of Environment and Forests.

Grandjean, P. 2008. Seven deadly sins of environmental epidemiology and the virtues of precaution. *Epidemiology* 19: 158–162.

Gurr, G.M., K.L. Heong, J.A. Cheng, and J. Catindig 2012. Ecological engineering against insect pests in Asian irrigated rice. In G.M. Gurr, S.D. Wratten, W.E. Snyder, and D.M.Y. Read (eds.), *Biodiversity and Insect Pests: Key Issues for Sustainable Management.* New York: John Wiley & Sons: 214–229.

Gurr, G.M., Z. Lu, X. Zheng, H. Xu, P. Zhu, G. Chen, X. Yao, J. Cheng, Z. Zhu, J.L. Catindig, S. Villareal, H. Van Chie, L.Q. Cuong, C. Channoo, N. Chengwattana, L.P. Lan, L.H. Hai, J. Chaiwong, H.I. Nicol, D.J. Perovic, S.D. Wratten, and K.L. Heong. 2016. Multi-country evidence that crop diversification promotes ecological intensification of agriculture. *Nature Plants* 2: 16014.

Hamilton, A.C. 2004. Medicinal plants, conservation and livelihoods. *Biodiversity and Conservation* 13: 1477–1517.

Hariramamurthi, G., P. Venkatasubramanian, P.M. Unnikrishnan, and D. Shankar. 2006. Home herbal gardens – a novel health security strategy based on local knowledge and resources. In G. Bodeker and G. Burford (eds.), *Traditional, Complementary and Alternative Medicine: Policy and Public Health Perspectives*. London: Imperial College Press: 167–184.

Hartig, T., R. Mitchell, S.D. Vries, and H. Frumkin. 2014. Nature and health. *Annual Review of Public Health* 35: 207–228.

Heong, K.L. 2009. Are planthopper problems due to breakdown in ecosystem services? In K.L. Heong and B. Hardy (eds.), *Planthoppers – New Threats to the Sustainability of Intensive Rice Production Systems in Asia*. Los Baños: International Rice Research Institute: 221–232.

Heong, K.L., L. Wong, and J.H. de los Reyes. 2015. Addressing planthopper threats to Asian rice farming and food security: fixing insecticide misuse. In K.L. Heong, J. Cheng, and M.M. Escalada (eds.), *Rice Planthoppers*. Amsterdam: Springer: 65–76.

Herbreteau, V., F. Bordes, S. Jittapalapong, Y. Supputamongkol, and S. Morand. 2012. Rodent-borne diseases in Thailand: targeting rodent carriers and risky habitats. *Infection Ecology and Epidemiology* 2: 18637.

Hoai, P.M., S. Sebesvari, M. Tu Binh, V. Pham Hung, and F.G. Renaud. 2011. Pesticide pollution in agricultural areas of northern Vietnam: case study in Hoang Liet and Minh Dai communes. *Environmental Pollution* 159: 3344–3350.

Hossain, M.S., L. Hein, F.I. Rip, and J.A. Dearing. 2013. Integrating ecosystem services and climate change responses in coastal wetlands development plans for Bangladesh. *Mitigation and Adaptation Strategies for Global Change* 20: 241–261.

Jax, K., D. Barton, K. Chan, R. de Groot, U. Doyle, U. Eser, C. Görg, E. Gómez-Baggethun, Y. Griewald, W. Haber, R. Haines-Young, U. Heink, T. Jahn, H. Joosten, L. Kerschbaumer, H. Korn, G. Luck, B. Matzdorf, B. Muraca, C. Neßhöver, B. Norton, K. Ott, M. Potschin, F. Rauschmayer, C. von Haaren, and S. Wichmann. 2013. Ecosystem services and ethics. *Ecological Economics* 93: 260–268.

Keesing, F. and R.S. Ostfeld. 2015. Is biodiversity good for your health? Disease incidence is often lower in more diverse communities of plants and animals. *Science* 349: 235–236.

Keune, H., C. Kretsch, G. De Blust, M. Gilbert, L. Flandroy, K. Van den Berge, V. Versteirt, T. Hartig, L. De Keersmaecker, H. Eggermont, D,. Brosens, J. Dessein, S. Vanwambeke, A. Prieur-Richard, H. Wittmer, A. Van Herzele, C. Linard, P. Martens, E. Mathijs, I. Simoens, P. Van Damme, F. Volckaert, P. Heyman, and T. Bauler. 2013. Science–policy challenges for biodiversity, public health and urbanization: examples from Belgium. *Environmental Research Letters* 8: 1–19.

Khan, M.M.H., P. Samuel, and O. Fund. 2008. *Protected Areas of Bangladesh: A Guide to Wildlife, Nishorgo Program, Wildlife Management and Nature Conservation Circle*. Dhaka: Bangladesh Forest Department.Laird, S. and R. Wynberg. 2008. *Access and Benefit-Sharing in Practice: Trends in Partnerships across Sectors*. Technical Series No. 38. Montreal: Secretariat of the Convention on Biological Diversity.

Lara, R.J., S. Neogi, M. Islam, Z. Mahmud, S. Islam, D. Paul, B. Demoz, S. Yamasaki, G. Nair, and G. Kattner. 2011. *Vibrio cholerae* in waters of the Sunderban mangrove: relationship with biogeochemical parameters and chitin in seston size fractions. *Wetlands Ecology and Management* 19: 109–119.

Lõhmus, M. and J. Balbus. 2015. Making green infrastructure healthier infrastructure. *Infection Ecology and Epidemiology* 5. www.ncbi.nlm.nih.gov/pmc/articles/PMC4663195/ (accessed 16 March 2017).

Long, N. 2001. *Development Sociology: Actor Perspectives*. London: Routledge.

Mcmichael, A.J. 2009. Human population health: sentinel criterion of environmental sustainability. *Current Opinion in Environmental Sustainability* 1: 101–106.

Ministry of Environment and Forest. n.d. *Convention and Treaty*. www.moef.gov.bd/html/protocol/protocol_main.html (accessed 16 March 2017).

Morand, S., J. Jittapalapong, Y. Suputtamongkol, M.T. Abdullah, and T.B. Huan. 2014. Infectious diseases and their outbreaks in Asia-Pacific: biodiversity and its regulation loss matter. *PLoS One* 9: e90032.

Mukul, S. A. 2007. Biodiversity conservation strategies in Bangladesh: the state of protected areas. *Tigerpaper* 34: 28–32.

Newman, D. and G. Cragg. 2007. Natural products as sources of new drugs over the last 25 years. *Journal of Natural Products* 70: 461–477.

NIPORT, Mitra and Associates, and ICF International. 2011. *Bangladesh Demographic and Health Survey*. Dhaka and Calverton, MD: NIPORT, Mitra and Associates and ICF International.

One World, One Health. n.d. *A Programme of the Wildlife Conservation Society*. www.wcs.org/conservation-challenges/wildlife-health/wildlife-humans-and-livestock/one-world-one-health.aspx (accessed 16 March 2017).

Parmesan, C., S.M. Skevington, J.-F. Guegan, P. Jutro, S.R. Kellert, A. Mazumder, M. Roue, and M. Sharma. 2009. Biodiversity and human health: the decision making process. In O.E. Sala, L.A. Meyerson, and C. Parmesan (eds.), *Biodiversity Change and Human Health: From Ecosystem Services to Spread of Disease*. Washington, D.C.: Island Press: 61–81.

Payyappallimana, U. and P.B. Nagendrappa. 2007. Traditional herbal medicines for malaria prevention. *Endogenous Development Magazine* 1: 16–18.

Payyappallimana, U. and S.M. Subramanian. 2012. *Biodiversity, Traditional Knowledge and Community Health: Strengthening Linkages*. Yokohama: UNU-IAS.

Prüss-Üstün, A. and C. Corvalán. 2006. *Preventing Disease through Healthy Environments: Towards an Estimate of the Environmental Burden of Disease*. Geneva: World Health Organization.

Santhanakrishnan, R., A. Hafeel, B.A. Hariramamurthi, and P.M. Unnikrishnan. 2008. Documentation and participatory rapid assessment of ethno veterinary practices. *Indian Journal of Traditional Knowledge* 7: 360–364.

Sarkki, S., J. Niemelä, R. Tinch, and the SPIRAL Team. 2012. *Criteria for Science–Policy Interfaces and their Linkages to Instruments and Mechanisms for Encouraging Behaviour that Reduces Negative Human Impacts on Biodiversity*. www.spiral-project.eu/sites/default/files/SPIRAL_3-1.pdf (accessed 16 March 2017).

Savary, S., F. Horgan, L. Willocquet, and K.L. Heong. 2012. A review of principles for sustainable pest management in rice. *Crop Protection* 32: 54–63.

Sebesvari, Z., T.T.H. Le, and F. Renaud. 2011. Climate change adaptation and agrichemicals in the Mekong Delta, Vietnam. In M. Steward and P. Coclanis (eds.), *The Mekong Delta: Environmental Change and Agricultural Sustainability*. Amsterdam: Springer: 219–239.

Sebesvari, Z., T.T.H. Le, P.V. Toan, U. Arnold, and F.G. Renaud. 2012. Agriculture and water quality in the Vietnamese Mekong Delta. In F.G. Renaud and C. Kuenzer (eds.), *The Mekong Delta System: The Mekong Delta System. Interdisciplinary Analyses of a River Delta*. Amsterdam: Springer: 331–362.

Settele, J., J.H. Spangenberg, K.L. Heong, B. Burkhard, J.V. Bustamante, J. Cabbigat, H. Van Chien, M.M. Escalada, V. Grescho, L.H. Hai, A. Harpke, F.G. Horgan, S. Hotes,

R. Jahn, I. Kühn, L. Marquez, M. Schädler, V. Tekken, D. Vetterlein, S.B. Villareal, C. Westphal, and M. Wiemers. 2015. Agricultural landscapes and ecosystem services in South-East Asia: the LEGATO Project. *Basic and Applied Ecology* 16: 661–664.

Sodhi, N.S., L.P. Koh, B.W. Brook, and P.K.L. Ng. 2014. Southeast Asian biodiversity: an impending disaster. *Trends in Ecology and Evolution* 19: 654–660.

Spangenberg J.H., J.-M. Douguet, J. Settele, and K.L. Heong. 2015. Escaping the lock-in of continuous insecticide spraying in rice: developing an integrated ecological and socio-political DPSIR analysis. *Ecological Modelling* 295: 188–195.

Sudmeier-Rieux, K., H. Masundire, I. Rizvi, and S. Rietbergen. 2006. *Ecosystems, Livelihoods and Disasters: An Integrated Approach to Disaster Risk Management*. Gland: IUCN.

Toan, P.V., Z. Sebesvari, M. Bläsing, I. Rosendahl, and F.G. Renaud. 2013. Pesticides in the Mekong Delta Vietnam: application practices and residues in sediment, surface and drinking water. *Science of the Total Environment* 452–453: 28–39.

Turnhout, E., B. Bloomfield, and M.B. Hulme. 2012. Conservation policy: listen to the voices of experience. *Nature* 488: 454–455.Van den Hove, S. 2007. A rationale for science–policy interfaces. *Futures* 39: 807–826.

Ved, D.K. and G.S. Goraya. 2008. *Demand and Supply of Medicinal Plants in India*. Dehra Dun: Bishen Singh, Mahendra Pal Singh.

World Health Organization (WHO). 2005. *Ecosystems and Human Well-being: Health Synthesis: A Report of the Millennium Ecosystem Assessment*. Geneva: WHO. www.maweb.org/documents/document.357.aspx.pdf (accessed 16 March 2017).

WHO. 2011. *The Use of DDT in Malaria Vector Control*. Geneva: WHO. http://whqlibdoc.who.int/hq/2011/WHO_HTM_GMP_2011_eng.pdf (accessed 16 March 2017).

WHO and CBD Secretariat. 2015. *Connecting Global Priorities: Biodiversity and Human Health: A State of Knowledge Review*. Geneva: WHO.

Wilcove, D.S., X. Giam, D.P. Edwards, B. Fisher, and L.P. Koh. 2013. Navjot's nightmare revisited: logging, agriculture and biodiversity in Southeast Asia. *Trends in Ecology and Evolution* 28: 531–540.

Yang, Y.-J., B.-Q. Dong, H.-X. Xu, X.-S. Zheng, K.L. Heong, and Z-X, Lu. 2014. Susceptibility to insecticides and ecological fitness in resistant rice varieties of field *Nilaparvata lugens* population free from insecticides in laboratory. *Rice Science* 21: 181–186.

14 International and regional governance in health and biodiversity

Claire Lajaunie and Serge Morand

> For the international environmental lawyer, law is a set of tools to help to solve problems. Just as a doctor seeks to diagnose a disease in order to know what cure to prescribe, understanding the causes of an environmental problem can help to identify the most appropriate policy responses.
>
> (Bodansky, 2011: 37)

The link between health and environment has been acknowledged progressively in international declarations and agreements, from the Stockholm Conference of 1972, which attempted to forge a "common outlook and common principles to inspire and guide the peoples of the world in the preservation and enhancement of the human environment"[1] to the recent Conference of the Parties of the Convention on Biological Diversity (CBD), which acknowledged the value of One Health to address the cross-cutting issue of biodiversity and human health,[2] and the 2015 state-of-knowledge review on biodiversity and human health published by the CBD and the World Health Organization (WHO and CBD, 2015).

In line with the growing commitment to sustainable development, numerous international conferences and declarations as well as diverse legal instruments give an international legal framework to the issues related to health and biodiversity.

In the meantime, as globalization contributes to the emergence or reemergence of diseases, the notion of global health governance appears along with the awareness that public health issues are linked to environmental factors.

Global governance can be defined as "the existing set of collective agreements and arrangements to set norms, make decisions, solve problems and monitor at the global level in the absence of a world government" (Carin et al., 2011: 6). As stated by Bodansky (2011: 39), "Disentangling the various factors that contribute to an international environmental problem thus constitutes an important step in determining the range of possible responses." In this respect, international environmental law might be appropriate in fostering action at the international and regional levels (Von Schirnding et al., 2002).

It is the nature of international environmental law to focus on states' responsibility to take into consideration their neighborhoods.[3] This fundamental aspect of international environmental law doctrine is expressed by Principle 21 of the Stockholm Declaration:

> States have, in accordance with the Charter of the United Nations and the principles of international law, the sovereign right to exploit their own resources pursuant to their own environmental policies, and the responsibility to ensure that activities within their jurisdiction or control do not cause damage to the environment of other States or of areas beyond the limits of national jurisdiction.[4]

In Southeast Asia, the national governments have become increasingly aware of the importance of biodiversity conservation for human development and ecosystem health, and the international documents are being steadily translated through various paths into regional initiatives, legal instruments and formal or informal action.

This chapter aims to proceed backwards from the present to determine the main steps leading to the elaboration of the international instruments related to health and biodiversity and to present the ways they have been integrated and implemented at the regional level. The goal is to unravel the milestones of international society's involvement in issues regarding the interrelations between health and biodiversity. Departing from a historical presentation (and its many details) helps to highlight how these turning points are captured by international governance and helps understanding of the progress of various concepts that are internationally debated from a legal perspective.

At the regional level, we will show the constant integration of those concepts in light of the involvement of the various Southeast Asian countries in the construction of the ASEAN community.

The One Health approach

The One Health approach has been presented as a necessary "paradigm shift" in the approach to global health by diverse voices located mainly in the arenas of veterinary science and biology (e.g. Kaplan and Echols, 2009; Zinsstag *et al.*, 2009; Atlas *et al.*, 2010; Bousfield and Brown, 2011). There are many definitions of the concept, but the American Veterinary Medical Association (AVMA, 2008: 13) defines One Health as "the integrative effort of multiple disciplines working locally, nationally, and globally to attain optimal health for people, animals, and the environment."

An expert consultation organized in Canada in 2009 called for action to implement the "One Health approach." The report underlines the role international agencies should play in promoting the harmonization and standardization of criteria, codes of practice and legislation, and in assisting countries with policy development and standards adoption (Public Health Agency of Canada, 2009).

A decisive step in the integration of the One Health approach into the international governance and legal arena came with an international conference co-organized in Hanoi in 2010 by the European Union and the United States in partnership with the FAO, WHO, World Organization for Animal Health (OIE), Asian Development Bank (ADB), World Bank (WB) and UNICEF, which assembled ministers and seniors officials from more than seventy countries with the aim of formulating practical, coordinated plans to combat pandemics (highly pathogenic avian influenza was specifically targeted) and control the spread of infectious diseases. The outcome of this conference was the Hanoi Declaration of 2010, which states that the participants are committed to increasing their efforts to review pandemic preparedness plans and jointly strengthen human and animal public health systems. It recognises that each country's strategies "should be aligned nationally and regionally to address the global 'One Health' challenges."[5] However, although the declaration refers to the animal–human–environment interface, it focuses on animal–human health interactions and provides no details on how the environment should be taken into account into the countries' strategies. Among all of the international agencies involved into the battle against infectious diseases, not one is specialized in biodiversity and environmental issues.

Interestingly, the link with the environmental side of the issue appeared with the concern about wildlife health at the 10th Conference of the Parties of the Convention on Conservation of Migratory Species,[6] which invited the CITES Secretariat to become a core affiliate member of the Scientific Task Force on Wildlife Disease. During the 26th meeting of the Animals Committee (2012), it was announced that the task force's purpose is to support "evidence-based decision processes and tools that consider disease dynamics in the broader context of sustainable biodiversity/ecosystem management, agricultural production and food security, socio-economic development, environmental protection and conservation of migratory species, their habitats and migration routes" through the promotion of an integrated scientific approach within One Health.[7] It aims to support the major conservation Multilateral Environment Agreements (MEAs) to integrate issues relating to the livestock–wildlife–human–ecosystem health interface into their activities, approaches and resolutions. Thus, the Conference of the Parties (COP) of the Ramsar Convention on Wetlands[8] and of the United Nations Convention to Combat Desertification[9] explicitly refer to that interface or to the One Health approach that is considered necessary for informed policy-making.

All the COPs of the main MEAs have passed resolutions regarding partnerships and synergies with MEAs and other institutions. There are thus undoubtedly strong influences between the different MEA mechanisms leading to the integration of the One Health concept into the resolutions adopted by the corresponding decision-making bodies. This effect is fostered by the existence of the Liaison Group of Biodiversity-Related Conventions, created in 2004 in order to increase coordination and enhance coherence and cooperation between the secretariats of the six biodiversity-related conventions.[10]

The pace of integration is accelerating with the multiplication of arrangements between international institutions, such as the agreement between the OIE and the of the CBD Secretariat (2013) to manage the risks presented by animal diseases and zoonoses at the animal–human–ecosystem interface[11] and the Workshop on the Interlinkages between Human Health and Biodiversity, organized jointly by the WHO and the CBD Secretariat (2012).

All of these documents mention the international, regional, national and local levels of implementation of the decisions and show how these levels are interlinked. One Health is now acknowledged by instruments of international environmental law (Lajaunie and Mazzega, 2016b), but there are still efforts to be made to integrate that recognition within a coordinated framework where evidence-based research informs policy strategy (Coker *et al.*, 2011), which could lead to practical and efficient implementation.

In Southeast Asia, several recent conferences and meetings have highlighted the need for an integrated understanding of the environment and health issues through the One Health approach. ASEAN decided to target the issues relating to the animal–human–environment interface by creating the ASEAN Secretariat Working Group for One Health, which is in charge of coordinating a series of diverse health-related initiatives, and strengthening its coordination of seven non-health sectors[12] that will be responsible for the delivery of essential services in the event of a pandemic (ASEAN +3, 2010). These coordination plans include a project to stockpile antiviral drugs and personal protective equipment while also ensuring continuity in the economic and social functions of each country. ASEAN focuses on emergency preparedness but does not consider the upstream issues arising from the interaction between animals and humans within the ecosystem, and there is no specific reference to the environment.[13]

ASEAN's coordination plans are different in nature from the type of coordination that is demanded by the One Health approach. ASEAN is formulating institutional coordination to meet the basic responsibilities of a state (or, in the case of ASEAN, a group of states committing to resolve issues in common) and maintain public order and security in the event of a pandemic, but failing to develop a comprehensive approach that also addresses human health, animal health and the environment according to a multidisciplinary perspective based on a shared wealth of knowledge. In this respect, regional initiatives led by researchers engaged in multidisciplinary work and representing a network of universities[14] could serve to open up the institutional world to a "common conceptualisation of the systems across which animals and human beings encounter each other" (Coker *et al.*, 2011: 327).

International environmental law could greatly contribute to define a holistic system approach as, "in this particular field, law intersects with many other fields such as physic, geology, biology and social sciences and it must cooperate with them" (Boisson de Chazournes, 2011: 11). It could help to build a bridge between the research framework and the regional and national institutions in charge of enacting the appropriate policies.

Intergovernmental Science-Policy Platform on Biodiversity and Ecosystem Services (IPBES)

The complexity of the issues related to biodiversity and ecosystem services and their cross-sectoral and cross-scale nature need to be addressed through strong scientific knowledge generated by collaboration among many different disciplines at various levels with the integration of best practices and knowledge of civil society, communities and/or beneficiaries of ecosystem services.

There is often inadequate coordination between policy sectors and jurisdictional levels in the policy-making process, and biodiversity governance can be partly ignored by policies on land use or economic activity.[15] Decision-makers often mention their difficulty in accessing scientific results, issues of reliability or independence with the scientific information they receive, or the fact that available data are inappropriate to their decision-making. The scientists, on their side, recognize their unfamiliarity with the needs and processes of policy-making and their ability to produce relevant work (Larigauderie and Mooney, 2010).

In light of the need for informed policy-making, there were calls for the creation of an intergovernmental platform to strengthen decision-making on biodiversity in order to tackle the lack of appropriate responses to the deterioration of biodiversity (Millennium Ecosystem Assessment, 2005). This marked a shift from a conservationist approach (taking into account species and habitats) to a holistic vision that encompassed conservation and the sustainable use of biodiversity and ecosystem services (Van den Hove and Chabason, 2009).

This IPBES is often referred to as the "Intergovernmental Panel on Climate Change for Nature," as it is modeled on the Intergovernmental Panel for Climate Change to be an efficient science–policy interface mechanism that will improve the quality of policy-relevant information from all relevant sources about the state, trends and outlook of human–environment interactions, with a focus on the impacts of ecosystem change on human well-being (UNEP and French Government, 2008). Its four main functions[16] are:

- knowledge generation;
- regular and timely assessments;
- support for policy formulation and implementation; and
- capacity building.

The process of establishing IPBES was lengthy,[17] but eventually a resolution of the United Nations General Assembly[18] asked the United Nations Environmental Programme to convene a plenary meeting in order to fully operationalize the platform at the earliest opportunity.[19] Consequently, IPBES was officially established in April 2012 as an independent intergovernmental body by a resolution adopted by ninety-four countries.[20]

The first meeting of the Plenary of the IPBES was held in 2013 in Bonn,[21] the city where the platform's secretariat is based. It agreed on the body's procedural rules and on the organization of the platform, and took the opportunity to elect an international group of renowned experts – the Multidisciplinary Expert Panel (MEP) – in order to ensure the scientific credibility and independence of the IPBES's work. The Plenary also decided that only governments and MEAs related to biodiversity and ecosystem services could make "requests" to the platform, with UN bodies and other stakeholders permitted to provide only "inputs and suggestions." However, the platform failed to define its relationship to the UN system.

The COP of the CBD has already asked IPBES to contribute to the preparation of the next global assessment on biodiversity and ecosystem services, to be launched in 2018, and to help countries to implement the Strategic Plan for Biodiversity 2011–2020 and achieve the Aichi Biodiversity Targets.

Reflections on health and human well-being were presented at the first plenary meeting after an informal multidisciplinary expert process. The experts identified the key features that might be included in an IPBES conceptual framework in order to address issues of biodiversity, ecosystem services and their impacts on human well-being.[22] Among them, we can highlight the need to integrate the views of indigenous and local knowledge experts, policy-makers and other relevant stakeholders (as in the One Health framework, the participatory approach is championed). The existence of various temporal and geographical scales should be explicitly considered for a better understanding of the multi- or cross-scale impact of changes. As with the One Health approach, it appears that the components of biodiversity, ecological functioning, ecosystem services and human well-being should be studied as parts of a socio-economic–ecological system. Any changes in those components will be scrutinized within that system. To provide appropriate strategies and interventions, the roles of institutions and decisions should also be examined as direct or indirect drivers of change in the state of biodiversity, ecosystem services and human well-being.[23]

The second plenary meeting focused on the work programme for 2014–2018, a draft of a "Stakeholder Engagement Strategy" and strategic partnership,[24] and the institutional arrangements that were developed during the intersessional process by the MEP and the Bureau of the IPBES (Figures 14.1 and 14.2).

The second objective of the IPBES 2014–2018 draft work programme is "to strengthen the science–policy interface on biodiversity and ecosystem services at regional and sub-regional levels"[25] and the programme specifies that the platform will ensure a bottom-up approach. In order to inform this objective, an Asia-Pacific workshop on regional interpretation of the IPBES conceptual framework and knowledge-sharing, organized in Seoul in 2013, acknowledged that the Asia-Pacific region expects to play an important role as it houses mega-biodiversity as well as large populations.[26] It proposed the creation of an IPBES regional hub to foster regional collaboration, policy coherence and the use of universal methods, and to address assessment shortfalls. The main idea is to unite the work of the region's various governments, organizations and stakeholders with a common repository of regional data gathered using standard methods, including the results

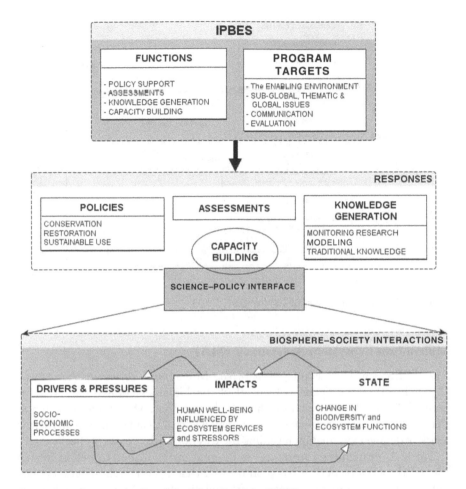

Figure 14.1 A conceptual model of the role of the IPBES

gathered by IUCN and AP-BON, for instance. The purpose is to coordinate regional interventions to take account of trade-off dynamics and institutions as well as to address geographic imbalance[27] and ensure policy coherence.

To create advanced knowledge systems, it seems necessary to integrate social science,[28] citizen, private sector, indigenous and local knowledge with contemporary science. In turn, practical methodologies to assess comparable synergies for trade-offs and co-benefits and in this respect tools such as TEEB (The Economics of Ecosystems and Biodiversity) illustrate useful typologies (ASEAN Center for Biodiversity *et al.*, 2013). Finally, regarding the impacts of policy,[29] the development of verifiable criteria might help with the reporting and monitoring of policy impacts on biodiversity and ecosystem services.

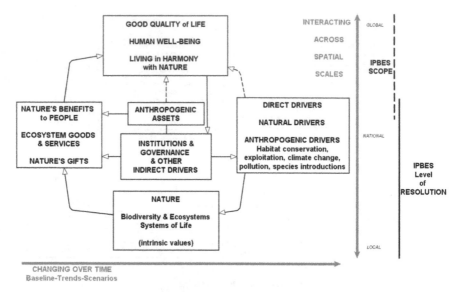

Figure 14.2 Analytical conceptual framework

Millennium Ecosystem Assessment (MA)

The IPBES is a direct consequence of the Millennium Ecosystem Assessment (MA). In 2000, the Secretary-General of the United Nations issued a report entitled *We the Peoples* which highlighted the fact that, while there is awareness that the natural ecosystems on which human life depends are under threat, there is a lack of detailed information regarding the extent and the causes of the damage.

The MA,[30] which is presented as "an international collaborative effort to map the health of our planet," is based on the fact that a good environmental policy must be based on reliable scientific data and on the need for a comprehensive global assessment of the world's major ecosystems to deliver the appropriate responses regarding their conservation and sustainable use. It focuses on ecosystem services and on how changes in them have affected and will impact upon human well-being. It was launched in 2001 and involved the work of more than 1,360 experts worldwide.

The MA provides an efficient conceptual framework that aids understanding of the dynamic interactions between people and other parts of their ecosystems as well as the direct and indirect drivers of change in ecosystems and their effects on human well-being. Ecosystem services are defined as the benefits people obtain from ecosystems. The MA distinguishes between four categories of ecosystem services:

- Provisioning services: goods, such as food or freshwater, that ecosystems provide and humans consume or use.

- Regulatory services: services such as flood reduction and water purification that healthy, natural systems, such as wetlands, can provide.
- Cultural services: intangible benefits, such as aesthetic enjoyment or religious inspiration, that nature often provides.
- Supporting services: basic processes and functions, such as soil formation and nutrient cycling, that are critical to the provision of the first three types of ecosystem services.

Norgaard (2010: 1219, 1227) has described this notion as an "eye-opening metaphor" which has "become integral to how we are addressing the future of humanity and the course of biological evolution."

The ecosystem approach[31] is an integrated approach that considers all ecosystem components. In this respect, the MA identifies ten categories of systems, which contain a number of ecosystems with boundaries that can overlap (for instance, marine and coastal systems). Echoing the concept of "land sickness," introduced by Leopold,[32] the ecosystem approach led to the idea of ecosystem health (Morand, 2010).

The results of the MA were presented in a synthesis report and five other reports that interpret the MA findings for specific audiences.[33] The general results show that, over the past fifty years, humans have changed ecosystems more rapidly and extensively than in any comparable period of time, and that almost 60 percent of the ecosystem services the MA examined were being degraded or used unsustainably – including fresh water, capture fisheries, air and water purification, and the regulation of regional and local climates, natural hazards and pests.

The report – *Ecosystems and Human Well-being: Health Synthesis* – was presented by the World Health Organization (WHO, 2005a) as a "call to the health sector, not only to cure the diseases that result from environmental degradation but also to ensure that the benefits that the natural environment provides to human health and well-being are preserved for future generations," in accordance with the principle of intergenerational equity, which is also a key element of sustainable development and a feature of international environmental law. This principle encompasses the concept of "wise stewardship [of natural resources] . . . and their conservation for the benefit of future generations."[34]

The report insists on the fundamental role of ecosystem services in the well-being and health of people everywhere, and on humanity's inherent need for food, water, clean air, shelter and relative climatic constancy (Figure 14.3). Among the ecosystem services influencing human health, we can notice:

- The critical role played by ecosystems in the recycling and redistribution of nutrients.
- The regulation of infectious diseases in animals and humans, as many ecosystem changes can alter the habitats and hence populations of disease-transmitting vectors.

The report also recognizes the health sector's responsibility for informing decision-makers about the health effects of ecosystem changes and potential interventions.[35]

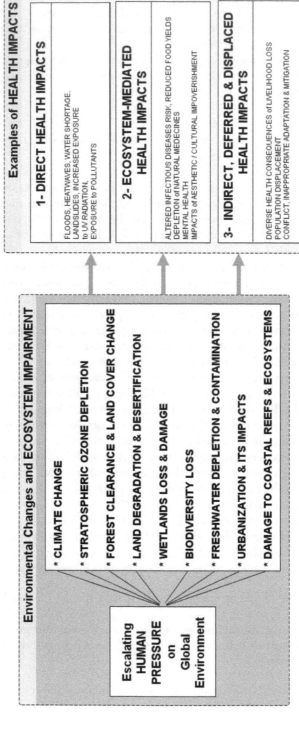

Figure 14.3 Ecosystem services and human well-being

Source: Adapted from WHO (2005a: 9).

The MA's sub-global assessments were designed to meet the needs of decision-makers at the scale at which they are undertaken, strengthen global findings with on-the-ground reality, and strengthen local findings with global perspectives, data and models. Eighteen MA-approved sub-global assessments and an additional fifteen with associated status have been conducted worldwide. In Southeast Asia they took place in Indonesia, the Philippines and the downstream Mekong River wetlands in Vietnam. The main outcome of the sub-global process has been the generation of new tools and methodologies as well as baseline information that have helped to empower stakeholders (Millennium Ecosystem Assessment, 2005).

A 2008 decision of the decision-making body (COP) of the CBD acknowledges the global strategy for follow-up to the MA[36] and invites parties, other governments, relevant organizations, indigenous and local communities and stakeholders to contribute actively to its implementation.[37] In 2012, the COP decided to invite the IPBES to include in its work programme the preparation of the next global assessment on biodiversity and ecosystem services, to be launched in 2018.[38]

New International Health Regulations and tools for surveillance and alerts

Together with awareness of the multi-level threats to biodiversity and ecosystem services, the traditional approach to the international spread of disease has been transformed by an acknowledgement that the globalization of trade and increased international travel have accelerated the risk of rapid worldwide spread of infectious diseases.

In 2003, the General Assembly of the United Nations recognized the need for greater international and regional cooperation to meet new and existing challenges to public health and the importance of active international cooperation in the control of infectious diseases, based on the principles of mutual respect and equality. It adopted a resolution urging member states to give high priority to work on the revision of the International Health Regulations (IHR).[39]

When adopted in 1969, the original IHR covered six diseases, but it was amended in 1973 to target only three (yellow fever, plague and cholera; WHO, 2005b: Foreword).[40] The only obligations of governments were to report outbreaks of these diseases to the WHO and to maintain minimal public health capabilities at ports and borders. Nevertheless, there was serious criticism of the lack of compliance with the IHR and the absence of coordination (Sing, 2009), and its ineffectiveness became a salient issue with the emergence and reemergence of infectious diseases in the 1980s and 1990s. In 2005, a serious outbreak of severe acute respiratory syndrome (SARS) led to the adoption of the revised International Health Regulations by the World Health Assembly (WHA, 2005).

The purpose of the new IHR is "to prevent, protect against, control and provide a public health response to the international spread of disease in ways that are commensurate with and restricted to public health risks, and which avoid unnecessary interference with international traffic and trade." Instead of being based on a list of diseases, the scope of IHR has been widened to cover any extraordinary event that could constitute a public health emergency of international concern.[41] The state parties must inform the WHO of any such event through an IHR National Focal Point,[42] connected to its corresponding IHR Regional Focal Point. The WHO is also now permitted to gather information from other sources, in addition to the traditional notifications and reports from states. It has an explicit mandate to obtain verification from states parties concerning unofficial reports or communications, and it must share any information it receives from non-governmental sources with all state parties and relevant intergovernmental organizations when these might be necessary to enable responses to public health risks.

As an international legal instrument, the IHR is binding on all state members of the WHO. It provides a framework of global and integrated governance in order to achieve global health security. This framework includes in its surveillance system governmental, intergovernmental and non-governmental actors. It is defined to take into account the continued evolution of diseases and the factors determining their emergence and transmission (WHO, 2005b: Foreword).

Many MEAs, whether they are related to the ozone layer, climate change, persistent pollutants, hazardous pesticides or movement of hazardous waste, have health implications. Thus, international environmental law plays an important role in the protection of health at the international level. As highlighted by Fidler (2005: 341), the adoption of those environmental agreements "demonstrated that cross-border threats to public health existed outside the infectious disease context," and the implementation of international environmental law can contribute – for instance, through the protection of natural resources or ecosystems – to mitigating the threat of disease.

The IHR states that the WHO shall cooperate and coordinate its activities with other competent intergovernmental organizations or international bodies,[43] notably the FAO and OIE (WHO, 2005b: Article 14). Hence, WHO, FAO and OIE launched a "Global Early Warning and Response System for Major Animal Diseases" (GLEWS) in 2006. This joint system combines and coordinates the alert and response mechanisms of the three organizations for the international community and stakeholders to assist in the prediction, prevention and control of animal disease threats, including zoonoses, by sharing information, epidemiological analysis and joint field missions to assess and control outbreaks, whenever needed.[44] Later, the same three organizations adopted a concept note regarding their collaboration, considered as a strategic alignment, to address gaps and strengthen collaboration in human and animal health. The aim is to improve national-, regional- and community-level pandemic preparedness and response through complementary work to develop normative standards and field programs and achieve One Health goals.

Also in pursuance of One Health objectives, in 2013 the three organizations established, as an addition to GLEWS, a cross-sectoral mechanism (GLEWS+) for conducting robust and timely joint risk assessments, with the aim of formulating risk management options for health events at the human–animal–ecosystem interface. It targets health events of potentially international concern that affect domestic or wild animal populations, humans or the food chain, and benefits from the network of expert resources within WHO, OIE and FAO and their collaborating centers.

At the regional level, various networks are supported by these three organizations. For instance, the ASEAN Animal Health Collaboration website,[45] supported by FAO, is a platform that facilitates dissemination and sharing of relevant documents and/or information relating to animal health initiatives in the region among the member states, stakeholders and relevant partners. Meanwhile, the ASEAN Regional Animal Health Information System (ARAHIS) is open to authorized representatives of ASEAN countries for the purpose of sharing timely information on livestock diseases and to improve regional disease control.

Other networks,[46] such as the Mekong Basin Disease Surveillance (MBDS) network, formally established in 2001 by the ministers of health of the countries in the Greater Mekong sub-region, participate in the improvement of cross-border infectious disease outbreak investigation and response by sharing surveillance data and best practices in disease recognition by jointly responding to outbreaks.

Convention on Biodiversity framework

Adopted in 1992 at the Rio Conference and ratified by more countries than any other international agreement, the Convention on Biodiversity (CBD) currently has 196 parties.[47] Considered a landmark agreement in the environmental field, its preamble affirms that the conservation of biodiversity is now a common concern of humankind.

The CBD is innovative in that it recognizes three levels of diversity and represents an all-encompassing agreement on conservation and the sustainable use of biodiversity and natural resources. It defines "biodiversity" as: "the variability among living organisms from all sources including, inter alia, terrestrial, marine and other aquatic ecosystems and the ecological complexes of which they are part; this includes diversity within species, between species and of ecosystems" (Article 2).

Its framework for action has three main objectives:

- the conservation of biological diversity;
- the sustainable use of its components; and
- the fair and equitable sharing of the benefits arising out of the utilization of genetic resources.

The CBD is a framework agreement, so it requires completion by additional international protocols. It establishes a body of flexible obligations that parties may apply through their own national laws and policies,[48] so decision-making is vested at the national level. Thus, individual countries can interpret the provisions of the CBD according to their own national or regional priorities in order to make them operational. However, the convention is legally binding and requires all parties to implement its provisions. The parties should develop

National Biodiversity Strategies and Action Plans (NBSAP) and report to the CBD on their efforts to implement their convention commitments.

Article 8 of the convention provides a list of measures that are required to protect the essential elements of in-situ conservation (including establishing protected areas; regulation and management of biological resources both inside and outside protected areas; and protection of ecosystems, natural habitats and species populations), while the Article 9 lists the necessary ex-situ measures (including the adoption of measures for the recovery and rehabilitation of threatened species and their reintroduction into their natural habitats, and the establishment of facilities for ex-situ conservation of and research into plants, animals and microorganisms).[49]

The Conference of the Parties (COP), the convention's decision-making-body, supervises the implementation of the convention. In this regard, it may consider and adopt protocols such as the Cartagena Protocol on Biosafety,[50] which entered into force in 2003, or the Nagoya Protocol on access to genetic resources and the fair and equitable sharing of benefits, which entered into force in October 2014.

In 2010, the Secretary-General of the United Nations, Ban Ki-Moon, declared the world's collective failure to achieve significant reduction in the rate of bio-diversity loss[51] and stressed that higher priority should be given to tackle biodiversity loss in all areas of decision-making (CBD, 2010: 5).

In response to the alarming conclusions of the Third Global Biodiversity Outlook, the COP decided to adopt[52] the Strategic Plan for Biodiversity 2011–2020 and the Aichi Biodiversity Targets. This plan includes five strategic goals, subdivided into twenty targets, with the commitment that "by 2050, biodiversity is valued, conserved, restored and wisely used, maintaining ecosystem services, sustaining a healthy planet and delivering benefits essential for all people."

At the regional level, the urgent need to conserve biodiversity, together with adherence to the CBD, led to the establishment of the ASEAN Center for Biodiversity, based in the Philippines, in 2005. This is an intergovernmental regional institution that facilitates cooperation and coordination among the ten ASEAN member states and with relevant regional and international organizations and the private sector on the conservation and sustainable use of biological diversity, and the fair and equitable sharing of benefits.

Nevertheless, the ASEAN Biodiversity Outlook of 2010 confirmed that the region, like the rest of the world, had continued to lose biodiversity and thus had failed to meet its biodiversity preservation targets. An ASEAN conference on the theme "Biodiversity in Focus: 2010 and Beyond" discussed the key biodiversity issues that are important to the region, including gaps and challenges, and issued various recommendations. Those regarding cross-cutting concerns[53] stressed the urgent need to forge links between policy and science, integrate different fields, such as the social sciences, biology and economics, and conduct transdisciplinary research to form the basis for decision-making support.[54]

The ASEAN Center for Biodiversity implemented a plan for 2014–2020 with the intention of strengthening the role of the center and its work with multisectoral partners to enable ASEAN member states to meet the Aichi Targets and their commitments to various multilateral environmental agreements (ASEAN Center for Biodiversity, 2014).

Global and regional strategies on health and the environment

In May 1992, the 45th World Health Assembly asked the Director-General of the WHO to "formulate a new global WHO strategy for environmental health based on the findings and recommendations of the WHO Commission on Health and Environment and on the outcome of the United Nations Conference on Environment and Development."[55] A year later, the WHO presented its Global Strategy for Health and the Environment, which built on the global objectives drafted by an independent commission established by the Director of the WHO in order to "bring to the forefront the health dimension of the environmental and development crisis" (WHO Commission on Health and Environment, 1992: ix).

Those global objectives are:

- achieve a sustainable basis for health for all;
- provide an environment that promotes health; and
- make all individuals and organizations aware of their responsibility for health and its environmental basis.

The Global Strategy calls for wider action on health and the environment across the whole health sector and for close cooperation with the other socio-economic development sectors. Among its elements, there is a call for a "broader action and collaboration throughout WHO on matters related to health and the environment" (Ozolins and Stober, 1994: 6), specifically by putting health at the heart of all programs and decision-making on matters relating to the environment and development, and promoting and improving the organization of health systems at national and local levels. The new strategy also calls for the WHO to play an increased role at the country level vis-à-vis other ministries (Ozolins and Stober, 1994).

The World Health Assembly endorsed the Global Strategy in Resolution WHA46.20 and suggested regional committees should use it to develop corresponding regional strategies and actions plans.[56] In the meantime a WHO inter-regional initiative was launched to promote the involvement of the health sector in national planning for sustainable development and to prepare action plans for health and the environment (Ozolins and Stober, 1994).

In 1993, following the Global Strategy, the WHO in the Southeast Asia Region (SEARO) adopted a Regional Strategic Plan for Health and the Environment. This identified four priority program areas:

- urban health management;
- water supply, hygiene and sanitation (including food safety) in rural areas;
- health and environmental aspects of water resources development; and
- chemical safety.

Important work on drafting subsequent national plans of action began in 1995. By 2004, nine Southeast Asian countries had prepared their National Health and Environment Action Plans.[57] In addition, the WHO SEARO insisted on the opportunity provided by preparation for the World Summit on Sustainable Development (2002) to the ministries of health to ensure that the place of health in sustainable development is fully appreciated by planners and decision-makers at the national level. In turn, it is assumed to enhance the position of health in regional and global developmental agendas.[58]

The WHO SEARO and Asia-Pacific, the UNEP and the ADB, responding to the concerns of the WEHAB initiative,[59] decided to join forces to tackle the various problems related to health and the environment in the region and held the first high-level meeting on "Environment and Health in Southeast and East Asian Countries" in Manila in 2004. This meeting marked the start of the "Initiative on Environment and Health in Southeast and East Asian Countries," which expressed the following objectives:

- to review and identify major and common environmental health issues and challenges facing ASEAN countries, China, Japan, Mongolia and the Republic of Korea; and
- to delineate actions by countries and partner agencies that would strengthen the effective collaboration between the health and environment sectors for sound environmental health policies and interventions.[60]

A Regional Charter on Environment and Health[61] and the Bangkok Declaration, endorsed by the First Ministerial Meeting in Bangkok in 2007, established the Regional Forum on Environment and Health. The general objective was to promote the implementation of integrated environmental health strategies and regulations at the national and regional level to address the need for policies that will protect and enhance the environment and so improve living conditions and quality of life "through enforceable legislation and other legal instruments."[62] The Bangkok Declaration refers explicitly to existing international agreements on protection of the ozone layer, climate change, biodiversity conservation, the management of chemicals and waste and other initiatives related to the environment and health as well as to international law principles, such as the precautionary principle, the polluter-pays principle and the norms of good governance.[63]

Reviews of national activities[64] are ensured through scheduled high-level meetings of officials: each country presents its progress on its national action plan and is encouraged to share any difficulties of implementation with the others.

Sustainable development and international environmental law

The United Nations Millennium Declaration, which was adopted unanimously in 2000, acknowledged the collective responsibility of member states to uphold the principles of human dignity, equality and equity at the global level. Among six essential values in international relations in the twenty-first century, it included "respect for nature," which required a change in the current unsustainable patterns of production and consumption, and prudence "in the management of all living species and natural resources, in accordance with the precepts of sustainable development." As one of the key objectives for international cooperation, it emphasized "the protection of our environment."[65]

In order to measure achievement of the key objectives contained in the Millennium Declaration, and to translate them into action, concrete targets called the Millennium Development Goals (MDGs) were established. Many of these MDGs were related to health (child mortality; maternal health; combat HIV/AIDS, malaria and other diseases; eradicate extreme poverty and hunger) and one concerned the need to ensure "environmental sustainability" through discrete targets (such as the need to reverse the loss of environmental resources through the integration of sustainable principles into policies or to improve sustainable access to safe drinking water). However, the linkage between health and environment was not endorsed by the MDGs, even though it had previously been clearly stated that the most critical health problems in the world would not be resolved without major improvements in environmental quality (Smith *et al.*, 1999).

Although the fragmentation of the MDGs and the lack of synergy between them (Waage *et al.*, 2010) have been criticized,[66] the goals are now recognized "as the most successful global anti-poverty push in history"[67] because the member states committed to a common agenda which led them to take concrete actions and improve coordination.

The notion of "sustainable development" that is integral to the MDG targets first became central to international environmental policy during the United Nations Conference on Environment and Development of 1992. In the 1970s, international law had focused on the protection of specific sectors related to the environment, but the emergence of new issues such as depletion of the ozone layer and the risk of nuclear accidents led the international community to adopt cross-sectoral consideration of environmental questions.

In the resolution that convened the Rio Conference, the General Assembly of the United Nations expressed its deep concern over the

> continuing deterioration of the state of the environment and the serious degradation of the global life-support systems . . . [Certain] trends, if allowed to continue, could disrupt the global ecological balance, jeopardize the life-sustaining qualities of the Earth and lead to an ecological catastrophe . . . [D]ecisive, urgent and global action is vital to protecting the ecological balance of the Earth.

It also acknowledged that the global character of environmental problems necessitates action at the global, regional and national levels, and the commitment and participation of all countries.[68]

The Rio Conference is considered as a major environmental legal landmark (Handl, 2012). Two important environmental conventions were opened for signature: the United Nations Framework Convention on Climate Change and the Convention on Biological Diversity. The conference also adopted the Forest Principles[69] as well as an action programme called Agenda 21, which contributed to the promotion of health sector involvement in addressing health and environment issues (WHO, 1997). Finally, the Rio Declaration led to the formulation (or reaffirmation) of principles of particular importance for the development of international environmental law (Kiss and Shelton, 2007). For instance, Principle 2 completes Principle 21 of the Stockholm Declaration by affirming that states' exploitation of their resources can be carried out according to their own environmental "and developmental" policies. It also restates rights of public information, participation and remedies; the need for the development of liability rules; and the requirement to notify other states about emergencies and projects that may affect their environments.

It also formulates new principles, such as the polluter-pays principle, the precautionary principle, the general requirement of environmental impact assessments for proposed activities, the principle of common but differentiated responsibilities and the obligation to enact effective environmental legislation. Principle 4 states that "in order to achieve sustainable development, environmental protection shall constitute an integral part of the development process and cannot be considered in isolation from it."

The Rio Conference considered health in a very broad way,[70] as is illustrated by Principle 1 of the Declaration: "Human beings are at the center of concerns for sustainable development. They are entitled to a healthy and productive life in harmony with nature." Nevertheless, sustainable development became a major principle of international environmental law during this conference.

The Stockholm Conference

The 1972 United Nations Conference on the Human Environment affirmed for the first time the link between environment and well-being. For instance, its first principle states: "Man has the fundamental right to freedom, equality and adequate conditions of life, in an environment of a quality that permits a life of dignity and well-being." This echoes the WHO's definition of health, as stated in 1946: "Health is a state of complete physical, mental and social well-being and not merely the absence of disease or infirmity."[71] The Stockholm Conference therefore led to international recognition of the health dimension in environmental issues.

Given the global extent of environmental problems and the fact that they affect the international common realm, the conference stressed that it would "require extensive cooperation among nations and action by international organizations in the common interest."[72] The Stockholm Declaration of Principles for the Preservation and Enhancement of Human Development was proclaimed in that respect. The texts adopted during the conference, notably this declaration, constitute the first comprehensive statements of international concern for environmental protection and thus they comprise one of the founding instruments of international environmental law. Principle 23 notably recommends that states should develop international environmental law further, taking into consideration the system of values prevailing in each country, especially developing countries.[73]

The Stockholm Conference recommended the creation of the United Nations Program for the Environment (UNEP) to address environmental issues. UNEP was expected to promote international cooperation on the environment and to recommend policies to this end. In fact, it went on to play a much larger role and has been the most active UN body in the development of MEAs and international environmental law in general.

Environmental law affects all human activities, falls within the domain of every governmental institution and level of lawmaking, and guides the conduct of non-state as well as state actors.[74] Therefore, it is an appropriate instrument for an integrated approach to resolve the issues regarding the relationship between health and biodiversity.

Conclusion

From the One Health approach to the Stockholm Conference, we have seen the main steps leading to an integrative approach of health linking human and animal health within their ecosystems and the evolution of international governance in the areas of health and biodiversity.

From a theoretical point of view, the One Health approach and international environmental law have common features: each is a multidisciplinary approach that takes into account the environment and enriches its reflection with the input of science. International environmental law became a tool of global public health protection (Onzivu, 2006) and it has developed principles of sustainable development which advocate the integration of environmental protection as an integral part of the development process.[75]

Since the Stockholm Declaration, the necessity of an integrated and coordinated approach to protect and improve the environment for the benefit of humanity has been affirmed.[76] The link between a healthy environment and the promotion of the human well-being is now acknowledged, and the notion of well-being has been placed at the heart of the International Science-Policy Platform on Biodiversity and Ecosystem Services framework. The second objective of the IPBES's draft work programme for 2014–2018 takes into consideration the necessity to adopt strategies at the relevant geographical, and thus ecological, scale, as illustrated by the Asian-Pacific Workshop's proposal to create an IPBES regional hub.

We should keep in mind that the IPBES is intended to follow a participatory, bottom-up approach, including the incorporation of indigenous and local knowledge in the conservation and sustainable use of biodiversity and ecosystems.[77] In this regard, the IPBES should consider biodiversity and ecosystem services beyond their economic value or their utility for humans and integrate the notions of co-viability, co-evolution and resilience of socio-ecosystems.

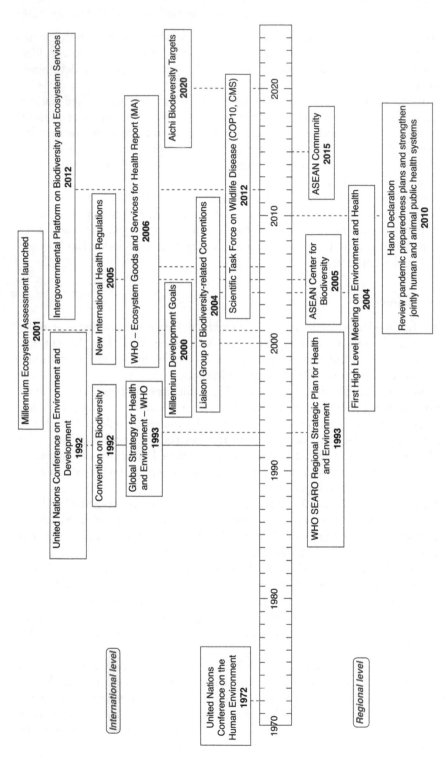

Figure 14.4 Timeline of the main international and regional instruments related to health and biodiversity

In order to respond to the pressing need to embody the fact that biodiversity is a common concern of humankind and to give shape to the notion of global environmental responsibility, we expect the IPBES to be an appropriate framework through which to study the linkage between health and biodiversity. It should provide findings encompassing the different dimensions of the well-being of society in interaction with its environment which ultimately might help to address the issue of environmental justice.

Notes

1　*Stockholm Declaration*, 16 June 1972, Preamble. (To aid readability, we have presented all legal references to COP decisions, international agreements and international conferences as notes in this chapter.)

2　CBD, COP 12, Decision XII/18, *Sustainable Use of Biodiversity: Bushmeat and Sustainable Wildlife Management*, Pyeongchang, Republic of Korea, 6–17 October 2014, §4.

3　"This neighbourhood is not only geographical – regional or hemispheric – but may also need to be considered in terms of ecosystems" (Doumbé-Billé, 2011: 119).

4　In ICJ Report 226 (1996), *Advisory Opinion on the Legality of the Threat or Use of Nuclear Weapons*, §19, the International Court of Justice affirmed that "the existence of the general obligation of states to ensure that activities within their jurisdiction and control respect the environment of other states or of areas beyond national control is now part of the corpus of international law relating to the environment."

5　International Ministerial Conference on Avian and Pandemic Influenza, *Hanoi Declaration*, Vietnam, 19–21 April 2010, §5 and §9.

6　UNEP/CMS Resolution 10.22, *Wildlife Disease and Migratory Species*, Bergen, 20–25 November 2011.

7　CITES Resolution 16.4, *Cooperation of CITES with Other Biodiversity-Related Conventions*, Bangkok, 3–14 March 2013: "Recommends that Parties further strengthen the cooperation, coordination and synergies among the focal points of the biodiversity-related conventions."

8　Ramsar Resolution XI.12, *Wetlands and Health: Taking an Ecosystem Approach*, Bucharest, Romania, 6–13 July 2012, notably §4. An interesting element is mentioned in Annex 2 in relation to the key messages for policy-makers: "Humans can be exposed to health risks in wetland ecosystems: toxic materials, water-borne or vector-borne diseases. While steps can be taken to ameliorate these risks, the risks can increase (sometimes dramatically) if disruption occurs to ecosystems and the services they provide" (Annex 2, §8).

9　UNCCD, COP(11)/CST/INF3/Corr1, *Final Outcome of the UNCCD 2nd Scientific Conference*, Windhoek, Namibia, 17–20 September 2013, §29.

10　Convention on Biological Diversity (CBD); Convention on International Trade in Endangered Species of Wild Fauna and Flora (CITES); Convention on the Conservation of Migratory Species of Wild Animals (CMS); Ramsar Convention on Wetlands (Ramsar); Convention Concerning the Protection of the World Cultural and Natural Heritage (WHC); and International Treaty on Plant Genetic Resources for Food and Agriculture (ITPGRFA). See *Modus Operandi for the Liaison Group of the Biodiversity-Related Conventions*, Geneva, 4 September 2011.

11　*Cooperation Agreement between the Secretariat of the Convention on Biological Diversity and the World Organization for Animal Health*, 15 February 2013, Article 1, §1.

12　Water and sanitation; food supply; utilities and energy; public transportation; communication; security and order; and finance and banking. See ASEAN +3, 2010: 125–126.

13 Even in the aforementioned non-health sectors. The same remark could be made about ASEAN, 2010.
14 On the capacity of a network of organizations to function as an information integration system, as required in the performance of complex common objectives such as the design of inter-sectoral policies in "Health and Environment," see Lajaunie and Mazzega, 2016c.
15 Policies remotely linked to biodiversity conservation can even have goals that are contradictory to safeguarding biodiversity: for instance, planning policies can hinder the enforcement of conservation measures or sustainable land use rules. See Paloniemi *et al.*, 2012.
16 See UNEP/IPBES/3/3, 2010, *Third ad hoc Intergovernmental and Multi-stakeholder Meeting on an Intergovernmental Science-Policy Platform on Biodiversity and Ecosystem Services* ("Busan Outcome"), Busan, Republic of Korea, 7–11 June 2010, §6.
17 For a history of the IPBES consultation process, see Larigauderie and Mooney, 2010.
18 UNGA Resolution 65/162, *Report of the Governing Council of the United Nations Environment Programme on its Eleventh Special Session*, 20 December, 2010,§17. It followed the COP of the Convention on Biodiversity, Decision X/11, *Science–Policy Interface on Biodiversity, Ecosystem Services and Human Well-being and Consideration of the Outcome of the Intergovernmental Meetings*, 29 October 2010, concluding that an intergovernmental science-policy platform on biodiversity and ecosystem services should be established.
19 The resolution specifies that the meeting should provide "for the full and effective participation of all Member States, in particular representatives from developing countries, to determine modalities and institutional arrangements for the platform."
20 UNEP/IPBES MI/2/9, 2012, *Plenary Meeting to Determine Modalities and Institutional Arrangements for an Intergovernmental Science-Policy Platform on Biodiversity and Ecosystem Services*, Second Session, Panama City, 16–21 April 2012.
21 The Plenary is the decision-making body of the platform, open to membership from all state members of the United Nations.
22 IPBES/1/INF/9, 2013, *Outcome of an Informal Expert Workshop on Main Issues Relating to the Development of a Conceptual Framework for the Intergovernmental Science-Policy Platform on Biodiversity and Ecosystem Services*, Bonn, Germany, 21–26 January 2013.
23 Examples of key questions coming from those integrated considerations – such as "What are the relationships between biodiversity and the health and well-being of humans, their domesticated animals and cultivated plants, and the persistence of wild populations?" – are given into Larigauderie *et al.*, 2012.
24 Those documents have been submitted to governments and stakeholders of the IPBES for comments.
25 IPBES/2/2, *Draft Work Programme for the Period 2014–2018*, Second Session, Antalya, Turkey, 9–14 December 2013, pp. 6–7.
26 United Nations University–Institute for Sustainability and Peace, *Asia-Pacific Workshop on Regional Interpretation of the IPBES Conceptual Framework and Knowledge Sharing*, Seoul, Republic of Korea, 2–4 September 2013.
27 For instance, it appears that current knowledge of western Asian biodiversity and ecosystem assessments are inadequate.
28 Recognizing that elements such as the perception, behavior and action of humans towards biodiversity and ecosystem services are missing in current assessments, the workshop insisted on the integration of social science issues, such as culture, language, local knowledge and history, to better understand socio-ecological relationships and human well-being.
29 On the participatory procedure of assessment of impacts of policies on health in Thailand and the rest of Southeast Asia, see Lajaunie and Morand, 2015.

30 Scientists had already identified the need for an international ecosystem assessment. For a history of the MA, see Millennium Ecosystem Assessment Steering Committee, 1999.

31 Endorsed by the CDB, Decision COP5/VI, *Ecosystem Approach*, Nairobi, Kenya, 15–26 May 2000, §7, A2: "an ecosystem approach is based on the application of appropriate scientific methodologies focused on levels of biological organization, which encompass the essential structure, processes, functions and interactions among organisms and their environment. It recognizes that humans, with their cultural diversity, are an integral component of many ecosystems."

32 "The land consists of soil, water, plants, and animals, but health is more than a sufficiency of these components. It is a state of vigorous self-renewal in each of them, and in all collectively. Such a collective functioning of interdependent parts of the maintenance of the whole is characteristic of an organism. In this sense land is an organism, and conservation deal with its functional integrity, or health" (Leopold 1938/1991: 310).

33 Biodiversity; Health; Desertification; Wetlands and Water; and Opportunities and Challenges for Business and Industry.

34 ICJ, Report 38, *Denmark v. Norway, Concerning Maritime Delimitation in the Area between Greenland and Jan Mayen* (1993), Separate Opinion of Judge Weeramantry.

35 Among the promising interventions to reduce ecosystem change's pressure on health services, the report (p. 44) mentions the importance of an integrated action for health, making use of tools such as health impact assessments of major development projects, policies, programs and indicators for health and sustainable development.

36 It recognizes that the MA has demonstrated more comprehensively than ever before the important links between ecosystems, ecosystem services and human well-being. See CBD/COP9/INF/26, 2008, *The Millennium Ecosystem Assessment Follow-up: A Global Strategy for Turning Knowledge into Action*, Preamble.

37 CBD, COP 9, Decision IX/15, *Follow-up to the Millennium Ecosystem Assessment*, Bonn, 19–30 May 2008, §6.

38 CBD, COP 11, Decision XI/2, *Review of Progress in Implementation of National Biodiversity Strategies and Action Plans and Related Capacity-Building Support to Parties*, §28.

39 UNGA Resolution 58/3, *Enhancing Capacity-Building in Global Public Health*, 27 October 2003.

40 The second amendment, in 1981, marked the global eradication of smallpox.

41 Defined as an event that constitutes "a public health risk to other States through the international spread of disease and to potentially require a coordinated international response" (Article 1). In this case, each state party is required to provide notification of such an event to the WHO within twenty-four hours (WHO, 2005).

42 The states are obliged to update their national legislation in order to comply with the IHR.

43 For the list of competent intergovernmental organizations or international bodies with which WHO is expected to cooperate and coordinate its activities, see WHA, 2005, §4.

44 The definition of GLEWS is in Article 3.2 (FAO, OIE, and WHO, 2006: 13).

45 *35th Meeting of the ASEAN Ministers on Agriculture and Forestry*, Kuala Lumpur, Malaysia, 26 September 2013, §9.

46 Among them, we can mention the Greater Mekong Subregion's Regional Communicable Diseases Control Project and the Information Center on Emerging Infectious Diseases in the ASEAN +3 countries.

47 That is, almost every member state of the United Nations, with the notable exception of the USA.

48 On the core characteristics of the CBD (comprehensiveness, complexity, compromise), see McGraw, 2002.

49 For further information on CBD requirements, see Birnie *et al.*, 2009.
50 Its objective is "to contribute to ensuring an adequate level of protection in the field of the safe transfer, handling and use of living modified organisms resulting from modern biotechnology that may have adverse effects on the conservation and sustainable use of biological diversity, taking also into account risks to human health, and specifically focusing on transboundary movements" (*Cartagena Protocol on Biosafety*, Montreal, 29 January 2000, Article 1).
51 CBD, COP 6, Decision VI/26, *Strategic Plan for the Convention on Biological Diversity*, The Hague, Netherlands, 7–19 April 2002, Annex, §11: "Parties commit themselves to a more effective and coherent implementation of the three objectives of the Convention, to achieve by 2010 a significant reduction of the current rate of biodiversity loss at the global, regional and national level as a contribution to poverty alleviation and to the benefit of all life on earth."
52 CBD, COP 10, Decision X/2, *Strategic Plan for Biodiversity 2011–2020 and the Aichi Targets*, 18–29 October 2010.
53 On the cross-cutting issues linked to health in the CBD, see Lajaunie and Mazzega, 2016a.
54 ASEAN Conference on Biodiversity, Singapore, 21–23 October 2009, in ASEAN Center for Biodiversity, 2010.
55 WHA45.31, *Health and Environment*, 14 May 1992, §1.
56 WHA46.20, *WHO Global Strategy for Health and the Environment*, 12 May 1993, §5.
57 WHO, SEA/RC61/13, *Regional Initiative on Environment and Health*, 23 July 2008, p. 1.
58 WHO, SEA/RC54/11, *Health and Environment in National Development: Regional Progress and Preparation for Rio+10 Conference*, 17 July 2001, p. 10.
59 "Well-developed health-and-environment information systems, based on relevant data sets, are essential if scientific monitoring information is to be provided in support of policy and decision-making, planning and evaluation. In general, knowledge of environment and health risks is segmented, and information is incomplete. Commonly, mechanisms to ensure co-ordination at the national, regional and local levels regarding health effects, impact assessment and the development of adequate reporting systems are lacking. In addition, mechanisms are frequently not in place to ensure that such information, once obtained, is transmitted to the various relevant sectors for action" (WEHAB, 2002: 16).
60 WHO/UNEP/ADB, RS/2004/GE/20(PHL), *High-Level Meeting on Health and Environment in ASEAN and East Asian Countries*, 2004, p. 5.
61 WHO/UNEP, *Second High-Level Meeting on Health and Environment in ASEAN and East Asian Countries*, Bangkok, Thailand, 12–13 December 2005.
62 With a regular review to take into account new knowledge and emerging technologies. See MF1/4, *Charter of the Regional Forum on Environment and Health in Southeast and East Asian Countries*, Bangkok, Thailand, 9 August 2007, p. 6.
63 "Including civic engagement and participation, efficiency, equity, transparency and accountability" (*Bangkok Declaration on Environment and Health*, Bangkok, Thailand, 9 August 2007).
64 Thematic Working Groups (TWGs) deal with seven prioritized issues: air quality; water supply, hygiene and sanitation; solid and hazardous waste; toxic chemicals and hazardous substances; climate change, ozone depletion and ecosystem changes; contingency planning, preparedness and response in environmental health emergencies; and health impact assessments. The members of each TWG are drawn from member countries concerned with a specific issue or have expertise which can be shared to benefit other members. Each TWG is responsible for knowledge management and technical support, drafting progress reports for the Regional Forum, coordination and advocacy, and resource mobilization.

65 Conforming with the principles of justice and international law. See UNGA Resolution 55/2, *United Nations Millennium Declaration*, 8 September 2000.
66 The new Sustainable Development Goals of 2015 have been designed to develop connections between the goals in order to address the lack of synergy in the MDGs.
67 Ki-Moon, 2013: 3.
68 UNGA Resolution 44/228, *United Nations Conference on Environment and Development*, 22 December 1989.
69 The title – "Non-legally Binding Authoritative Statement of Principles for a Global Consensus on the Management, Conservation and Sustainable Development of All Types of Forests" – reflects the divergence of views during the negotiations.
70 Except for the actions included in Chapter 6 of Agenda 21, "Protecting and promoting human health."
71 WHO, *Constitution of the World Health Organization as Adopted by the International Health Conference*, New York, 19–22 June 1946, Preamble.
72 UNCED, *Declaration of the United Nations Conference on the Human Environment*, 16 June 1972, Preamble, §7.
73 In line with the principle of common but differentiated responsibilities. See Kiss and Shelton, 2007: 37.
74 Elements of international law given in the Foreword of Kiss and Shelton, 2007.
75 Principle 4 of the Stockholm Declaration.
76 Principle 13 of the Stockholm Declaration.
77 UNEP/IPBES/3/3, 2010, *Third ad hoc Intergovernmental and Multi-stakeholder Meeting on an Intergovernmental Science-Policy Platform on Biodiversity and Ecosystem Services* ("Busan Outcome"), Busan, Republic of Korea, 7–11 June 2010, p. 6, §7(d).

References

American Veterinary Medical Association (AVMA). 2008. *One Health: A New Professional Imperative*. AVMA, Schaumberg, IL.

ASEAN. 2010. *The Roadmap for an HPAI-Free ASEAN by 2020.* www.asean.org/wp-content/uploads/images/2012/publications/HPAI%20in%20ASEAN,%20Strategies%20and%20Success%20Stories.pdf (accessed 16 March 2017).

ASEAN. 2013. *35th Meeting of the ASEAN Ministers on Agriculture and Forestry*. ASEAN, Kuala Lumpur.

ASEAN +3. 2010. *Good Practices in Responding to Emerging Infectious Diseases: Experiences from ASEAN Plus Three Countries*. ASEAN, Bangkok.

ASEAN Center for Biodiversity. 2010. *ASEAN Biodiversity Outlook*. ASEAN, Philippines.

ASEAN Center for Biodiversity. 2014. *Annual Report 2013*. ASEAN, Philippines.

ASEAN Center for Biodiversity, UNEP, GIZ, and British Foreign and Commonwealth Office. 2013. *The ASEAN TEEB Scoping Study: Valuing Ecosystem Services in Southeast Asia*. www.teebweb.org/wp-content/uploads/2013/02/FLYER-ASEAN-TEEB-Side-Event-v01-Oct-12.pdf (accessed 16 March 2017).

Atlas R., C. Rubin, and S. Maloy. 2010. One Health: attaining optimal health for people, animals, and the environment. *Microbe Magazine* 5: 383–389.

Birnie P., A. Boyle, and C.Redgwell. 2009. *International Law and the Environment*. 3rd edn. Oxford University Press, New York.

Bodansky, D. 2011. *The Art and Craft of International Environmental Law*. Harvard University Press, Cambridge, MA.

Boisson de Chazournes, L. 2011. Features and trends in international environmental law. In Y. Kerbrat and S. Maljean-Dubois (eds.), *The Transformation of International Environmental Law*. Pedone and Hart, Oxford: 9–20.

Bousfield, B. and R. Brown. 2011. One world, one health. *Veterinary Bulletin* 1(7). www.afcd.gov.hk/tc_chi/quarantine/qua_vb/files/OWOH2.pdf (accessed 30 March 2017).

Carin, B., P. Heinbecker, P. Jenkins, and D. Runnalls. 2011. *An Unfinished House: Filling the Gaps in International Governance*. Centre for International Governance Innovation, Waterloo, Canada.

Coker R., J. Rushton, S. Mounier-Jack, E. Karimuribo, P. Lutumba, D. Kambarage, D.U. Pfeiffer, K. Stärk, and M. Rweyemamu. 2011. Towards a conceptual framework to support one-health research for policy on emerging zoonoses. *Lancet Infectious Diseases* 11: 326–331.

Convention on Biological Diversity (CBD). 2010. *Global Biodiversity Outlook 3*. Secretariat of the Convention on Biological Diversity, Montreal.

CBD, CITES, CMS, Ramsar, WHC, and TPGRFA. 2011. *Modus Operandi for the Liaison Group of the Biodiversity-Related Conventions*. Geneva.

Doumbé-Billé, S. 2011. Regionalism and universalism in the production of environmental law. In Y. Kerbrat and S. Maljean-Dubois. 2011. *The Transformation of International Environmental Law*. Pedone and Hart, Oxford: 117–129.

FAO, OIE, and WHO. 2006. *Global Early Warning and Response System for Major Animal Diseases (GLEWS), Including Zoonoses*. www.oie.int/doc/ged/D11304.PDF (accessed 16 March 2017).

FAO, OIE, and WHO. 2013. GLEWS+: *The Joint FAO–OIE–WHO Global Early Warning System for Health Threats and Emerging Risks at the Human–Animal–Ecosystems Interface*. www.fao.org/docrep/019/i3579e/i3579e.pdf (accessed 16 March 2017).

Fidler, D.P. 2005. From international sanitary conventions to global health security: the new International Health Regulations. *Chinese Journal of International Law* 4: 325–392.

Handl, G. 2012. *Declaration of the United Nations Conference on the Human Environment (Stockholm Declaration) 1972 and the Rio Declaration on Environment and Development, 1992*. United Nations Audiovisual Library of International Law. http://legal.un.org/avl/ha/dunche/dunche.html (accessed 16 March 2017).

Kaplan, B. and M. Echols M. 2009. One Health: the Rosetta Stone for 21st century health and health providers, *Veterinaria Italiana* 45(3): 377–382.

Ki-Moon, B. 2013. Foreword. In *The Millennium Development Goals Report 2013*. United Nations, New York.

Kiss, A. and D. Shelton. 2007. *Guide to International Environmental Law*. Martinus Nijhoff, Boston.

Lajaunie, C. and P. Mazzega. 2016a. Mining CBD. *Revista de Direito Internacional* 13(2): 277–292.

Lajaunie, C. and P. Mazzega. 2016b. One Health and biodiversity conventions: the emergence of health issues in biodiversity conventions. *IUCN Academy of Environmental Law eJournal* 7: 105–121.

Lajaunie, C. and P. Mazzega. 2016c. Organization networks as information integration system: case study on environment and health in Southeast Asia. *Advances in Computer Science: an International Journal* 5(2): 28–39.

Lajaunie, C. and S. Morand. 2015. A legal tool for participatory methods in land systems science: the Thai model of health impact assessment and the consideration of zoonotic diseases concerns into policies. *Global Land Project News* 11: 30–33.

Larigauderie, A. and H.A. Mooney. 2010. The Intergovernmental Science-Policy Platform on Biodiversity and Ecosystem Services: moving a step closer to an IPCC-like mechanism for biodiversity. *Current Opinion in Environmental Sustainability* 2: 9–14.

Larigauderie A., A.H. Prieur-Richard, G.M. Mace, M. Lonsdale, H.A. Mooney, L. Brussaard, D. Cooper, W. Cramer, P. Daszak, S. Díaz, A. Duraiappah, T. Elmqvist, D.P. Faith, L.E. Jackson, C. Krug, P.W. Leadley, P. Le Prestre, H. Matsuda, M. Palmer, C. Perrings, M. Pulleman, B. Reyers, E.A. Rosa, R. J. Scholes, E. Spehn, B.L. Turner, and T. Yahara. 2012. Biodiversity and ecosystem services science for a sustainable planet: the DIVERSITAS vision for 2012–20. *Current Opinion in Environmental Sustainability* 4: 101–105.

Leopold, A. 1938/1991. Conservation: in whole or in part. In S.L. Flader and J.B. Callicott (eds.), *The River of the Mother of God and Other Essays*. University of Wisconsin Press, Madison: 310–319.

McGraw, D.M. 2002. The CBD: key characteristics and implications for implementation. *Review of European Community and International Environmental Law* 11: 17–28.

Millennium Ecosystem Assessment. 2005. *Ecosystems and Human Well-being: Synthesis*. Island Press, Washington, D.C.

Millennium Ecosystem Assessment. 2006. *Ecosystems and Human Well-being: Multiscale Assessments*. Island Press, Washington, D.C.

Millennium Ecosystem Assessment Steering Committee. 1999. *Groundswell: The Newsletter of the Millennium Assessment of Global Ecosystems*. www.millenniumassessment. org/documents/document.33.aspx.pdf (accessed 17 March 2017).

Morand, S. 2010. Biodiversity: an international perspective. *Revue Scientifique et Technique Office International des Epizooties* 29: 65–72.

Norgaard, R. 2010. Ecosystem services: from eye-opening metaphor to complexity blinder. *Ecological Economics* 69: 1219–1227.

Onzivu, W. 2006. International environmental law, the public's health, and domestic environmental governance in developing countries. *American University International Law Review* 21(4): 597–684.

Ozolins, G. and J. Stober. 1994. *WHO Global Strategy for Health and Environment and Related Events*. WHO/EOS/94.41. WHO, Geneva.

Paloniemi R., E. Apostolopoulou, E. Primmer, M. Grodzinska-Jurczak, K. Henle, I. Ring, M. Kettunen, J. Tzanopoulos, S.G. Potts, S. van den Hove, P. Marty, A. McConville, and J. Similä. 2012. Biodiversity conservation across scales: lessons from a science–policy dialogue. *Nature Conservation* 2: 10.

Public Health Agency of Canada. 2009. *One World One Health: From Ideas to Action: Report of the Expert Consultation*. Public Health Agency of Canada, Ottawa.

Shelton, D. 2002. *Human Rights, Health & Environmental Protection: Linkages in Law & Practice: A Background Paper for the World Health Organization*. Health and Human Rights Working Paper No. 1. WHO, Geneva.

Sing, S.K. 2009. International health regulations: a major paradigm shift from 1969 to 2005. *Journal of Communicable Diseases* 41: 113–116.

Smith, K.R., C.F. Corvalán, and T. Kjellström. 1999. How much global ill health is attributable to environmental factors. *Epidemiology* 10: 573–584.

UNEP and French Government. 2008. *An Intergovernmental Science-Policy Platform on Biodiversity and Ecosystems Services, Building on the Global Strategy for Follow-Up to the Millennium Ecosystem Assessment (MA) and the Consultative Process towards an International Mechanism of Scientific Expertise on Biodiversity (IMoSEB)*. www.iddri.org/ Iddri/08_IPBES_concept-note_July.pdf (accessed 16 March 2017).

United Nations. 2013. *The Millennium Development Goals Report 2013*. United Nations, New York.

Van den Hove, S. and L. Chabason. 2009. The debate on an Intergovernmental Science-Policy Platform on Biodiversity and Ecosystem Services (IPBES): exploring gaps and needs. *Idées pour le débat, Institut du Développement Durable et des Relations Internationales* 1: 1–24.

Von Schirnding, Y., W. Onzivu, and A.O. Adede. 2002. International environmental law and global public health. *Bulletin of the World Health Organisation* 80: 970–974.

Waage, J., R. Banerji, O. Campbell, E. Chirwa, G. Collender, V. Dieltiens, A. Dorward, P. Godfrey-Faussett, P. Hanvoravongchai, G. Kingdon, A. Little, A. Mills, K. Mulholland, A. Mwinga, A. North, W. Patcharanarumol, C. Poulton, V. Tangcharoensathien, and E. Unterhalter. 2010. The Millennium Development Goals: a cross-sectoral analysis and principles for goal setting after 2015. *Lancet* 376: 991–1023.

WEHAB Working Group. 2002. *A Framework for Action on Health and the Environment*. World Summit on Sustainable Development, Johannesburg.

World Health Assembly (WHA). 2005. *Revision of the International Health Regulations*. WHA58.3. WHA, Geneva.

World Health Organization (WHO). 1997. *Health and Environment in Sustainable Development: Five Years after the Earth Summit*. WHO/EHG/97.12.E. WHO, Geneva.

WHO. 2005a. *Ecosystems and Human Well-being: Health Synthesis: A Report of the Millennium Ecosystem Assessment*. WHO, Geneva. www.millenniumassessment.org/documents/document.357.aspx.pdf (accessed 30 March 2017).

WHO. 2005b. *International Health Regulations*. 2nd edn. WHO, Geneva.

WHO Commission on Health and Environment, 1992. *Our Planet, Our Health*. WHO, Geneva.WHO and CBD. 2015. *Connecting Global Priorities: Biodiversity and Human Health: A State of Knowledge Review*. WHO and CBD, Geneva.

Zinsstag, J., E. Schelling, B. Bonfoh, A.R. Fooks, J. Kasymbekov, D. Waltner-Toews, and M. Tanner. 2009. Towards a "One Health" research and application tool box. *Veterinaria Italiana* 45(1): 121–133.

Index